Practical Ha... Oil and St...

Stationary, Marine, Tr... ...urners, Oil Burners, Etc.; Farm, ...tion, Automobile, Locomotive; A simple, practical and comprehensive book on the construction, operation and repair of all kinds of engines. Dealing with the various parts in detail and the various types of engines and also the use of different kinds of fuel

John B. Rathbun

Alpha Editions

This edition published in 2024

ISBN 9789361478376

Design and Setting By

Alpha Editions

www.alphaedis.com

Email - info@alphaedis.com

As per information held with us this book is in Public Domain.
This book is a reproduction of an important historical work.
Alpha Editions uses the best technology to reproduce historical work
in the same manner it was first published to preserve its original nature.
Any marks or number seen are left intentionally to preserve.

Contents

CHAPTER I HEAT AND POWER (1) The Heat Engine.	- 1 -
(3) Combustion In the Cylinder.	- 9 -
(4) Compression.	- 13 -
(5) Efficiency and Heat Losses.	- 16 -
(6) Expansion of the Charge.	- 18 -
CHAPTER II FUELS AND COMBUSTION (7) Combustion.	- 24 -
(8) Gaseous Fuels.	- 27 -
(9) Gasifying Coal.	- 30 -
(10) Water Gas.	- 31 -
(11) Blast Furnace Gas.	- 32 -
(12) Producer Gas.	- 33 -
(13) Producer Gas From Peat.	- 38 -
(14) Crude Oil Producers.	- 41 -
(15) Operation of Producers.	- 42 -
(16) Coal.	- 45 -
(17) Fuel Oils.	- 48 -
(18) Tar for Fuel.	- 51 -
(19) Residual Oils.	- 53 -
(20) Gasoline.	- 54 -
(21) Benzol.	- 58 -
(22) Alcohol.	- 59 -
(23) Kerosene Oil.	- 60 -
CHAPTER III WORKING CYCLES (24) Requirements of the Engine.	- 64 -
(25) Four Stroke Cycle Engine.	- 66 -

(26) Two Stroke Cycle Engine. — 70 —

(27) Three Port—Two Stroke Cycle Engine. — 73 —

(28) Reversing Two Cycle Motors. — 77 —

(29) Scavenging Engines. — 80 —

CHAPTER IV INDICATOR DIAGRAMS (35)
General Description. — 81 —

(36) Diagram of Four Stroke Cycle Engine. — 83 —

(37) Detecting Faults With the Indicator. — 88 —

(38) Two Stroke Cycle Diagram. — 91 —

(39) Diagram of Diesel Engine. — 92 —

(40) Gas Turbine Development. — 93 —

CHAPTER V TYPICAL FOUR STROKE CYCLE
ENGINES (41) Essential Parts of the Gas Engine. — 98 —

(42) Application of the Four Stroke Principle. — 101 —

(43) Horizontal Single Cylinder Engine. — 103

(44) Multiple Cylinder Engines. — 107 —

(45) Four Cylinder Vertical Auto Motor. — 111 —

(46) Stationary Four Cylinder Engine. — 118 —

(47) The "V" Type Motor. — 122 —

(48) Mesta Gas Engines. — 129 —

(49) Knight Sliding Sleeve Motor. — 132 —

(50) Reeves Slide Sleeve Valve. — 139 —

(51) Argyll Single Sleeve Motor. — 141 —

(53) Sturtevant Aeronautical Motor. — 144 —

(54) The Rotating Cylinder Motor. — 147 —

(55) The Gyro Rotary Motor. — 148 —

(56) Gnome Rotary Motor. — 151 —

CHAPTER VI TWO STROKE CYCLE ENGINES (30)

The Junker Two Stroke Cycle Engine.	- 158 -
(34) Koerting Two Stroke Cycle Engine.	- 162 -
(57) Two Stroke Cycle Rail Motor Cars.	- 165 -
(58) Rotating Cylinder Two Stroke Cycle Motor.	- 167 -
(60) Gnome Radial Two Stroke Motor.	- 170 -
(62) Variable Speed Two Stroke Motor.	- 175 -

CHAPTER VII OIL ENGINES (31) Diesel Oil Engine.

	- 178 -
(63) Diesel Engine (Marine Type).	- 186 -
(64) The M.A.N. Diesel Engine.	- 190 -
(65) Mirlees-Diesel Engines.	- 192 -
(66) Willans-Diesel Engines.	- 193 -
(67) Installation and Consumption of Diesel Plant.	- 194 -
(32) Semi-Diesel Type Engine.	- 197 -
(68) De La Vergne Oil Engines.	- 199
The De La Vergne Oil Engine (Type HA)	- 204 -
(69) Operating Costs of the Semi-Diesel Type.	- 206 -
(70) Elyria Semi-Diesel Type.	- 208 -
(71) Remington Oil Engine.	- 212 -

CHAPTER VIII IGNITION SYSTEMS (73)

Principles of Ignition.	- 218 -
(74) Advance and Retard.	- 220 -
(75) Preignition.	- 222 -
(76) Misfiring.	- 223 -
(77) Hot Tube Ignition.	- 224 -
(78) Electrical Ignition.	- 227 -
(79) Sources of Current.	- 229 -

(80) Primary Batteries. - 231 -

(81) Dry Batteries. - 234 -

(82) Series and Multiple Connections. - 239 -

(83) Multiple-Series Connections. - 241 -

(84) Operation of Dry Cells. - 243 -

(85) Storage Batteries. - 244 -

(86) Care of the Storage Cell. - 247 -

(87) Make and Break System (Low Tension). - 250 -

(88) Operation of the Make and Break Igniter. - 256 -

(89) Jump Spark System (High Tension System). - 258 -

(90) Vibrator Construction. - 261 -

(91) Operation of the Jump Spark Coil. - 263 -

(92) Primary Timer. - 267 -

(93) Timer Construction. - 268 -

(94) Operation of Timers. - 271 -

(95) High Tension Spark Plug. - 273 -

(96) Care of Spark Plug. - 278 -

(97) Magnetos. - 281 -

(98) Low Tension Magneto. - 283 -

(99) Care of Low Tension Magnetos. - 288 -

(100) High Tension Magnetos. - 289 -

(101) Bosch Oscillating High Tension Magneto. - 295 -

(102) The Mea High Tension Magneto. - 297 -

(103) The Wico High Tension Igniter. - 302 -

(104) Starting On Magneto Spark. - 306 -

CHAPTER IX CARBURETORS (105) Principles of Carburetion. - 309 -

(106) Schebler Carburetor. - 311 -

(107) Two Cycle Carburetors. — 313 —

(108) Kingston Carburetors. — 314 —

(109) The Feps Carburetor. — 316 —

(111) Gasoline Strainers. — 317 —

(112) Installing Gasoline Carburetors. — 318 —

(113) Installing the Carburetor. — 319 —

(114) Kerosene Vaporizer for Motorcycles. — 324 —

CHAPTER X LUBRICATION (116)
General Notes on Lubrication. — 326 —

(117) Force Feed Lubricating System. — 331 —

(118) Bosch Force Feed Oiler. — 332 —

(119) Castor Oil for Aero Engines. — 335 —

(120) Force Feed Troubles. — 337 —

(121) Oil Cup Failure. — 338 —

(122) Hot Bearings. — 339 —

(123) Cold Weather Lubrication. — 340 —

(124) Plug Oil Holes When Painting. — 342 —

(125) Oiling the Magneto. — 343 —

CHAPTER XI COOLING SYSTEMS — 344 —

(126) Cooling System Troubles. — 349 —

CHAPTER XII GOVERNORS AND VALVE GEAR (127)
Hit and Miss Governing. — 353 —

(128) The Throttling System. — 354 —

(129) The Controlling Governor. — 356 —

(130) Types of Governors. — 359 —

(131) Governor Troubles. — 360 —

(132) Throttling Governor Troubles. — 361 —

(133) Valve Gear Arrangement. - 362 -

(134) Cam Shaft Speeds. - 363 -

(135) Valve Gear Troubles. - 364 -

(136) Valve Timing. - 367 -

(137) Valve Setting on Stationary Engines. - 368 -

(138) High Speed Engine Valve Timing. - 369 -

(139) Timing Offset Cylinders. - 370 -

(140) Auxiliary Exhaust Ports. - 371 -

(141) Valves and Compression Leaks—Misfiring. - 372 -

CHAPTER XIII TRACTORS AND FARM POWER - 375 -

(142) The Gas Tractor. - 379 -

(143) Construction of Gas Tractors. - 384 -

(144) Fairbanks-Morse Oil Tractor. - 388 -

(145) The Rumely "Oil Pull" Tractor. - 391 -

(146) The "Big Four" Tractor. - 396 -

CHAPTER XIV THE STEAM TRACTOR (147) The Steam Tractor. - 402 -

(148) The Cylinder and Slide Valve. - 403 -

(149) Expansion of Steam. - 406 -

(150) Speed Regulation. - 407 -

(151) Reverse Gear. - 408 -

(152) Feed Pump. - 411 -

(153) The Boiler. - 412 -

(154) Oil-Burning Steam Tractors. - 413 -

(155) Care of the Steam Tractor. - 414 -

CHAPTER XV. OIL BURNERS. (156) Combustion. - 418 -

CHAPTER I
HEAT AND POWER
(1) The Heat Engine.

Heat engines, of which the steam engine and gas engine are the most prominent examples, are devices by which heat energy is transformed into mechanical power or motion. In all heat engines, this transformation of energy is accomplished by that property of heat known as "expansion," by which an increase or decrease of temperature causes a corresponding increase or decrease in volume of the material subjected to the varying temperatures. The substance whose expansion and contraction actuates the heat engine is known as the "working medium," and may be either a solid, liquid, or a gas. The extent to which the working medium is expanded depends not only upon the change of temperature but also on its composition.

In all practical heat engines, the heat energy is developed by the process of combustion, which is a chemical combination of the oxygen of the air with certain substances, such as coal or gasoline, known as "fuels." The heat producing elements of the fuels are generally compounds of carbon and hydrogen, which when oxydized or burnt by the oxygen form products that are unlike either of the original components. It is due to this chemical change that heat energy is evolved, for the heat represents the energy expended by the sun in building up the fuel in its original form, and as energy can neither be created nor destroyed, heat energy is liberated when the fuel is decomposed. The heat energy thus liberated is applied to the expansion of the working medium to obtain its equivalent in the form of mechanical power.

During the period of expansion, the heat obtained by the combustion is absorbed by the working medium in proportion to its increase in volume, and as this increase is proportional to the mechanical effort exerted by the engine, it will be seen that the output of the engine in work is a measure of the heat applied to the medium. The quantity of heat absorbed by the medium represents the energy required to set the molecules of the medium into their new positions in the greater volume, or to increase their paths of travel. In the conversion of heat, each heat unit applied to the medium results in the production of 778 foot pounds of energy, providing that there are no heat or frictional losses.

In explanation of these terms or units, we wish to say, that the unit of heat quantity, called the **BRITISH THERMAL UNIT** is the quantity of heat required to raise one pound of water, one degree Fahrenheit, and the **FOOT POUND** is the work required to raise one pound through the vertical distance of one foot. As the British Thermal Unit = 778 foot pounds it is equivalent to the work required to raise 778 pounds one foot or one pound 778 feet, or any other product of feet and pounds equal to the figure 778.

As liquids expand more than solids with a given temperature, and gases more than either, the mechanical work returned for a given amount of thermal energy (the **EFFICIENCY**) will be greater with an engine using gas as a working medium than one using a solid or liquid working medium. The steam engine and the gas engine are both examples of heat engines using gaseous working mediums, the medium in the steam engine being water vapor and in the gas engine, air and the gaseous products of combustion. For this reason the working medium will be considered as a gas in the succeeding chapters.

Practically the only way of obtaining mechanical effort from an expanding gas is to enclose it in a cylinder (c) fitted with a freely sliding plunger or piston (p) as shown in Fig. 1. Two positions of the piston are shown, one at M indicated by the dotted lines, and one at N indicated by the full lines. It will be assumed that the space between the cylinder head P and the piston at M represents the volume of the gas before it is heated and expanded, and that the volume between O and N represents the volume after heating and expansion have occurred. The vessel B represents a chamber containing air that is periodically heated by the lamp L, and which is connected to the working cylinder C by the pipe O.

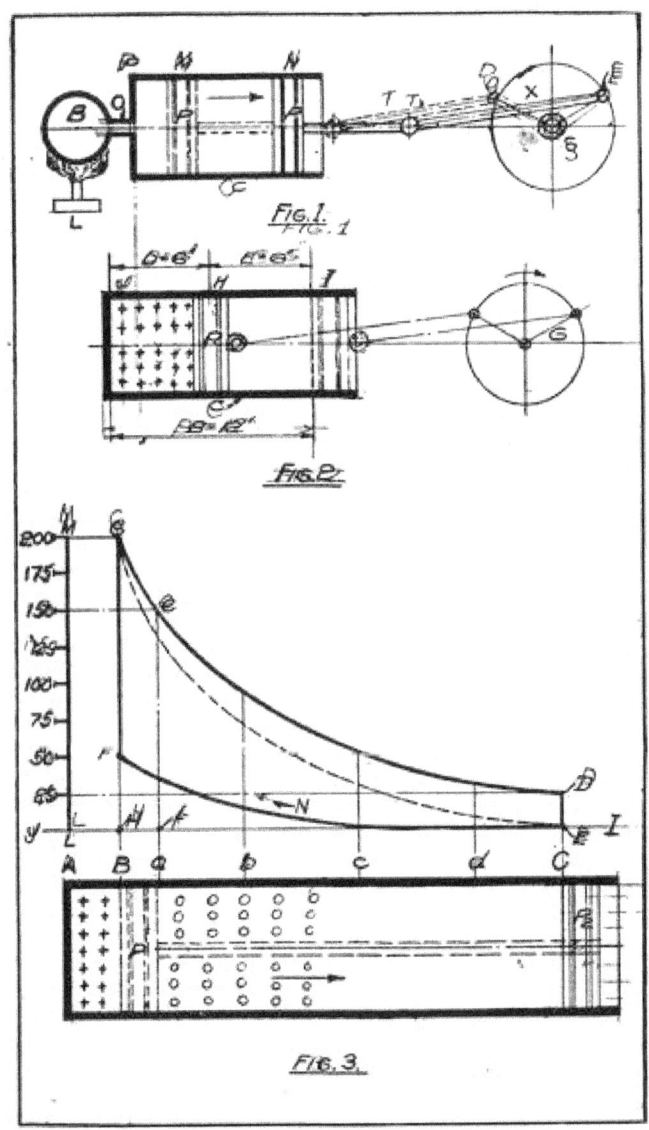

Figs. 1–2–3. Showing Expansion in an External Combustion Engine, the Cycle of Operations in an Internal Combustion Engine, and the Pressure Diagram of the Latter Engine Giving the Pressures at Various Points in the Stroke.

With the piston at M, the lamp L is lighted and placed under the retort B which results in the immediate expansion of the air in B. The expanded air passes through O into the cylinder, and if sufficient heat is supplied, exerts pressure against the piston since it occupies much more than its original volume. Providing that the friction of the device and the load on the shaft S are low enough the pressure on the piston will, move it to the position N in the direction of the arrow, thus accomplishing mechanical work. The motion of the piston revolves the crank to which it is connected by the rod X from D to E. During the trip from M to N the volume of gas has greatly increased being supplied continuously with heat from the lamp. As a considerable amount of heat has been radiated from the cylinder during the piston travel, and a considerable portion of the mechanical work lost through the friction of the piston on the cylinder walls, and by the crank, not all of the heat units are represented at the crank as mechanical effort.

Because of the limiting length of the cylinder, and the temperature limits of the lamp it is not possible to expand the working medium and increase the temperature indefinitely, therefore there must be a point where the application of heat must cease and the temperature be reduced in order to bring the gas back to its original volume and the piston to its original position so that the expansion may be repeated. This condition results in a very considerable loss of heat and power in addition to the losses previously mentioned, as the heat taken from the medium to reduce it to its original volume is thrown away as far as the production of power is concerned. To return the piston to its former position without expending energy on the engine, the volume and pressure may be reduced either by allowing the gas to escape to the atmosphere by means of a valve, or by removing the lamp and cooling the air by the application of water, but in any case the heat of the air is lost and the efficiency of the engine reduced.

To increase the efficiency of the engine and reduce the loss just mentioned, nearly all heat engines, either steam or gas, have the working medium at the highest temperature for only a small portion of the stroke, after which no heat is supplied to the cylinder. As the pressure forces the piston forward the volume increases, and as no more heat is supplied, both the pressure and the temperature continue to decrease until the end of the stroke is reached, thus utilizing the greater part of the heat in the expansion. Since the temperature at the end of the stroke is comparatively low, very little heat is rejected when the valve is opened for the return stroke. This loss would be the least when the temperature of the gas at the end of the stroke was equal to the temperature of the surrounding air. With both the internal and external temperatures equal, there would be no difference between the pressure of the gas in the cylinder and that of the surrounding air.

Fig. 1-a. Fairbanks-Morse Two Cylinder, Type "R E" Stationary Engine Direct Connected to a Dynamo.

It will be seen from the example just given that the heat engine performs mechanical work by dropping the working medium from a high to a low temperature, as it receives the medium at a high temperature from the lamp and rejects it at atmospheric temperature after delivering a small percentage of useful work. This may be compared to a water wheel which receives the working medium (water) at a high pressure and rejects it at a lower pressure. Carrying this comparison still further, it is evident that an increase in the range of the working temperatures (high and low) would increase the output of the heat engine in the same way that an increase in the range of pressures would increase the output of the water wheel. The temperature at which the engine receives the working medium and the temperature at which it is rejected determines the number of heat units that are available for conversion into mechanical energy, and therefore, if the range be increased by either raising the upper limit of temperature or by reducing the lower limit, or by the combined increase and decrease of the limits, the available heat will be increased.

Based on the temperature range, the maximum possible efficiency of the heat engine may be expressed by the ratio—

$$E = \frac{\text{Reception Temperature} - \text{Rejection Temperature}}{\text{Reception Temperature}}$$

Reception Temperature

This maximum defined by Carnot establishes a limit that can be exceeded by no engine, whatever the construction or working medium.

According to the methods adopted in applying the heat of combustion to the working medium, heat engines are divided into two general classes, (1) External combustion engines, (2) Internal combustion engines. The expressions "Internal" or "External" refer to the point at which combustion takes place in regard to the working cylinder, thus an internal combustion engine is one in which the combustion takes place in the working cylinder, and an external combustion engine is one in which the combustion takes place outside of the working cylinder. The steam engine is an example of an external combustion engine, the fuel being burned in the furnace of a boiler which is independent of the engine cylinder proper. As the fuel is burned directly in the cylinder of a gas engine it is commonly known as an internal combustion engine.

An external combustion engine, such as the steam engine is subject to many serious heat losses because of the indirect method by which the heat is supplied to the working cylinder, aside from the losses in the cylinder. Much of the heat goes up the smoke stack and much is radiated from the boiler settings and the steam pipes that lead to the engine. The greatest loss however is due to the fact that the range of temperatures in the working cylinder is very low compared to the temperatures attained in the boiler furnace, for it is practically impossible to have a greater range than 350°F to 100°F with a steam engine, while the furnace temperatures may run up to 2500°F and even beyond.

High temperatures with a steam engine result in the development of enormous pressures, a temperature of 547°F corresponding to an absolute pressure of 1000 pounds per square inch. This pressure would require an extremely heavy and inefficient engine because of the terrific strains set up in the moving parts. The pressures established by air as a working medium are very much lower than those produced by air or any permanent gas at the same temperature, and for this reason it is possible to exceed a working temperature of over 3000°F in the cylinder of a gas engine without meeting with excessive pressures. This high working temperature is one of the reasons of the extremely high efficiency of the gas engine.

In order to compete with the gas engine from the standpoint of efficiency, the steam engine builders have resorted to super-heating the steam after it has left the boiler in order to increase the temperature range in the cylinder. By applying additional heat to the steam after it has passed out of contact with the water it is possible to obtain up to 600°F without material increase in the pressure, but the practical gains have not been great enough to approach the gas engine with its 3000°F. After reaching his maximum temperature at this comparatively low pressure, the steam engineer has still to eliminate a number of other losses that do not obtain with the gas engine.

Since the radiation losses of a burning fuel are proportional to the time required for burning, it is evident that the rate of combustion has much to do with the efficient development of the heat contained in it, and it is true that rapid combustion develops more useful heat from a given fuel than slow. In the gas engine the combustion is practically instantaneous with a low radiation loss, but in the steam engine the rate is slow, and with the excess of air that must necessarily be supplied, a great part of the value of the fuel is lost before reaching the water in the boiler. The temperature of the medium determines the efficiency of the engine and as rapid combustion increases the temperature it is evident that the gas engine again has the best of the problem.

In the case of the gas engine where the fuel (in gaseous form) is drawn directly into the working cylinder in intimate contact with the working medium (air) and in the correct proportions for complete combustion, each particle of fuel, when ignited, applies its heat to the adjacent particle of air instantly and increases its volume with a minimum loss by radiation.

A gas engine is practically a steam engine with the furnace placed directly in the working cylinder with all intervening working mediums removed, the gases of combustion acting as the working medium. It derives its power from the instantaneous combustion of a mixture of fuel and air in the cylinder, the expansion of which causes pressure on the piston. Under the influence of the pressure on the piston, the crank is turned through the connecting rod and delivers power to the belt wheel where it is available for driving machinery. Whether the fuel be of solid, liquid, or gaseous origin it is always introduced into the cylinder in the form of a gas.

Fig. 1-b. The English Adams Automobile Motor (End View), Showing the Magneto Driven by Spiral Gears at Right Angles to the Crank-Shaft.

(3) Combustion In the Cylinder.

As the working medium in an internal combustion engine is in direct contact with the fuel it must not only be uninflammable but it must also be capable of sustaining combustion and must have a great expansion for a given temperature range. Since atmospheric air possesses all of these qualifications in addition to being present in all places in unlimited quantities it is natural that it should be used exclusively as the working medium for gas engines. Unlike the vapor working medium in a steam engine the medium in the gas engine not only acts in an expansive capacity but also as an oxydizing agent for burning the fuel, and therefore must bear a definite relation to the quantity of the fuel in the cylinder to insure complete combustion.

In the gas engine the use of gaseous fuel is imperative since there must be no solid residue existing in the cylinder after combustion and also for the reason that the fuel must be in a very finely subdivided state in order that the combustion shall proceed with the greatest possible rapidity. In addition to the above requirements the introduction of a solid fuel into the cylinder would involve almost unsurmountable mechanical problems in regard to fuel measurement for the varying loads on the engine. This limits the fuel to certain hydrocarbon or compounds of hydrogen and carbon in gaseous form of which the following are the most common examples:

(a) **CARBURETED AIR** consisting of a mixture of atmospheric air and the vapor of some hydrocarbon (liquid) such as gasoline, kerosene or alcohol.

(b) **OIL GAS** formed by the distillation of some heavy, nonvolatile oil, or the distillation of tar or paraffine.

(c) **NATURAL GAS** obtained from natural accumulations occurring in subterranean pockets in various parts of the country.

(d) **COAL GAS**, made artificially by the distillation of coal, commonly called "illuminating" gas.

(e) **PRODUCER GAS**, some times known as "fuel gas," produced by the incomplete combustion of coal in a form of furnace called a "producer."

(f) **BLAST FURNACE GAS**, the unconsumed gas from the furnaces used in smelting iron, somewhat similar to producer gas but lower in heat value.

It should be noted that there is no essential difference between engines using a permanent gas or an oil as in either case the fuel is sent into the cylinder in the form of a vapor. In the case of oil fuel, the vapor is formed by an appliance external to the engine proper. In this book, the heat action of an engine using one form of fuel applies equally to the engine using another. The selection of a particular fuel for use with a gas engine depends not only upon its value in producing heat, but also upon its cost, the ease with which it meets the peculiar conditions under which the engine is to work, and its accessibility.

Neglecting for the moment, all of the items that do not affect the operation of the engine from a power producing standpoint, the principal requirement of a fuel is the production of a high temperature in the cylinder since the output is directly proportional to the temperature range. Since a very considerable mass of air is to be raised to this high temperature, the heat value, or **CALORIFIC VALUE** of the fuel in British Thermal units is of as much importance as the temperature attained in the combustion. The calorific value of different fuels vary widely when based either on the cubic foot or pound, and a considerable variation exists even among fuels of the same class owing to the different methods of production or to the natural conditions existing at the mine or well from which they originated. The principal elements of gas engine fuels, carbon and hydrogen, exist in many different combinations and proportions, and require different quantities of air as oxygen for their combustion because of this difference in chemical structure.

Since complete combustion is never obtained under practical working conditions, the actual evolution of heat and the actual temperatures are always much lower than those indicated by the **CALORIMETER** or heat measuring device. Besides the loss of heat due to imperfect combustion, there are many other losses such as the loss by radiation, connection, and slow burning, the latter being the principal cause of low combustion temperatures. From the statements in the foregoing paragraphs it will be seen that the theoretical or absolute calorific value of a fuel is not always a true index to its efficiency in the engine.

Complete combustion results in the carbon of the fuel being reduced to carbon dioxide (CO_2) and the hydrogen to water (H_2O), with the liberation of atmospheric nitrogen that was previously combined with the fuel, and some oxygen. The reduction of the fuel to carbon dioxide and water produces every heat unit available since the latter compounds represent the lowest state to which the fuel can be burned. Carbon however may be burned to an intermediate state without the production of its entire calorific contents when there is not sufficient oxygen present to thoroughly consume the fuel. Incompletely consumed carbon produces a gas, carbon monoxide, as a product of combustion, and a quantity of solid carbon in a finely subdivided state known as "soot." Unlike the products of complete combustion, both the carbon monoxide and soot may be burned to a lower state with a production of additional heat when furnished with sufficient oxygen, both the soot and the monoxide being reduced to carbon dioxide during the process.

Fig. F-2. Sunbeam Engine with Six Cylinders Cast "En Bloc" (in one piece). At the Right and Under the Exhaust Pipe is the Compressed Air Starting Motor that Starts the Motor Through the Gear Teeth Shown on Flywheel. *From "Internal Combustion."*

As the soot and monoxide have a calorific value it is evident that much of the heat of the fuel is wasted if they are exhausted from the cylinder without further burning at the end of the stroke. To gain every possible heat unit it is necessary to furnish sufficient oxygen or air to reduce the fuel to its lowest state. As the free oxygen and nitrogen contained in the fuel are without fuel value, their rejection from the cylinder occasions no loss except for that heat which they take from the cylinder by virtue of their high temperature.

With complete combustion the **TEMPERATURE** attained increases with the rate of burning, while the number of heat units developed remain the same with any rate of combustion. Because of the conditions under which the fuel is burned in the gas engine the fuel is burned almost instantaneously with the result that high temperatures are reached with fuels of comparatively low calorific value. With a given gas the rate of combustion is increased with an increase in the temperature of the gas before ignition and remains constant for all mixtures of this gas in the same proportion when the initial temperature is the same. The rate of combustion also varies with the composition of the gas, hydrogen burning more rapidly than methane. As a rule it might be stated that the rate of burning decreases with the specific gravity of the gas, the light gases such as hydrogen burn with almost explosive rapidity, while the heavier gases such as carbon dioxide are incombustible or have a zero rate of combustion. In practice an increased rate of burning is obtained by heating the charge before ignition by a process that will be explained later.

Another factor governing the output of an engine with a given size cylinder is the amount of air required to burn the fuel. The quantity of air necessary for the combustion of the fuel determines the amount of fuel that can be drawn into a given cylinder volume, and as we are dependent upon the fuel for the expansion it is evident that with two fuels of the same calorific value, the one requiring the least air will develop the most power. Since the air required to burn hydrogen gas is only one fourth of that required to burn the same amount of methane it is clear that more hydrogen can be burned in the cylinder than methane. This great increase in output due to the hydrogen charge is however, considerably offset by the greater calorific value of the methane.

Should the air be in excess of that required for complete combustion, or should a great quantity of incombustible gas, such as nitrogen be present in the mixture, the fuel will be completely burned, but the speed of burning will be reduced owing to the dilution. As the air is increased beyond the proper proportions the explosions become weaker and weaker as the gas becomes leaner until the engine stops entirely. Because of the fact that it is impossible in practice to so thoroughly mix the gas and air that each particle of gas is in contact with a particle of air, the volume of air used for the combustion is much greater than that theoretically required. A **SLIGHT** excess of air, making a lean mixture, increases the efficiency of combustion although it reduces the temperature and pressure attained in the cylinder. This is due to the fact that while the temperature of the mixture is lower than with the theoretical mixture the temperature of the burning gas itself is much higher. A mixture that is too lean to burn at ordinary temperatures will respond readily to the ignition spark if the temperature or pressure is raised.

(4) Compression.

In the practical gas engine the gas is not ignited at the beginning of the suction stroke by which it is drawn into the cylinder, but is compressed in the front end of the cylinder by the return stroke of the piston, and then ignited. The process of compression adds greatly to the power output of a given sized cylinder and increases the efficiency of the fuel and expansion. In order to understand the relation that the compression bears to the expansion let us refer to Fig. 2 in which C is the working cylinder, P the piston and G the crank. While the piston is moving towards the crank in the direction of the arrow A it draws the mixture, indicated by the marks x x x x x, into the cylinder, the quantity being proportional to the position of the piston. In this particular case let us assume that the area of the piston is 50 square inches and that the entire stroke (B) of the piston is 12 inches. To prevent confusion due to considerations of heat loss we will further assume that the cylinder is constructed of non-conducting material.

With the piston at the position H, midway between J and I, the volume D is filled with the explosive mixture at atmospheric pressure and a temperature of 500° absolute. Since D = 6 inches and the area of the piston is 50 square inches, the volume D is equal to 6 × 50 = 300 cubic inches, and the entire volume is 2 × 300 = 600 cubic inches. On igniting this mixture (at atmospheric pressure) the temperature will rise immediately, say to 1000°F with the piston at H. According to a law governing the expansion of gases, known as Gay-Lussac's Law, the expansion $\frac{v \times T}{t}$ = V where v = the initial volume of the gas before ignition = 300 cubic inches; t = the temperature before ignition 500° absolute; V = the volume of the gas after expansion; and T = temperature after ignition = 1000° absolute. Inserting the values in numerical form we have as the final volume:—$\frac{300 \times 1000}{500}$ = 600 cubic inches = the volume after expansion, or twice the original volume of gas. This means that the expansion is capable of driving the piston from H to I before the pressure is reduced again to atmospheric pressure. As the volume is expanded to twice that of the original volume at atmospheric pressure (14.7 pounds per square inch), the pressure against the piston before it starts moving will be 2 × 14.7 = 29.4 pounds per square inch.

Let us now consider the case in which the charge is compressed before ignition occurs and compare the expansion and pressure established with that produced by ignition at atmospheric pressure. To produce the compression the piston will travel through the entire stroke to the position I on the suction stroke filling the entire cylinder volumes of 600 cubic inches with the mixture. On the return stroke the piston stops at H, reducing the original volume of 600 cubic inches to 300 cubic inches, doubling the pressure of the gas. The initial and final temperatures will be considered as being the same as those in the first example, 500° and 1000°. From Gay-Lussac's Law— $v \times T \over t = V$ and substituting the numerical values $600 \times 1000 \over 500$ = 1200 cubic inches, or the expanded volume will be four times the compressed volume, or four times the initial volume of the first case where the gas was ignited at atmospheric pressure.

It should be noted however, that while the expansion has been greatly increased by the compression, that this is not all gain, as equivalent work has been expended in compressing the charge. With the exception of doubling the fuel taken into the cylinder, and consequently doubling the output for a certain cylinder capacity, there has been no increase in fuel efficiency except that due to conditions other than the mere reduction in volume. In the second case the volume was increased four fold which resulted in a piston pressure of $4 \times 14.7 = 58.8$ pounds per square inch before the piston increased the volume by moving from H to I.

The work done by the engine on the charge in compressing is converted into heat energy causing a rise in the temperature of the gas. This would not be a loss as it would reappear as mechanical energy on the return stroke of the piston through its expanding effect on the gas. This heat would, in effect, be added to the temperature due to ignition, and the sum would produce its equivalent expansion. The temperature due to the combustion may be determined by reversing Gay-Lussac's Law—

$$\frac{t}{p} = \frac{I}{P} \text{ or } T = \frac{Pt}{p}$$

Where t = initial temperature; T = temperature combustion; P = pressure after combustion; p = pressure before combustion.

Because of the fact that the act of compressing the charge in the cylinder before ignition increases the temperature of the working medium, the compression will increase the speed of combustion and efficiency of the fuel as the rate of combustion increases with the initial temperature. This increased temperature due to initial compression of course results in a greater temperature range and output due to the increased rate of burning, and this rate of combustion may be varied for different fuels by changing the compression pressure. In a previous paragraph it was explained that the fuel efficiency was increased by a slight dilution or excess of air, and that while the temperature and pressure of the mixture were reduced by the dilution the temperature of the fuel was increased, provided that the inflammability was not decreased.

Compression affords a means of using dilute mixtures without loss of inflammability, as the heat gained by the compression restores the inflammability lost by the effects of dilution. Increased compression pressures increases the possible range of dilution, so that extremely lean gases and mixtures may be used with success with appropriately high compression. As an example of this fact we can refer to the engine using blast furnace gas, a fuel that is so lean that it cannot be ignited under atmospheric pressure. By increasing the piston speed, the heat of the compression can be made more effective as the gas lies in contact with the cylinder walls for a shorter time which of course reduces the heat to the jacket water.

(5) Efficiency and Heat Losses.

Up to the present time we have considered an engine in which there is no heat loss or loss from friction, but in the actual engine such losses are large and tend to materially reduce the values of heat and pressure to be obtained from a fuel with a given calorific content. Applying the rule for heat engines given in a previous section where the efficiency is—

$$E = \frac{T - t}{T}$$

We have the theoretical efficiency of a gas engine, neglecting friction, loss to the cylinder walls, and loss through the rejection of heat with the exhaust gas, equal to—

$$E = \frac{1960 - 520}{1960} = 73.5 \text{ percent.}$$

In substituting the numerical values in the above calculation it was assumed that the temperature of the burning mixture would be 1500° F above zero, and that the exhaust temperature would be as low as 60. Since the calculation is made from absolute zero, which is 460° below the zero marked on our thermometers, the temperature of the burning charge, $T = 1500 + 460° = 1960°$ above absolute zero. Similarly the absolute temperature of the exhaust would be, $t = 60 + 460 = 520°$ absolute. The application of the absolute temperatures will be seen from the calculation for efficiency. The value given, 73.5 per cent, it should be understood is the theoretical efficiency and is at least 20 per cent above the best results obtained in practice. The best record that we have had to date, is that established by a Diesel engine which returned 48.2 per cent of the calorific value of the fuel in the form of mechanical energy. In order that the reader may have some idea of the losses that occur in the engine, and their extent we submit the following table. These are the results of actual tests obtained from different sources and represent engines built for different services and of various capacities:

LOSSES—DATA	Automobile Motor	Stationary Engine	Stationary Engine
Horse-power	30.	200	1000

Heat lost to jacket water	35.8%	31.0%	2970 B.T.U.	
Heat lost in exhaust	24.6%	30.0%	2835 B.T.U.	Loss at per Horse-power in B.T.U.'s
Friction loss	8.6%	6.5%	810 B.T.U.	
Heat lost by radiation	15.4%	8.2	540 B.T.U.	
Heat available as power			2700 B.T.U.	
Efficiency (per cent)	15.6%	24.3		
Fuel	Gasoline		Producer Gas	

The remarkable efficiency of the Diesel engine is due principally to the extremely high compression pressure, which was from 500 to 600 pounds per square inch. When this is compared to the 60 to 70 pounds compression pressure used with automobile engines it is easy to see where the Diesel gains its efficiency. It is evident that as much depends on the manner in which the fuel is used in the engine as on the calorific value of the fuel.

(6) Expansion of the Charge.

When an explosive mixture is ignited in the cylinder with the piston fixed in one position thus making the volume constant, the increase of temperature is accompanied by an increase of pressure. If the piston is now allowed to move forward increasing the volume, the increase of volume decreases the pressure. Since in the operation of the gas engine the piston continuously expands the volume on the working stroke it is evident that there is no point in the stroke where the pressures are equal, and that the pressure is the least at the end of the stroke, it being understood of course that no additional heat is supplied to the medium after the piston begins its stroke.

This distribution of pressure in the cylinder in relation to the piston position is best represented graphically by means of a diagram as shown by Fig. 3, in which K is the cylinder and P the piston. Above the cylinder is shown the diagram HGDE the length of which (HE) is equal to the stroke of the piston shown by (BC). Intersecting the line HI are vertical lines, A, a, b, c, C, which represent certain positions of the piston in its stroke. The height of the diagram H G represents to scale the maximum explosion pressure in pounds per square inch, and the line HG is drawn immediately above the piston position B which is at the inner end of the stroke. To the left of the line HS is drawn a scale of pressures ML divided in pounds per square inch so that the pressures may be read off of the pressure curve GD. The line JI represents atmospheric pressure, and the divisions on ML, of course, begin from this line and increase as we go up the column. As an example in the use of the scale we find that the point F is at 50 pounds pressure above the atmospheric line JI.

We will consider that the clearance space AB is full of mixture at the point B, and that it is moved toward the left to the point C filling the space AC full of mixture at atmospheric pressure. The location of the piston on the diagram is shown by D and E. The opening through which the gas was supplied to the cylinder is now closed, and the piston starts on its compression stroke, moving from C to A. As the volume is reduced from AC to AB, there is an increase of pressure which is shown graphically by the rising line EF. This line rises gradually from the line JI in proportion to the reduction in volume until the piston reaches the end of the compression stroke at B, at which point the compression is at a maximum. The extent of this pressure is shown by the length of HF which on referring to the scale of pressure at the left will be found to be 50 pounds per square inch.

Ignition now occurs and the pressure increases instantly from the compression pressure at F to the maximum pressure at G which on referring to the scale will be found to equal 200 pounds. The actual increase of pressure due to ignition above the compression pressure will be shown by the length of the line FG which is equal to 150 pounds. As the pressure is now established against the piston it will begin to move forward with an increase of volume and a corresponding decrease in pressure, until it reaches the point C. This point at the end of the stroke is indicated on the diagram by D which by reference to the scale will be found equal to 25 pounds above atmosphere. An exhaust valve is now opened allowing the gas to escape to the atmosphere which reduces the pressure instantly from D to E on the atmospheric line. Expansion along the line GD is not complete as the pressure is not decreased to atmospheric pressure in the cylinder which means that there is a considerable loss of heat in the exhaust. In practice the expansion is never complete, but ends considerably above atmospheric pressure as shown.

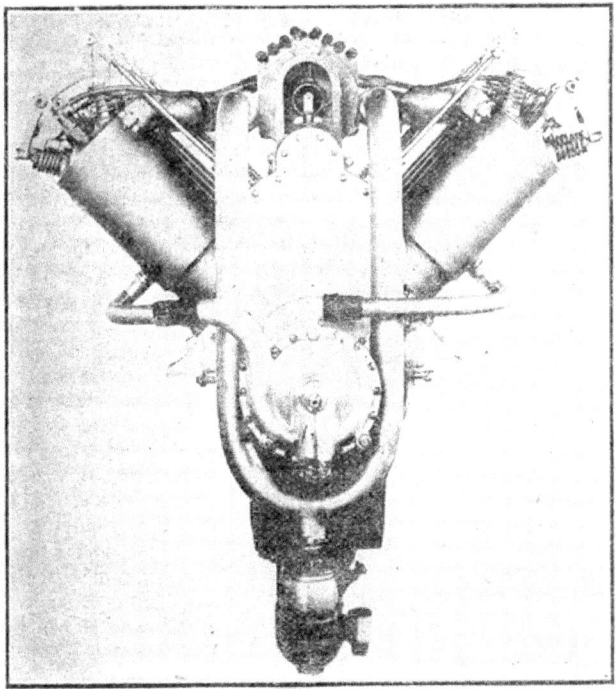

Fig. F-3. Front Elevation of Curtiss "V" Type Aeronautical Motor. This is the Front View of the Motor Shown in the Frontispiece. See Chapter V for Description of this Type of Motor.

Complete expansion is shown by the dotted line GE which terminates at E on the atmospheric line. By following the vertical lines up from the points a, b, c, and d, the pressures corresponding to these piston positions can be found by measuring the distance of the curve from the atmospheric line, on the given lines a, b, c or d. To find the pressure at the position a, for instance, follow upwards along the line a to the point c on the curve, the length of the line ef from the curve to the atmospheric represents the pressure, which by reference to the scale ML will be found equal to 125 pounds. The pressure at any other point can be found in a like manner. Compression pressures may be found at any point by measuring from the atmospheric line to the compression curve FE along the given line. It will be noted that the combustion is so quick that the pressure rises in a straight line along GH, indicating that combustion was complete before the piston had time to start on the outward stroke. The expansion curves GE and GD are similar to the compression curve FE. With the actual engine the shape of the ideal card as shown by Fig. 3 is sometimes considerably deformed owing to the effects of defective valves, leaks, or improperly timed ignition.

Pressure curves of actual engines are of the greatest value as they show the conditions within the cylinder at a glance and make it possible to detect losses due to leaks, poor valve settings, etc. These curves are traced by means of the **INDICATOR** which is an instrument consisting of a small cylinder which is connected to the cylinder of the engine, and an oscillating drum that is driven to and fro by the engine piston. The piston in the indicator cylinder is provided with a spring that governs its movements and communicates its motion to a recording pencil through a system of levers. The spring is of such strength that a pressure of so many pounds per square inch in the cylinder causes the pencil to draw a line of a definite length, this line being equivalent to the pressure line GH in Fig. 3. A piece of paper is wrapped about the indicator drum, and the drum is attached to the piston in such a manner that it turns a certain amount for every piston position, the complete stroke of the piston turning the drum through about three-quarters of a revolution. Rotation of the drum traces the horizontal lines of the diagram and the movement of the piston draws the vertical lines, so the combined movements of the drum and piston records the pressures and piston positions as shown by Fig. 3.

Since the movement of the indicator piston represents the pressures in the cylinder to scale it is possible to compute the power developed in the cylinder as the output in mechanical units is equal to the product of the average force acting on the piston multiplied by the speed of the piston in feet per minute. This product of the force and velocity (known as "foot pounds per minute") divided by 33,000 (one horse-power = 33,000 foot pounds) gives the output of the engine, in horse-power.

As the pressure on the piston fluctuates throughout the stroke, it would be wrong to consider the force, in the calculation for power as being equal to the explosion pressure, and so the effective pressure is taken as being the average of all the pressures from the point of explosion to the exhaust. The average pressure or "mean effective pressure" as it is called is computed from the indicator diagram by dividing it into a number of equal parts along the horizontal line, adding the lengths of the pressure lines such as CH, CF, etc., and dividing the total length by the number of the lines. After the average height of the diagram is thus determined, the average length is multiplied by the scale of the indicator or the pressure that is shown by it per inch.

Fairbanks-Morse Gasoline Pumping Engine. Pump is Gear Driven From the Engine Crank-Shaft at Reduced Speed.

Knowing the mean effective pressure, the total pressure on the piston, or the force is found by multiplying the area of the piston in square inches by the average pressure per square inch. This product is multiplied by the piston speed in feet per minute and is divided by the product of the number of strokes to the explosion and the quantity 33,000. Should there be more than one cylinder the result is multiplied by the number of cylinders, and this is multiplied by 2 in the case of a double acting engine. Stated as a formula this rule becomes:

$$\text{H.P.} = \frac{A \times P \times 2R \times L \times N \times O}{33000 \times C}$$

When A = Area of piston in square inches.

P = Average or mean effective pressure per square inch. About 75 pounds for Gasoline Engines. See Table on Page 31.

R = Revolutions per minute.

L = Stroke of piston in feet.

N = Number of cylinders.

O = 2 when engine is double acting, that is when explosions occur on both sides of the piston.

C = Number of strokes per explosion. C = 4 in a four cycle engine, and 2 in a two cycle.

It should be specially noted that the area of the piston is given in square inches and the stroke of the piston in feet. The number of revolutions per minute, R, is multiplied by two in order to obtain the number of strokes, as there are two strokes per revolution. When the engine governs its speed by dropping explosions to meet varying loads, the quantity C should be omitted and the explosions counted.

Due to the fact that the incoming charge of the mixture is expanded by the heat of the passages, a full charge computed at atmospheric temperature is never obtained in the cylinder and for this reason the gas should be kept as cold as possible before entering the passages in order to obtain the maximum output. Friction due to restricted passages and valve openings also reduces the amount of mixture available. Small exhaust valves and pipes prevent the gases from escaping freely to the atmosphere and produces a back pressure on the piston which cuts down the effective pressures. All of these items are recorded by the indicator and makes it possible to make alterations that will increase the output of the engine.

Because of the reduced atmospheric pressures at high altitudes the output and compression are reduced for every foot of elevation above sea level. As the weight of the atmosphere is reduced, less mixture is drawn into the cylinder. Taking the output of the engine as 100 per cent at sea level, it is reduced to less than 62 per cent at an elevation of 15,000 feet.

CHAPTER II
FUELS AND COMBUSTION
(7) Combustion.

The phenomenon called combustion by which we obtain the heat energy necessary for the operation of the internal combustion engine is a chemical combination of the air with the fuel. This process results in heat and some light which is equal in quantity to the energy required to separate the fuel compound into its elements or to build it up in its present form from the original elements. If the process is comparatively slow, the compound is called a fuel, if it is instantaneous it is called an explosive. Some substances produce mechanical force through an instant, without the evolution of much heat, due to the disintegration of an unstable compound. The effect of the latter type of which dynamite is an example is static, that is to say, it is not capable of producing power, but only pressure. For this reason, compounds having an instantaneous effect without the ability to produce the pressure through a distance, or an expansion, are not considered as suitable fuels for a heat engine.

A fuel is essentially a substance which is capable of generating heat, which is a form of energy, and not static pressure. The heat engine is an instrument which transforms this energy into power which is again dissipated into heat through the friction of the engine itself and by the load that it drives. This is an illustration of the physical law that "energy can neither be created nor destroyed," that is, the heat energy developed by the fuel is converted into mechanical energy which is again transformed into heat energy through friction.

It should be understood that fuel belongs to that class of substances that will not burn nor evolve energy under any temperature, pressure, or shock, without an outside supply of oxygen. This is the characteristic property of all fuels used with the internal combustion engine. Each element, such as carbon and hydrogen, in a compound fuel, develops a certain definite amount of heat during their complete combustion, and at the close of the process certain compounds are formed that represent the lowest chemical form of the compound. To restore the products of combustion to their original form as fuel would require an expenditure of energy equal to that given out in the combustion.

While all substances that are capable of oxydization or combustion can be made to liberate heat energy, it does not follow that all of them can be successfully used as fuels. A fuel suitable for the production of power must be cheap, accessible and of small bulk, and must burn rapidly. Such fuels must also be products of nature that require no expenditure of energy in their preparation or completion.

Fig. F-4. Fairbanks-Morse Producer Plant and Engine, Connected for Operation.

In practical work, the natural fuels are coal, mineral oils, natural gas, and wood, which are compounds of the elements carbon and hydrogen. When these fuels are burned to their lowest forms the products of combustion consist of carbon dioxide and water, the first being the result of the oxydization of carbon, and the latter a compound of oxygen and hydrogen. In solid fuels, such as coal, a portion of the compound consists of free carbon and the remainder of a compound of carbon and hydrogen known as a **HYDROCARBON**. In liquid fuels there is little, if any, free carbon, the greater proportion being in the form of a hydrocarbon compound. Natural gas is a hydrocarbon compound.

It should be noted that a definite amount of oxygen is required for the complete combustion of the fuel elements, and that a smaller amount of oxygen than that called for by the fuel element results in incomplete combustion, which produces a product of higher form than that produced by the complete reduction. The product of incomplete combustion represents a smaller evolution of heat than that of the complete process, but if reburned in a fresh supply of oxygen the sum of the second combustion together with that of the first will equal the heat of the complete oxydization. When pure carbon is incompletely burned the product is carbon monoxide (CO) instead of carbon dioxide (CO_2).

Carbon completely burned to carbon dioxide produces 14,500 British thermal units per pound of carbon, while the incomplete combustion to carbon monoxide evolves only 4,452 British thermal units, or less than one-third of the heat produced by the complete combustion. Theoretically one pound of carbon requires 2.66 pounds of oxygen to burn it to carbon dioxide. On supplying additional oxygen, the carbon monoxide may be burned to carbon dioxide and the remainder of the heat may be recovered, or 10,048 British thermal units. When a hydrocarbon, either solid, liquid or gaseous is burned with insufficient oxygen, solid carbon is precipitated together with lower hydrocarbons, and tar. In an internal combustion engine the precipitated solid carbon is evident in the form of smoke.

Since the carbon and hydrogen elements of a fuel exist in many different proportions and conditions in coal and oil, different amounts of oxygen are required for the consumption of different fuels. It should also be borne in mind that a greater quantity of air is required for the combustion of a fuel than oxygen, as the air is greatly diluted by an inert gas, nitrogen, which will not support combustion. Because of the impossibility of obtaining perfectly homogenous mixtures of air and the fuel, a greater quantity of air is used in practice than is theoretically required.

In a steam engine the fuel can be used in any form, solid, liquid, or gaseous, but in an internal combustion, it must be in the form of a gas no matter what may have been the form of the primary fuel. Fortunately there is no fuel which may not be transformed into a gas by some process if not already in a gaseous state. The petroleum products are vaporized by either the heat of the atmosphere or by spraying them on a hot surface. Coal is converted into a gas by distilling it in a retort or by incomplete combustion. The heat energy developed by a gas when burning in the open air depends on its chemical combustion, but its mechanical equivalent in power when burned in the cylinder of the engine depends not only upon its composition but upon the conditions under which it is burned as stated in the chapter devoted to the subject of heat engines.

(8) Gaseous Fuels.

While the calorific values of the different gases given in the accompanying table are approximately correct for gases burning in the open air at atmospheric pressure they develop widely different values in the cylinder of an engine because of the effects of compression and preheating. The table serves, however, as an index to the relative values of the fuels under ordinary conditions without compression. While natural gas has nearly eight times the calorific value of producer gas in the open air, its actual heat value in the cylinder is only about 45 per cent greater. While acetylene has an exceedingly high calorific value and explodes five times as fast as gasoline gas, it develops only 20 per cent more power in the same cylinder. Another item affecting the value of a gas is the rate at which it burns, which is in part a characteristic of the fuel and partly a factor of the conditions under which it is burnt. This subject is treated of in the chapter devoted to the heat engine.

The calorific value of a gas may either be computed from its chemical composition or by burning it in an instrument known as a **calorimeter**. A gas calorimeter consists of a small boiler or heating tank which is carefully covered with some non-conducting material so as to prevent a loss of heat to the atmosphere. The gas under test is burned in the boiler whose extended surface catches as much of the heat as possible and transfers it to the water in the boiler. The weight of the water heated and its temperature are taken when a certain amount of the gas has been burned (say 100 cubic feet), and from this data, the heat units per cubic foot of gas are computed.

FUEL GASES.

GAS	B.T.U. per Cubic Foot	Cubic Feet of Air Required to Burn 1 Cubic Foot of Gas Actual	Theoretical	Usual Compression Lbs. per Sq. Inch	Ratio of Gas to Air	Explosion Pressure in Lbs. per Sq. In.	Temperature of Combustion F°	Ignition Temperature F°	Weight per Cubic Foot, Lbs.	Candle Power	Mean Effective Pressure
Natural Gas	1000.	12.60	9	130	1–12.6	375		1100	.0459		94.00
Natural Gas	1000.			110	1–6	245		1000			72.00
Coal Gas	650.	9.00	5.85	80	1–9	285		1200	.035	18.00	85.00
Producer Anthracite	140.	1.20	1.85	160	1–1.2	360		1450	.065		88.00

Fuel										
Producer Bituminous	160.	3.20	2.20					1350		
Water Gas (Uncarb.)	290.	3.60	2.20						.044	
Water Gas (Carb.)	500.	8.50	5.15		1.8					22.00
Blast Furnace Gas	94.	1.10	0.70	170				1560	.080	77.5
Acetylene	1500.	20.00	12.60							
Gasolene Vapor	520.			70	1.12	245	1865	1550		79.00
Gasolene Vapor	520.		See Table of Liquid Fuels	70	1.8	360	2950	925		82.00
Gasolene Vapor	520.			70	1.6	410	3160	910		84.50
Kerosene Vapor				60	1.8	285		945		85.00
Coke Oven Gas	520.	7.	5.4						.042	
Alcohol				180		450				

The values given above are approximate, and vary not only with the engine used but also with the method used in producing the gas, and with the character of the fuel used. The figures will give an idea of the relative value of the gases in a rough way.

As a British thermal unit is the amount of heat required to raise the temperature of one pound of water through one Fahrenheit degree (at about 39.1° F.), the total heat per cubic foot of gas as observed by the calorimeter is equal to the weight of the water multiplied by its rise in temperature in degrees, divided by the number of cubic feet of gas burned in the calorimeter. Since a British thermal unit is equal to 778 foot pounds in mechanical energy, its mechanical equivalent is equal to the number of British thermal units multiplied by 778.

Another difference between the actual and theoretical results obtained is that due the perfect combustion in the calorimeter and the imperfect combustion in the engine. Since some gases require more air for their combustion than others, less of the first gas will be taken into the cylinder on a charge than the latter, which tends still further to balance the heating effect of rich and lean gases in the cylinder.

(9) Gasifying Coal.

Coal Gas or **Illuminating Gas** is generated by baking the coal in a closed retort or chamber out of contact with the air so that no combustion takes place either complete or incomplete. The hydrocarbon gases and tars are set free from the coal as permanent gases and are then piped to a gas holder after going through various purifying processes to remove the tars, oils, moisture and dust. The free or solid part of the coal remains in the retort in the form of **coke**, which is again burned for fuel.

Because of its high carbon content, coal gas burns with a yellowish-white flame and is extensively used for lighting purposes, hence the name **illuminating gas**. In many ways coal gas is an ideal fuel for power purposes as it has a high calorific value (650–750 B.T.U. per cubic ft.), is supplied by the illuminating company at practically a constant pressure, and is uniform in quality. Its only drawback is its comparatively high cost.

This gas is always obtained from the city service mains as its preparation is too expensive and complicated for the gas engine owner. Because of its cost, the use of coal gas is restricted to small engines.

(10) Water Gas.

Water gas is made by blowing air through a thick bed of some coal that is low in hydrocarbons until the coal becomes incandescent, the gases that are formed are allowed to escape to the atmosphere. At this point a jet of steam is blown into the incandescent bed, which is broken up into its elements, oxygen and hydrogen, by the heat of the fuel. As there is no air present the oxygen combines with the carbon of the fuel to form carbon monoxide while the hydrogen goes free. Both of these gases, carbon monoxide and hydrogen, are collected and supplied to the engine. The production of water gas is intermittent, as the steam blast cools down the fuel bed, and requires further blowing before more steam can be passed. While this gas has a lower heating value than coal gas, it is much cheaper to make and all of the coal is consumed in the process.

Water gas is high in hydrogen and is too "snappy" for gas engines; the hydrogen places a limit on the allowable compression.

For each thousand feet of **water gas** generated, approximately 24 pounds of water are required.

By the introduction of hydrocarbons or vaporized oil, illuminating value is given to water gas, this process is called **carburetion**. Carbureted gas is not usually used for power, as it is expensive, and is not proportionately high in heating value.

(11) Blast Furnace Gas.

Many steel companies are utilizing the unconsumed gas of the blast furnaces for power.

Blast furnace gas is of very low calorific value, rarely if ever, exceeding 85 B.T.U. per cubic foot. This allows of very high compression, which greatly increases the actual power delivered by the engine.

A smelter produces approximately 88,000 cubic feet of gas per ton of iron smelted.

Blast furnace gas is so lean that it cannot be burned satisfactorily under a boiler; the high compression of the gas engine makes its use possible.

(12) Producer Gas.

Producer gas which is generated by the incomplete combustion of fuels in a deep bed is the most commonly used gas for engines having a capacity of 50 horsepower and over, because of the simplicity and economy of its production. While producer gas has been obtained from practically every solid fuel, of which coal, coke, wood, lignite, peat, and charcoal are examples, the fuel most generally used is either coal or coke. While producer gas is much lower in calorific value than either natural or illuminating gas it gives admirable results in the gas engine and is a much cheaper fuel than coal gas in units above 50 horse-power capacity. The fuel is completely burned to ash in the producer without the intermediate coke product that exists in the manufacture of coke.

A producer consists of three independent elements as shown by Fig. F-6; the **PRODUCER** or generator (A), the steam boiler (B), and the **SCRUBBER** or purifier (C). The incandescent fuel (F) in the form of a cone lies on the grate bars (G) at the lower end of the producer. Above the burning fuel is a deep bed of coal (D) which reaches to the top of the producer at which point it is admitted to the bed through the charging valve or gate (H). The gas resulting from the combustion in the producer is drawn out of the tank through the gas outlet pipe (E) by the suction of the engine. The air for the combustion is drawn up through an opening in the ash pit (J) by the engine.

When the oxygen of the air strikes the incandescent fuel on the grate it combines with a portion of it forming carbon dioxide (CO_2) which is an incombustible gas, but on passing through the burning fuel above this point, one atom of the oxygen in the CO_2 recombines with the fuel forming the combustible gas—carbon monoxide (CO). Because of the distilling effect of the heat in the bed, the volatile hydrocarbons of the coal are set free and mingle with the CO formed by the combustion. The producer gas consists, therefore, principally of CO, with a certain proportion of the volatile hydrocarbons of the coal such as marsh gas, ethylene, and some oil vapor.

Since the hydrocarbons are easily condensed on coming into contact with the coal walls of the piping, to form trouble making tars and oils, they must either be washed out of the gas in the purifier or passed again through the high temperature zone to convert them into permanent gases. In the usual producer, the hydrocarbons are reheated, as they form a considerable percentage of the heat of the gas. After the volatile constituents are reheated, the gases pass through the boiler (B), which absorbs the heat of the gas in generating steam, and from this point the gases enter the scrubber where the dust and the residual tars are removed. The scrubber, which is a sort of filter, is an important factor in the generating plant, for if the dust and dirt were allowed to pass into the cylinder of the engine it would only be a question of a short time until the valves and cylinder would be ground to pieces.

When the steam from the boiler is allowed to flow into the ash pit of the producer and up through the incandescent fuel, the heat separates the water vapor into its two elements, oxygen and hydrogen. The oxygen set free combines with the carbon in the coal forming more carbon monoxide, while the hydrogen which is unaffected by the combustion adds to the heat value of the gas. The last additions to the combustion due to the disassociation of the steam are really what is known as "water gas." A limited amount of steam may be admitted continuously in this manner without lowering the temperature of the fuel below the gasifying point, and its presence is beneficial for it not only provides more CO and hydrogen but produces it without introducing atmospheric nitrogen. The steam is also a great aid in preventing the formation of clinkers on the grate bars. Since the air used in burning the fuel in the first reaction contains about 79 per cent of nitrogen, which is an inert gas, the producer gas is greatly diluted by this unavoidable admixture, which accounts for its low calorific value.

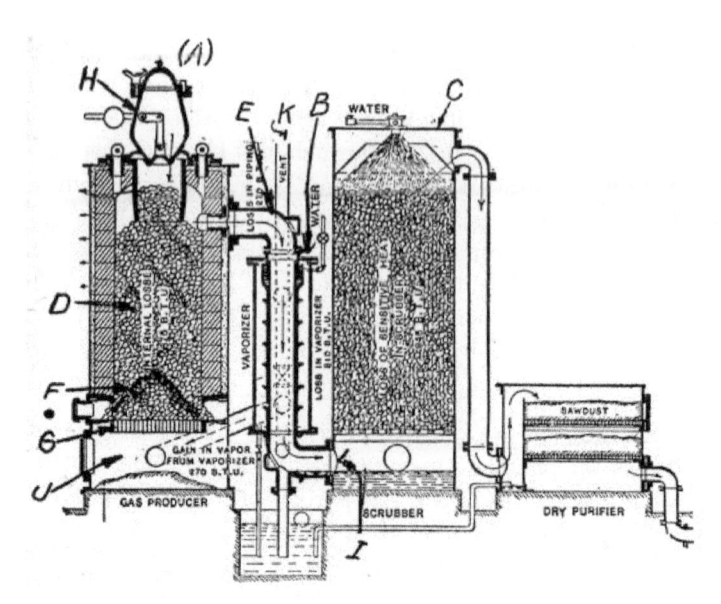

Fig. F-6. Diagram of Suction Gas Producer Showing the Generator, Boiler and Washer.

While the air required for the combustion of the fuel is drawn through the producer by the suction of the engine in the example shown (**SUCTION PRODUCER**), there is a type in common use called a **PRESSURE PRODUCER** in which the air is supplied under pressure to the ash pit by a small blower, which causes a continuous flow of gas above atmospheric pressure.

Gas producers are divided into two classes: suction producers and pressure producers. The suction producer presents the following advantages:

1. The pipe line is always less than atmospheric pressure, hence no leaks of gas to the air are possible.

2. The regulation of the gas supply is automatic.

3. No gas storage tank is required.

4. The production of gas begins and stops with the engine.

5. Uniform quality of gas.

The suction producer is limited to power application and cannot be used where the gas is to be used for heating, as in furnaces, ovens, etc., or where the engine is at a distance from the producer, unless pumped to its destination.

The pressure producer does not yield a uniform quality of gas, hence requires a storage tank where low quality gas will blend with gas of higher calorific values and produce a gas of fairly uniform quality.

The pressure producer is adapted to the use of all grades of fuels, such as bituminous coal and lignite.

Anthracite coal contains little volatile matter and is an ideal fuel for the manufacture of producer gas, while bituminous coal with its high percentage of volatile matter and tar, requires more efficient scrubbing, as these substances must be removed from the gas.

On starting the producer shown by Fig. 6, the producer is filled with the proper amount of kindling and coal, and a blast of air is sent into the ash pit by a small blower, the products of combustion being sent through the by-pass stack (K) until the escaping gas becomes of the quality required for the operation of the engine. The by-pass valve is now closed, and the gas is forced through the scrubber to the engine until the entire system is filled with gas. When good gas appears at the engine test cock the engine is started, and the blower stopped, the gas now being circulated by the engine piston. The volume of gas generated by the producer is always equal to that required by the engine so that no gas receiver or reservoir is required. Because of the friction of the gas in passing through the fuel, scrubber and piping, its pressure at the engine is always considerably below that of the atmosphere, which of course reduces the amount of charge taken into the cylinder. Because of the weak gas and the low pressure in the piping, it is necessary to carry a much higher compression with producer gas than with natural or illuminating gas.

The efficiency of a producer is from 75 to 85 per cent, that is, the producer will furnish gas that has a calorific value of an average of 80 per cent of the calorific value of the fuel from which it is made, the remaining 15 to 20 per cent being consumed in performing the combustion. This is far above the efficiency of the furnace in a steam boiler, as an almost theoretically exact amount of air can be supplied in the producer to effect the combustion, while in the boiler furnace about ten times the theoretical amount is passed through the fuel bed to burn it. Heating up this enormous volume of air to the temperature of the products of combustion consumes a large amount of fuel and reduces the efficiency of the furnace considerably. Because of the reduction in the air supply, a gas fired furnace is always more efficient than one fired with coal. Producer gas with 300,000 British thermal units per thousand cubic feet, and oil having 130,000 British thermal units per gallon will result in 1,000 cubic feet of gas being equal to about 2.30 gallons of fuel oil.

If the gas is to be used for heating ovens or furnaces in connection with the generation of power, the character of the fuel will be determined to a great extent by the requirements of the ovens and by the type of producer used, as each fuel will give the gas certain properties. Thus gas used for firing crockery will not be suitable for use in open hearth steel furnaces, as the impurities in the various fuels may have an injurious effect on the manufactured product. The cost of the fuel, cost of transportation, heat value, purity, and ease of handling are all factors in the selection of a fuel.

The size and condition of a fuel is also of importance. Exceedingly large lumps and fine dust are both objectionable.

Wet fuel reduces the efficiency of the producer, as the water must be evaporated, this causing a serious heat loss.

With careful attention a producer gas engine will develop a horse-power hour on from 1 to 1¼ pounds of anthracite pea coal, and in many instances the consumption has been less than this figure. The efficiency in dropping from full load to half load varies by little, one test showing a consumption of 1.1 pounds of coal per horse-power hour at full load and 1.6 pounds of coal at half load. Producer gas power is nearly as cheap as water power, in fact the producer gas engine has displaced at least two water plants to the writer's knowledge. According to an estimate made by a well known authority, Mr. Bingham, it is possible for a producer gas engine to generate power for only .1 of one cent more per K.W. hour than it is generated at Niagara Falls.

According to the United States Bureau of Mines,

"The tests in the gas producer have shown that many fuels of so low grade as to be practically valueless for steaming purposes, such as slack coal, bone coal and lignite, may be economically converted into producer gas and may thus generate sufficient power to render them of high commercial value.

"It is estimated that on an average each coal tested in the producer-gas plant developed two and one-half times the power that it would develop in the ordinary steam-boiler plant.

"It was found that the low-grade lignite of North Dakota developed as much power when converted into producer gas as did the best West Virginia bituminous coals burned under the steam boiler.

"Investigations into the waste of coal in mining have shown that it probably aggregates 250,000,000 to 300,000,000 tons yearly, of which at least one-half might be saved. It has been demonstrated that the low-grade coals, high in sulphur and ash, now left underground, can be used economically in the gas producer for the ultimate production of power, heat and light, and should, therefore, be mined at the same time as the high-grade coal.

"As a smoke preventer, the gas producer is one of the most efficient devices on the market, and furthermore, it reduces the fuel consumption not 10 to 15 per cent, as claimed for the ordinary smoke preventing device offered for use in steam plants, but 50 to 60 per cent."

(13) Producer Gas From Peat.

The production of gas from peat having a low water content (up to about 20 per cent) for use in suction gas engines has already met with considerable success in Germany, but for a number of years efforts have been made to utilize peat with a water content as high as 50 to 60 per cent and thus eliminate the costly process of drying the raw material.

Difficulties have been encountered in preventing a loss of heat through radiation and other causes, and in getting rid of the dust and tar vapors carried over by the gases to the engine; but great strides have been made recently in overcoming these obstacles. Peat with a water content up to 60 per cent has been found to be a suitable fuel. Owing to its great porosity and low specific gravity it presents a large combustion surface in the generator, so that the oxygen in the air used as a draft can easily unite with the carbon of the peat.

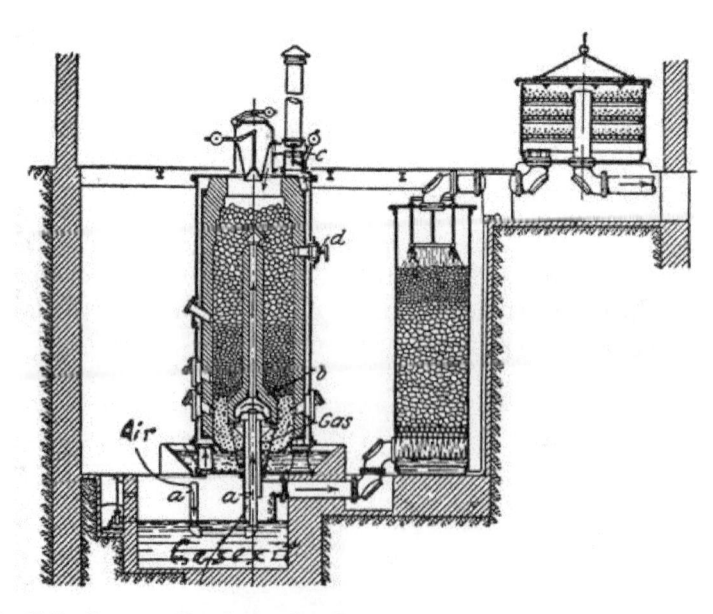

Fig. F-7. German Producer for Generating Producer Gas from Peat.

One of the great difficulties is to eliminate the tar vapors that clog up many of the working parts of the engine. The passing of the gas through the wet coke washers and dry sawdust cleansers does not appear to have thoroughly remedied the evil. Efforts were therefore made to remove the tar-forming particles of the gas in the generator itself or to render them harmless. That of the Aktien-Gesellschaft Gorlitzer Maschinenbau Ansalt und

Eissengiesserei of Gorlitz, was displayed at the exposition at Posen in 1911. The gas from the generating plant was employed in a gas suction engine of 300 horse-power used to drive a dynamo for developing the electric energy for the exposition. The fuel used was peat with a water content of about 40 per cent. The efficiency and economy results obtained were very promising.

The advantages claimed for the Gorlitz engine are that the sulphurous gases and those containing great quantities of tar products are drawn down by the suction of the engine through burning masses of peat and thus rid of their deleterious constituents. The air for the combustion purposes is well heated before entering the combustion chamber, thereby producing economical results. It is claimed also that the gas produced by its system is so free from impurities that the cleaning and drying apparatus may be of the simplest kind.

In *Stahl und Eisen*, an abstract is given of a paper by Carl Heinz describing a peat gas producer, built by the Goerlitzer Maschinenbauanstalt. We are indebted to *Metallurgical and Chemical Engineering* for the translation of this paper:

Air and fuel enter the producer at the top, and the gas exit is in the center of the bottom so that the air is forced to pass through the center of the producer, decomposing the volatile matter into gases of calorific value. The moisture which is present in the peat fuel in considerable quantities must be taken into consideration. For its decomposition which passing through the hot-fire zone only a certain amount of heat is available. It is, therefore, important that the heat from the gasification be fully utilized.

There are two kinds of heat losses in a gas producer, due to radiation and to the sensible heat of escaping gases. Both these amounts of heat, however, are utilized according to the special design of this producer. The air circulates first through the lower conduit and comes so in contact with the warm scrubber water. A part of the air which has been preheated is carried upwards through the pipe **A** in the center of the producer where it is thoroughly preheated by the hot gases and enters then the air superheater **B** in which the temperature rises to a still higher degree.

The other part of the air passes through the feet of the producer into an air jacket which envelops the whole shell of the producer and enters finally the producer by the reversing valve **C** on top of the producer. In this way the outer surface of the producer is maintained at a temperature hardly higher than that of the surrounding air. The escaping gases are cooled down so far that the gas outlet into the scrubber may be touched by hand. All ordinary heat losses are thus made use of in the gasification process.

If there is a large excess of moisture in peat, the process is somewhat modified by regulating both air supplies in such a way that the gasification in the upper part of the fuel-bed takes place in two directions, one downwards and the other upwards.

It seems that a content of 80 per cent moisture and 20 per cent dry fuel in the peat is about the limit permitting evaporation of the water, but it is, of course, impossible to obtain in this case a gas of calorific value.

The modification of the process for very wet fuel is as follows:

When the fire on top of the fuel bed appears to disappear, the heater opens the stack and valve **D**. Valve **C** is then closed, to prevent air from entering on top. The preheated air enters by **D** causing a down draft combustion due to the suction of the gas engine and an upward combustion due to the draft in the stack. The moisture is evaporated and escapes through the stack. When the fire has burned through at the top, the valve is switched over. The bad smelling gases rising from the scrubber enter the producer together with air and are there consumed.

In commercial use at the exhibition in Posen the whole plant worked continuously day and night and cleaning of the gas engines was necessary only every three months. Slagging of ashes is done during the operation of the producer, without any nuisance from dust.

The highest percentage of moisture in peat gasified was 50 per cent. The fuel consumption per horse-power hour is 2.2 lb. (1 kg.) of peat. Careful tests made by Prof. Baer, of Breslau, showed that with a cost of peat of $1 per ton the kw-hour at the switchboard costs 0.15 cent.

(14) Crude Oil Producers.

The development of the crude oil gas producer, for which there is great demand, in oil regions remote from the coal field, has been exceedingly slow but it is believed that definite progress has recently been made along this line. The most recent notes on this subject relate to the Grine oil producer. In this type steam spray is used for atomizing the oil which is introduced into the upper part of the generator where partial combustion takes place. The downdraft principle is then applied and the hydrocarbon broken up and the tar fixed by passing through a bed of incandescent coke. Mr. Grine reports that a power plant using one of these producers has been in operation a year in California. With crude oil as a fuel costing 95 cents per barrel, or 2.3 cents per gallon, the plant is reported to develop the same amount of power per gallon of crude as is ordinarily developed by the standard internal combustion engine operating on distillates at 7 cents per gallon. Including the cost of fuel, labor, supplies, interest, depreciation and taxes, Mr. Grine states the cost per b.h.p. hour to be 0.76 cents for a plant of 100 h.p. rating.

(15) Operation of Producers.

A good producer operator is simply a good fireman, he must know how to keep a uniform bed of coal and how to draw the fire. While there are many thousands of men running producer plants without previous mechanical training, there are now but few steam engineers running steam engines of the same capacity but what have had at least two years' training and sufficient mechanical knowledge to pass an examination and obtain a license. While a considerable amount of skill is necessary to obtain the best efficiency from a producer, it is a knack that is easily acquired in a short time by "sticking around" the plant. Skill in operating a producer consists chiefly in keeping the right sort of a fire without damage to the lining by poking down ashes and clinkers. When a new plant is installed, the manufacturer generally sends an instructor to operate the plant for a short time so that with a few days running in his hands any man with ordinary intelligence can overcome the difficulties which arise from time to time.

While there are many types of producers, the main difference will be found in the character of the draft, that is whether it is up, down, or crossways. Down draft producers are generally used with bituminous coals, as the tars and oils that emanate from the coal are drawn through the fire which converts them into a permanent gas, and avoids the difficulty of removing great quantities of the tar from the producer. An up draft producer will not do this as the gas is drawn directly into the mains without coming into contact with the fire. This would result in considerable expense due to the frequent cleaning. Anthracite coal which does not contain much tar can be used successfully in an up draft producer.

A compromise between the up draft and down draft producer is had in the **DOUBLE ZONE** producer, which "burns the candle at both ends" as it were, a fire being at both the top and bottom of the producer. Nearly any class of fuel may be used with this type.

It should be remembered that a hot fire and fuel are required for the manufacture of gas, and that the ash pit and grate must be kept clear of the ashes and clinkers that not only reduce the temperature of the fire, but also reduce the gas available at the cylinder by increasing the friction. Shaking down and cleaning out will in nearly every instance start a bucking producer into operation.

When operating under full load a much hotter fire is required than when operating under a reduced load, or the producer will not furnish the necessary gas. According to the size of the producer, the depth of the incandescent fuel will run from 30 inches in the large sizes to 15 inches in the smaller. After being charged up, suction producers will continue to give gas in sufficient quantities with the bed at half this depth. This is only possible with a hot producer, and when no fuel is being fed, as the feeding of a cold charge will reduce the output. A steady depth of fire should be kept to maintain a uniform quality of gas.

In suction producers careful watch should be kept for leaks, as the gas being below atmospheric pressure gives no outward signs of dilution. If water seals are used in the system they should be given careful attention. When using coals that are rich in tar or hydrocarbons, or with fuels that have much fine dust, considerable trouble is had with some types of producers due to "caking" or to the adhesion of the coal particles to the walls of the producers or to their adhesion to one another. In the latter case the "stickiness" of the fuel prevent the proper feed. This difficulty may often be overcome by a change in the rate of feeding or by regulating the depth of the incandescent bed.

Porosity of the fuel, and the rate at which the air is supplied to the producer determines the depth of the incandescent bed. Particular care should be taken that the blast or draft occurs evenly over the fire surface, and that no holes occur in the fire which will cause more rapid combustion in one spot than in another. Neglect of this precaution not only causes a waste of fuel but often results in the fuel "arching" and preventing further feed. The producer should be so proportioned that at full load, the rate of combustion does not exceed 24 pounds of fuel per square foot of producer area per hour.

In his researches, Professor Bone (Iron and Steel Institute, May, 1907) has shown that up to 0.32 lbs. of steam per lb. of coal can be completely decomposed in a producer, but that, from 0.45 lbs. to 0.55 lbs. should be used, approximately 80% more.

Now, in considering the question of the proper proportion of steam for the production of gas for power purposes we must bear in mind that as much heat as possible should be utilized in the producer itself. Some manufacturers of plant go so far as to state that as much as 1 lb. of steam per lb. of coal should be used, but we are safe in saying that 0.5 lb. to 0.7 lb. should be the figure for a power plant. The common practice is to use a blast saturation of 55% whenever the clinkering character of the coal renders it possible. This figure corresponds to about .57 of steam per lb. of coal gasified.

It is of the utmost importance that the proportion of steam and air should be constant, and the best figure being determined, it should not be varied to any degree. It is equally important that the fuel depth should be left constant. By this I mean that not only should the coal in the producer be kept at a specific level, but the position of the fire on the ash bed should be kept as near as possible a fixed point. Ashes should be drawn at regular intervals, or, if desired, continuously by mechanical means.

Further, the supply of air and steam should be regularly distributed, so that the velocity of the gases through the fuel shall be as nearly as possible regular across its whole area.

In some cases the by-products of a producer, such as ammonia, tar, etc., have a commercial value, and if a large amount of gas is generated it will sometimes pay to select a fuel that is rich in these particular substances.

(16) Coal.

Coal which is the basis of producer gas, is composed generally speaking of the combustible matter, moisture, ash and sulphur. The combustible element may be subdivided into the **HYDROCARBONS, OR VOLATILES**, and the solid fixed carbon. The exact composition of coal is generally given by what is known as **PROXIMATE** analysis, which analysis divides the constituents of the coal into five groups, viz.: **MOISTURE, VOLATILES, FIXED CARBON, ASH,** and **SULPHUR**. Ultimate analysis resolves the coal into its ultimate chemical elements, such as hydrogen, carbon, nitrogen, sulphur, etc., and being a difficult and tedious process it is not much used.

The proximate analysis gives all the necessary information and takes less time to perform.

VALUES OF COAL

Location of Mine	Moisture	Volatile Matter	Fixed Carbon	Ash	Sulphur	Calorific Value in B.T.U. per Lb. of Coal
ANTHRACITE						
Northern Pa.	3.39	4.41	83.30	8.17	.73	13,200
Eastern Pa.	3.70	3.07	86.42	6.18	.63	13,440
Western Pa.	3.12	3.76	81.60	10.61	.53	12,875
SEMI-ANTHRACITE	1.25	8.15	83.30	6.27	1.63	13,900
SEMI-BITUMINOUS						
Pennsylvania	.80	15.60	77.40	5.35	.85	14,900
Pennsylvania	1.55	16.45	71.50	8.63	1.87	14,200
Pocahontas Va.	1.00	21.00	24.40	3.02	.58	15,100
West Virginia	.90	17.83	77.70	3.30	.27	15,230
BITUMINOUS						

Youghiogheny Pa.	1.00	36.50	59.00	2.59	.86	14,400
Sample No. 2	1.20	30.18	59.00	8.84	.78	14,400
Hocking Valley	6.5	35.06	48.80	8.05	1.59	12,100
Kentucky	4.00	34.00	54.70	7.00	.03	12,800
Indiana	8.00	30.20	54.20	7.60		12,500
Illinois	10.50	36.15	37.00	12.90	3.45	10,500
Colorado	6.00	38.01	47.90	8.09		12,200
LIGNITE	9.00	42.26	44.30	3.27	1.18	11,000

The **CALORIFIC VALUE** of a fuel may be calculated from its analysis, or may be determined by means of the **CALORIMETER** from a sample of the coal; the latter method is the most reliable. Table gives approximately the calorific values, and the proximate analysis of several representative coals from various sections of the country. The values given in the table are not exact, as the coal from each locality varies considerably in quality, but the figures will indicate what may be expected from each type of coal.

Connellsville, Pa., Coke has a calorific value of approximately 13,000 B.T.U.'s per pound, contains no volatile matter, and has an approximate content of 10% ash. Coke is a valuable fuel for the gas producer, but is rather expensive. It is clean and the absence of volatile matter reduces the "scrubbing" problem to a minimum.

Small coal such as buckwheat and pea contain a much higher percentage of moisture than given in the table, running from 5% to 10% higher than the given values.

Bituminous coal is high in hydrocarbons or volatiles which condense easily and form tar. If the tar is not removed or converted into a permanent gas, it will clog the passages of the producer and the engine and cause trouble.

The removal of the tar and ash from a gas is called **SCRUBBING**, and is performed by a device much resembling a filter. Anthracite coal and coke are low in volatiles or hydrocarbons, and therefore do not cause trouble with tar deposits.

A high percentage of volatile matter also causes trouble by the tar cementing the particles of fuel together. This interferes with the proper action of the producer.

Fuels having a high percentage of ash call for perfect filtering or "scrubbing" as such fuels will fill the gas passages with dust. Dust should be kept out of the engine at all costs, for the dust even in a quantity will cause wear in the cylinder.

Depending on the quality of the fuel, bituminous coal will produce about 4½ pounds of ammonia and 12 gallons of tar with about 5% of sulphur.

Anthracite coal will produce approximately six pounds of tar, and two pounds of ammonia with traces of sulphur.

Loose Anthracite coal requires approximately 40 cubic feet of storage space per ton of 2240 pounds and weighs about 56 pounds per cubic foot (market sizes).

Loose Bituminous coal requires approximately 45 cubic feet of storage space per ton of 2240 pounds, and weighs about 52 pounds per cubic foot in market sizes.

Dry coke requires approximately 85 cubic feet of storage space per ton of 2240 pounds, and weighs about 26 pounds per cubic foot.

(17) Fuel Oils.

Crude oil, a natural product, is the base of the fuels most commonly used in internal combustion engines, especially in the smaller sizes. From this compound the following derivatives are obtained by the process of distillation, a separation possible because of the different boiling points of the various oils. As each derivative or **DISTILLATE** has a different boiling point, the temperature of the crude oil is maintained at the boiling point of that product that is desired, and the resulting vapor is condensed. The following list is not anywhere near complete for there are several hundred distinctly different distillates, but it contains those that are of the most interest to the engine man.

- 1. Crude Oil.
- 2. Gasoline.
- 3. Naphtha.
- 4. Solar Oil.
- 5. Kerosene.

The specific gravity of the crude oil as obtained in the field will range from 12° to 56° Beaumé scale. The crude from Pennsylvania will average 40° Beaumé while that from Texas will average 20°. The accompanying table will give the calorific values and general properties of the principle liquid fuels. It should be noted that the weight or density of the liquids is given in terms of specific gravity or Beaumé scale, in which the **SPECIFIC GRAVITY** of the fuel is the ratio of its weight per unit volume to the weight of an equivalent volume of water. The specific gravity of a liquid is generally determined by an instrument known as a **HYDROMETER** which consists of a glass tube sealed at both ends carrying a graduated scale on the upper portion of the stem, and a ballast weight of shot or mercury at the bottom.

The hydrometer is floated in the liquid to be tested, and the lower the specific gravity, the lower the hydrometer sinks, and vice versa. The specific gravity of the liquid is read directly from the graduation on the stem that are on a level with the surface of the liquid under test. As in the case of thermometers, hydrometers are all graduated in two different scales, the specific gravity scale and the Beaumé scale. The specific gravity scale reads at 1.00 when floated on distilled water, and the Beaumé at 10.00 when floated on the same liquid.

A difference in temperature affects the density of a liquid, hence all hydrometers are graduated for a standard temperature of 60°F unless otherwise specified. For a difference of 10°F there is a variation of one degree gravity in the Beaumé scale, and for a difference of 20°F in temperature there is a change of one degree on the specific gravity scale. If the temperature differs from 60°F, the corresponding correction should be made in the reading.

To convert the Beaumé reading (B) to terms of the specific gravity scale (S) use the following formula:

$$S = \frac{140}{130 + B} = \text{specific gravity.}$$

$$B = \frac{140}{S} = \text{Beaumé scale.}$$

PROPERTIES OF OILS

	Degrees Beaumé	Specific Gravity	Weight per gal.	B.T.U.'S per lb.	B.T.U.'S per gal.
Gasoline	67.2	.7125	5.932	21120	125,284
Heavy naphtha	64.6	.7216	6.011	20527	123,388
Kerosene	48.8	.7848	6.538	20018	130,877
W. Virginia crude	40.0	.8251	6.874	19766	135,871
Penn. fuel oil	31.9	.8660	7.215	19656	141,818
Kansas crude	29.0	.8816	7.345	19435	142,750
Fuel oil	22.7	.9176	7.645	19103	146,042
California crude	22.5	.9248	7.710	18779	144,786
California crude	15.2	.9646	8.036	18589	149,381
Alcohol, 95%	41.9	.816	6.798	10500	71,380

It will be noted that the petroleum products contain an enormous amount of heat energy, nearly 25% more than that of the same weight of pure carbon. It will also be noted that the lighter products such as gasoline, kerosene, etc., have more heat per pound but less per gallon than the heavier oils. This is rather confusing at first, but as will be seen after deliberation that the heavier fuel is the most economical since the least is used per horse-power, and is bought by the gallon. The calorific values given in the table are obtained by a calorimeter, and are burnt in the open air, and consequently have a different heating value when under compression in the cylinder of the engine.

In all cases the liquids are vaporized before being introduced in the cylinder, the more volatile liquids such as gasoline being converted into vapor at atmospheric temperature, and the heavier non-volatiles by being sprayed into a heated vessel or preheated air. The percentage of liquid fuel contained in a cubic foot of air vapor mixture depends on the temperature, the boiling point of the liquid and upon the pressure and humidity.

Gasoline consists principally of compounds of the methane series, the one representative of gasoline being Hexane (C_6H_{14}). It requires 15.5 pounds of air for combustion theoretically and about 10 per cent. more in practice. The formation of gasoline vapor produces a drop in temperature of 50°F, and should be heated 100°F above the atmosphere for the best results. The volume of air required for the combustion is about 192 cubic feet. With alcohol at 20 cents per gallon and gasoline at 12½ cents the number of B.T.U.'s for one cent in the case of alcohol is 3594 and 9265 in the case of gasoline. In the engine the difference is not so great owing to the difference in compression pressures.

(18) Tar for Fuel.

Because of the increasing interest in the Diesel type engine and the low grade fuels that it has made possible, we quote the specifications laid down by Dr. Rudolph Diesel, the inventor, before the English Institution of Engineers.

(1.) Tar-oils should not contain more than a trace of constituents insoluble in xylol. The test on this is performed as follows:—25 grammes (0.88 oz. av.) of oil are mixed with 25 cm.^{3} (1.525 cub. in.) of xylol, shaken and filtered. The filter-paper before being used is dried and weighed, and after filtration has taken place it is thoroughly washed with hot xylol. After re-drying the weight should not be increased by more than 0.1 gr.

(2.) The water contents should not exceed 1 per cent. The testing of the water contents is made by the well-known xylol method.

(3.) The residue of the coke should not exceed 3 per cent.

(4.) When performing the boiling analysis, at least 60 per cent. by volume of the oil should be distilled on heating up to 300° C. The boiling and analysis should be carried out according to the rules laid down by the Trust. (German Tar Production Trust on Essen-Ruhr.)

(5.) The minimum calorific power must not be less than 8,800 cal. per kg. For oils of less calorific power the purchaser has the right of deducting 2 per cent of the net price of the delivered oil, for each 100 cal. below this minimum.

(6.) The flash-point, as determined in an open crucible by Von Holde's method for lubricating oils, must not be below 65° C.

(7.) The oil must be quite fluid at 15° C. The purchaser has not the right to reject oils on the ground that emulsions appear after five minutes' stirring when the oil is cooled to 8°.

Purchasers should be urged to fit their oil-storing tanks and oil-pipes with warming arrangements to redissolve emulsions by the temperature falling below 15° C.

(8.) If emulsions have been caused by the cooling of the oils in the tank during transport, the purchaser must redissolve them by means of this apparatus.

Insoluble residues may be deducted from the weight of oil supplied.

Coal tar oil is the distillate of the tar obtained from gas works, from which all valuable commercial materials such as aniline have been removed. Coal oil tar is also known as creosote oil and anthracene oil, the heat value of which is not quite 16,000 B.T.U. per pound.

(19) Residual Oils.

Residual oil is the residue left after the lighter oils have been distilled from the petroleum, which before the advent of the Diesel engine were useless. Residual oil which was hardly fluid at ordinary temperatures has been successfully used in the Diesel and semi-Diesel types of engines, by preheating it before admission to the inlet valves. The enormously increased demand for gasoline has resulted in a great increase of the formerly useless residual oil so that it is possible that the demand for gasoline will make the production of the residual great enough so that it can be seriously considered as a fuel.

(20) Gasoline.

Gasoline is by the far the most widely used fuel for internal combustion engines because of its great volatility and the ease with which it forms inflammable mixtures with the air at ordinary temperatures. Another point in its favor is the fact that it burns with a minimum of sooty or tarry deposits, without a disagreeable smell with moderate compression pressures and without preheating through a wide range of air ratios. Gasoline is a product of crude oil from which it is obtained by a process of distillation, and as it forms but a small percentage of the crude oil it is rapidly becoming more and more expensive as the demand increases. Some Pennsylvania crude oils will yield as much as 20 per cent of their weight in gasoline, while the low grade Texas and California crudes very seldom contain more than 3 per cent.

When considered as a term applying to some specific product, the word "Gasoline" is a very flexible expression as it covers a wide range of specific gravities, boiling points, and compositions, the latter items depending on the demand for the fuel and the taste of the manufacturer. Since the specific gravity of gasoline is a factor that determines its suitability for the engine, at least in regard to its evaporating power or volatility, it is graded according to its density in Beaumé degrees as determined by the hydrometer. According to this scale gasoline will range from 85° to 60° Beaumé, and even lower, although 60° is supposed to mark the lowest limit and to form the dividing line between gasoline and naphtha.

The density of the gasoline in Beaumé degrees is an index to the volatility, for the higher the degree as indicated on the hydrometer, the higher is the volatility at a given temperature, consequently a high degree gasoline will give a better mixture at a low temperature than one of a low degree. In cold weather all gasoline should be tested with a hydrometer when purchased to insure a grade that will be volatile enough for easy starting when the engine is cold. In cold weather the gasoline should not be lower than 68°, and for the best results should be above 72°, at least for starting the engine. Good gasoline should evaporate rapidly and should produce quite a degree of cold when a small amount is spread on the palm of the hand, and it should leave neither a greasy feeling nor a disagreeable odor after its evaporation.

The high gravity gasoline is of course the most expensive, as there is less of it in a gallon of the crude oil from which it is made; gasoline of 76° Beaumé being approximately 15c. per gallon in carload lots, while naphtha of 58° Beaumé brings 8½c. per gallon.

The calorific value of gasoline increases as the gravity Beaumé decreases per gallon; 85° gasoline having approximately 113,000 B.T.U. per gallon while 58° naphtha has an approximate value of 122,000 B.T.U. per gallon. The calorific value remains nearly constant per pound for all gravities.

It should be remembered that heat is absorbed in evaporating gasoline as well as in evaporating water, and that effects of cold weather are greatly increased by the amount of heat absorbed, (or cold produced) by the vaporization of the fuel. While the heat absorbed by evaporating a given quantity of gasoline is only .45 per cent of that absorbed by an equal amount of water, it is a fact that this heat must be supplied from some source to prevent a reduction in the vapor density. In starting the engine, the heat of evaporation is supplied by the atmosphere, and should the temperature of the air be below that required for a given vapor density, the engine will refuse to start.

By the use of two tanks and a three way valve, it is possible to use two grades of fuel: one tank containing high gravity gasoline, and the other low gravity; the high gravity being used for starting the engine in cold weather, and the cheaper, low gravity, being used for continuous running after the engine is warmed up—the change of fuels being made by throwing over the three way valve.

The **VAPOR DENSITY** of gasoline vapor is the ratio of the weight of the vapor compared with the weight of an equal volume of dry air at the same temperature. If the weight of a cubic foot of gasoline vapor is divided by the weight of a cubic foot of air at the same temperature the result will be the vapor density of the gasoline vapor. Compared to air, the gasoline vapor is quite heavy so that if a small quantity of gasoline is poured on the top of a table, the vapor will flow over the edge of the table and drop to the floor where it will remain until it has united with the air by the process of diffusion. Experiments have shown that pure, dry gasoline vapor has a density of about 3.28, or in other words weighs 3.28 times as much as an equal volume of dry air. This weight of course is the weight of pure vapor which is considerably heavier than the mixture of vapor and air that is used in the cylinder of the engine.

Dampness, or the presence of water vapor in the air reduces the quantity of gasoline vapor taken up by the air, but only by a small amount, the maximum difference being only about 2 per cent. Since it is very likely that the water vapor is broken up into its original elements, oxygen and hydrogen, by the heat of the combustion it is likely that there is no heat loss due to the vapor passing out through the exhaust. The principal trouble due to dampness is the mixture of water and liquid gasoline caused by the condensation of the water vapor.

All gasolines and oils contain water to a more or less degree, hence provision should be made for the draining of the water which collects in the bottom of the tank. Water in liquid fuels is the cause of much trouble.

Water in gasoline may be detected by dropping scrapings from an indelible pencil into a sample of the suspected fluid. If water is present in any quantity the gasoline will assume a violet color.

In filling a supply tank with gasoline, a chamois filter or chamois lined funnel should always be used, as the chamois skin allows the gasoline to pass but retains the water and impurities contained therein. There are many funnels of this type now on the market.

The rate at which gasoline burns depends on the amount of surface presented to the air by the fluid, for a given quantity of gasoline burns faster in a wide shallow vessel than in a deep jar. Since a spray of minute particles presents an enormously greater surface than the liquid its burning speed is correspondingly greater, and as a true vapor has an almost limitless area, its speed is much greater than that of the spray, the combustion under the latter condition being almost instantaneous. Besides the question of subdivision of the liquid, the rate of combustion also depends on the intimacy of contact of the vapor with the air and on the pressure applied to the vapor as previously explained under the head of "**COMPRESSION**" in another chapter.

CARBURETING AIR, or producing an explosive mixture of gasoline vapor and air is accomplished by two different methods, first by passing the air over the surface of the liquid, or by passing it through the liquid in bubbles; second by spraying the liquid into the air. The latter is the method most generally in use at the present time, the spray being formed by the suction of the intake air upon the open end of the spray nozzle. The vapor density of the mixture thus formed depends on the suction of the air and upon the nozzle opening, either of which may be varied in the modern carburetor to vary the richness of the mixture.

As a suggestion to the users of gasoline we append the following remarks.

Gasoline vapor will readily combine with air to form explosive mixtures, at ordinary temperature. This property at once makes it the most suitable fuel and the most dangerous to handle.

Never fill tanks or expose gasoline to the air in the presence of an open flame, or do not attempt to determine the amount of gasoline in a tank with the aid of a match. There are a number of people who have successfully accomplished this feat, and a very great number who have not.

Be very sparing in the use of matches around a gasoline engine; there are such things as **leaks**.

Always carefully replace the stopper or filler cap in a gasoline tank after filling. Never use the same funnel for water and gasoline, and avoid any possibility of water finding its way into the tank.

If you do succeed in igniting a quantity of free gasoline, do not attempt to extinguish the fire with water. Pouring water on burning gasoline spreads the fire. Extinguish it with earth or sand, or by the use of one of the dry powder extinguishers now on the market.

Water may be removed from gasoline by placing a few lumps of desiccated calcium chloride in the tank, the amount depending on the quantity of water.

Calcium chloride, has a great capacity for absorbing water, and in a short space of time will absorb all of the moisture contained in the tank.

The best way to introduce the chloride is to wrap the lumps in a sheet of wire gauze and lower into tank with a wire, the wire allowing it to be easily removed when saturated with water.

(21) Benzol.

Benzol has been used to some extent in Europe as a fuel, its use being due to the rapidly increasing cost of gasoline.

Benzol is a distillate of coal tar, and is a by-product of the coke industry. In England benzol brings approximately the same price as gasoline (called petrol), but benzol proves economical for the reason that it develops more power per gallon.

Benzol is not as volatile as gasoline, but is sufficiently volatile to allow of easy motor starting.

Benzol is also used for denaturing alcohol.

(22) Alcohol.

Alcohol is of vegetable origin, being the result of the destructive distillation of various kinds of starchy plants or vegetables. Starch is the base of alcohol.

As a fuel, alcohol has much in its favor, as it causes no carbon deposit, has smokeless and odorless exhaust, can stand high compression, and requires less cooling water than gasoline, as the heat loss is less through the cylinder walls, and for this reason it is more efficient fuel than gasoline.

At the present time the price of alcohol prohibits its general use. In order that alcohol equal gasoline in price per horse-power hour, it should sell for 10c. per gallon, the price of gasoline being 15c. per gallon.

Alcohol can be used in any ordinary gasoline engine with readjustment of carburetor and the compression.

The nozzle in the carburetor has to be of larger bore for alcohol than for gasoline, and the compression for alcohol in the neighborhood of 180 pounds per square inch.

The inlet air should be heated to about 280°F for alcohol fuel; approximately 6% of the heat of the alcohol is required for its vaporization. Alcohol is much safer to handle than gasoline owing to its low volatility.

90% alcohol has a calorific value of 10,100 B.T.U. per pound, its specific gravity being .815.

WOOD, or **METHYL** alcohol is made by distilling the starch contained in the fibres of some species of wood (Poisonous).

GRAIN, or **ETHYL** alcohol is the result of the distillation of the starch contained in grains, potatoes, molasses, etc. **ETHYL**, or **GRAIN** alcohol rendered unfit for drinking by the addition of certain substances, is called **DENATURED ALCOHOL**. The process of denaturing does not affect the calorific value of alcohol to any extent.

(23) Kerosene Oil.

Kerosene is a fractional distillate of crude oil which has a considerably higher vaporizing temperature than gasoline. It does not form an inflammable mixture with the air at ordinary temperatures, but is vaporized in practice by spraying it into a chamber heated to above 200°F. Kerosene forms a greater percentage of crude oil than gasoline and as there has been less demand for it up to the present time it is much cheaper. Pennsylvania crude oil produces only 20 per cent of gasoline while the kerosene contents will average nearly 42 per cent according to figures at hand.

Kerosene has a very high calorific value per gallon, 8.5 gallons of kerosene having the same heating effect as 10 gallons of gasoline. Because of its high calorific value and its low cost per gallon, many types of engines have been developed for its use during the last few years, several of which have been very successful. Before the advent of the modern kerosene engine much difficulty was experienced with the fuel because of its high vaporizing temperature and its tendency to carbonize in the cylinder, but as the price of gasoline continued to rise, the inventive genius of the gas engine builder overcame these troubles so that the kerosene engine is now as reliable as any form of prime mover.

Kerosene Vaporizer on Fairbanks-Morse Engine. The Engine is Started on Gasoline and When Hot, the Kerosene Feed is Turned on.

Any gasoline engine will run on kerosene, after a manner, if the engine is thoroughly heated to insure the vaporization of the kerosene, and if the fuel is heated in the carburetor. Such an arrangement is make-shift, however, and is not productive of good results in continuous service. If kerosene is to be used as a regular fuel, a kerosene engine should be used to avoid vaporizing and carbonizing difficulties as well as the sooty, offensive exhaust, and the loss of fuel represented by the soot.

Many kerosene engines are arranged to start on gasoline, and, after becoming heated, have the running feed of kerosene admitted through a three way valve. The gasoline feed is then stopped.

The above arrangement admits of easy starting in all weathers and temperatures.

In the Diesel engine there is no evaporating of fuel, and no deposits of carbon because of the high temperature of the combustion chamber. With engines that draw the mixture of vapor and air into the cylinder there are several methods of applying heat to the liquid, and the combustion of the vapor thus formed is perfected by the injection of water into the combustion chamber. It has been found by experiment that a small amount of water vapor introduced into the cylinder of a kerosene engine makes the engine run more smoothly and prevents a smoky exhaust and carbon deposits in the cylinder. The water is introduced into the cylinder through an atomizer in the form of a mist or fog, the particles of water being in a very finely subdivided state.

Kerosene Vaporizer on Fairbanks-Morse Vertical Engine. Started on Kerosene Directly by Heating Vaporizer with Torch.

The deposits of free carbon (soot) caused by the "cracking" or decomposition of the kerosene vapor before ignition, due to the high temperature of the cylinder, are burnt to carbon dioxide by the oxygen of the water which is also set free by the heat of the cylinder. This produces an odorless gas (CO_2) which indicates complete combustion. Besides the increase of fuel efficiency due to the water vapor, the cylinder is more thoroughly cooled and is more efficiently lubricated because of the reduction in temperature.

CHAPTER III
WORKING CYCLES
(24) Requirements of the Engine.

In order that an internal combustion engine shall operate and develop power continuously the following routine of events must occur in the cylinder in the following order, no matter what the type of engine.

(1) The cylinder must be filled with a combustible mixture of air and gaseous fuel at as nearly atmospheric pressure as possible.

(2) The mixture must be compressed in order to develop the value of the fuel.

(3) Ignition must take place at the end of the compression stroke or at the highest point of compression.

(4) Complete combustion of the fuel must follow the ignition of the charge, with an increase of temperature and pressure which will act on the piston to the end of the power stroke.

(5) After the piston has completed the working stroke the products of combustion must be ejected from the cylinder completely to make way for the admission of the new combustible mixture.

With the exception of the Diesel engine which (1) fills the cylinder with pure air without the fuel, and (2) injects the fuel after compression, all internal combustion engines not only perform each of these operations but proceed with events in the order given as well. The accomplishment of the five acts is called a "cycle of events," or a **"CYCLE,"** and the series is performed in different ways in different types of engines. In the operation of the engine, the series of events occur over and over again, always in the same order, 1–2–3–4–5, 1–2–3–4–5, 1–2–3–4–5, etc. The five events are generally given in terms of the number of strokes of the piston taken to accomplish the complete routine, thus a two stroke cycle engine performs the series in two strokes, and a four stroke cycle engine in four strokes, and so on.

In order to obtain the benefits of high compression, perfect scavenging of the products of combustion from the cylinder and perfect mixtures, a great variety of engines have been developed in which the number of strokes taken to accomplish the five events varies. In some engines the cycle is accomplished in two strokes, in other engines it is accomplished in six strokes, but in the great majority of cases the cycle is performed in either two or four strokes, and as these are by far the most common routines, we will confine our description to engines of these types.

(25) Four Stroke Cycle Engine.

The four stroke cycle engine, some times improperly called the "four cycle" engine is the most widely used type for all classes of service, except possibly for marine work. Its extended use is due to its superior scavenging, high efficiency and reliability, although it is somewhat more complicated than the two stroke cycle type. Its ability to function properly under a wide variation of speed has driven the two stroke cycle type out of the automobile field, and its many admirable characteristics have cut a wide swath in the marine field, the stronghold of the two stroke cycle type.

A four stroke cycle engine performs the cycle of events in four strokes or two revolutions, only one of the strokes being a power of working stroke. In a single cylinder engine the explosion in the working strokes supplies enough power to the fly-wheel to carry the engine and its load through the remaining three strokes. Thus the energy stored in the fly wheel is sufficient to carry not only the load during the idle strokes but to "inhale" and compress the charge as well. Due to the long interval that exists between explosions, they are corresponding heavy and are productive of heavy strains in the engine and are the cause of considerable vibration.

To reduce the ill effects of the heavy intermittent blows, the majority of automobile and stationary engines are provided with two or more cylinders, the power being equally divided among them. In a four cylinder engine, there are four times as many impulses as in a single cylinder engine and the blow dealt by the individual cylinder is only one-quarter as great. While a single cylinder engine has an impulse only once in every other revolution, the four cylinder has two impulses in one revolution. Besides the advantages gained by increasing the impulses, the mechanical balance of a multiple cylinder engine is always better than that of a single and is also much lighter in weight since less material is required to resist shocks of the explosions.

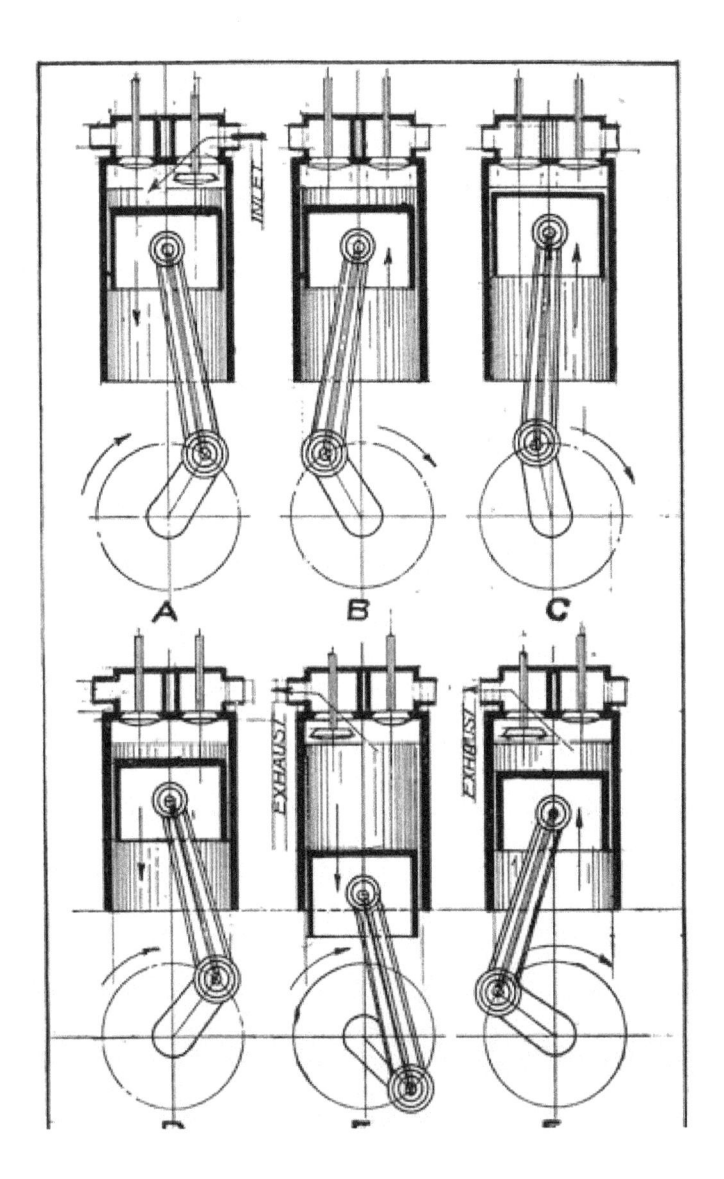

Fig. 4. Diagrammatic View of Four Stroke Cycle Engine with the Piston in Various Positions Corresponding with the Five Events. Diagram A—Suction. Diagram B—Compression. Diagram C—Ignition. Diagram D—Working Stroke. Diagram E—Release. Diagram F—Scavenging Stroke.

Engines with more than four cylinders have "overlapping" impulses, that is some cylinder on the engine is always delivering power, for before one cylinder reaches the end of the stroke, another has fired its charge and has started to deliver power. Thus the impulses "overlap" one another, and the result is an even and smooth application of power and a minimum of strain is imposed on the engine.

Aeronautical and speed boat engine builders have carried the multiple cylinder idea to an extreme because of the nature of their work. Eight cylinder aeronautical engines are very common and there are several built having sixteen cylinders. The latter type of engine gives eight impulses per revolution. To avoid a great multiplicity of cylinders, and to save on floor space, the great majority of heavy duty stationary engines are built double acting, that is an explosion occurs alternately in either end of the cylinder. In effect, a double acting cylinder is the same thing as a two cylinder single acting engine, as it gives twice the number of impulses obtained with a single acting cylinder.

The order in which the events occur in a four stroke cycle engine is as follows:

STROKE 1. First outward stroke of the piston causes a partial vacuum in the combustion chamber thus drawing a charge of combustible gas into the cylinder through the open inlet valve. The exhaust valve is closed. See diagram A in Fig. 4. (Suction Stroke.)

STROKE 2. Inlet valve closes at the end of the suction stroke and the piston starts on the inward stroke compressing the charge in the combustion chamber. See diagram B. (Compression Stroke.) At the end of the compression stroke, or a little before, the spark "S" occurs causing the ignition of the charge. See diagram C.

STROKE 3. Working Stroke. As the pressure is now established in the cylinder, the piston moves down on the working stroke forcing the crank around against the load and supplying sufficient energy to the fly wheel to carry the engine through the three idle strokes. See diagram D. When the piston reaches the end of the working stroke, or a little before, the exhaust valve opens to reduce the pressure and to allow the greater part of the burnt gas to escape. See diagram E.

STROKE 4. Scavenging Stroke. The exhaust valve remains open and the inwardly moving piston expels the remainder of the burnt gas through the exhaust valve, clearing the cylinder for the next fresh charge of mixture. See diagram F. The next stroke is the suction stroke explained under "Stroke 1."

In all of the diagrams the crank is supposed to turn in a right handed direction as indicated by the arrow, the piston moving in the direction shown by the arrow under the piston head. The valves are operated by cams on an intermediate shaft known as the "cam shaft." As the valves go through their series of movements in two revolutions of the crank shaft, and as the cam shaft must perform all of these operations in one revolution, it is evident that the cam shaft must run at exactly one-half the crank-shaft speed. This change of speed is accomplished by means of gearing between the cam shaft and crank-shaft from which the cam shaft is driven.

In some engines, notably the Diesel engine, pure air is drawn into the cylinder on stroke No. 1 instead of the entire mixture. Fuel is supplied in this type immediately after the end of the compression stroke.

While an electric spark is shown as the igniting medium in the diagrams, the ignition is sometimes performed by a hot tube, or simply by the heat of the compression as in the Diesel engine.

In the sliding sleeve type of four stroke cycle motor, the poppet or lifting type of valve as shown in Fig. 4, is replaced by a peculiar type of slide valve similar in action to the slide valves used on steam engines, except that it is cylindrical in form and entirely surrounds the piston. While there is a change in the form of the valve, and in a number of small details, the gases are drawn into the cylinder, compressed, ignited, and released in exactly the same way and in the same rotation, as in the poppet valve engine just described. A description of the Knight engine which is the most prominent example of the slide sleeve motor will be found in a succeeding chapter. Since the success of the slide valve type has been acknowledged by many prominent automobile manufacturers, there have been several similar types placed on the market, some with two sleeves and some with one, but in all cases the designers have had but two points in view, that is quiet running and free passages.

(26) Two Stroke Cycle Engine.

Two stroke cycle engines perform the five events of aspiration (suction), compression, ignition, expansion and release in two strokes or one revolution. Providing that these events are performed as efficiently as in the four stroke cycle engine, it is evident that with equal cylinder capacity, the two stroke cycle engine would have twice the output of a four stroke cycle since it gives twice the number of impulses per revolution. Unfortunately it is impossible to attain twice the output of the four stroke cycle type with the small two stroke engines built at the present time because of their imperfect scavenging and poor fuel economy. In the larger two stroke engines, the pumps and blowers used for scavenging the cylinders consume a considerable percentage of the output.

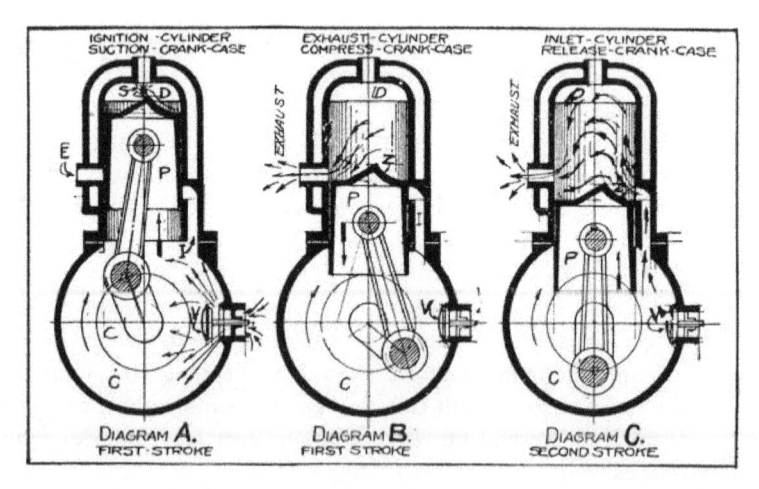

Fig. 5. Diagram of Two Port—Two Stroke Cycle Engine, Showing the Events in the Crank-Case and Cylinder.

A general classification of the two stroke cycle engine is not so simple a matter as that of the four stroke because of the differences in construction of large and small sizes. This difference between the large stationary engine and the small type commonly used on boats is due to the efforts of the builders of the large engine to obtain great fuel economy, while the chief endeavors of the builders of small engines is to build a simple and reliable engine for the use of inexperienced persons. While the smaller type of two stroke engine (less than 25 horse-power) has not been used in stationary practice to any extent, owing to the defects just named, or on automobiles, it has been widely used on motor boats, a service for which it is peculiarly adapted. Its extended use on boats is due to the fact that in such service it runs at practically a constant speed and works against a steady load, the conditions that are most favorable to the type. With automobiles where the motor speed is constantly varying, as well as the load, this type of motor is not flexible enough to meet the continually varying conditions.

The small two stroke motors are divided into two principal classes, the two port and three port type, depending on the method by which the charge is transferred to the cylinder. No valves are used in the cylinders of either type for the admission or release of the gases. As the two strokes of the cycle are the compression stroke and working stroke, it is evident that the charge must be introduced into the cylinder by means other than by the suction of the piston and at a time when there is no pressure in the cylinder. This is accomplished by a preliminary compression of the charge in the crank case which places the mixture under sufficient pressure to force it into the cylinder at the end of the working stroke and at the same time to displace the burnt gases left from the previous explosion. It should be noted that the incoming mixture is a substitute for both the suction and scavenging strokes of the four stroke cycle engine.

A diagrammatic view of a two port, two stroke cycle engine is shown by Fig. 5, in which P is the piston, C the crank case, I the transfer port, V the inlet valve, E the exhaust, and S the spark plug for igniting the charge. It should be noted that there are no valves in the cylinder and only three moving parts. The cycle of events for the two port type is as follows:

STROKE 1. We will consider the piston to be moving up on the compression stroke as shown in view (A), compressing the mixture in the combustion chamber D. While moving upwards in the direction of the arrow, the piston creates a vacuum in the crank case C drawing fresh mixture into the crank case. The piston at this time is covering the opening of the transfer port I and the exhaust port E so that the compressed mixture in the cylinder cannot escape. On reaching the end of the compression stroke, a spark occurs at S which drives the piston down and turns the crank towards the right as shown by the arrow.

STROKE 2. When the piston uncovers the exhaust port E on its downward working stroke as shown by view B, the exhaust gases being under pressure rush out into the atmosphere as shown by the arrows, and relieve the pressure in the cylinder. Some of the burnt gas remains in the cylinder at atmospheric pressure as there is no scavenging action up to this point. While the piston has moved down on the working stroke it has compressed the mixture in the crank case ready for admission to the cylinder. The valve V prevents the escape of the gas during the compression.

On reaching the end of the stroke the piston uncovers the transfer port which allows the compressed mixture in the crank case to rush into the cylinder through I, as shown by view C. Owing to the shape of the deflector plate Z on the piston head, the stream of mixture issuing from I is thrown up toward the top of the cylinder, as shown by the arrows, and consequently sweeps the remainder of the burnt gas before it through the exhaust port E. In this way the fresh mixture from the crank case scavenges the cylinder and fills it in one operation. Being filled with gas, the piston now moves up on the compression stroke for the next explosion as shown by view A.

Unfortunately the scavenging action of the incoming gas is not complete for the whirling motion of the charge causes it to mix with the residual gas to a certain extent which, of course, reduces the heating effect of the fuel and reduces the power output. Another factor that reduces the output of this type of engine is the loss of explosive mixture through the exhaust port at low engine speeds with an open throttle. In this case, the piston speed being low, part of the mixture has time to pass over the deflector plate and through the exhaust opening before the piston closes the exhaust port. At very high speeds the charge is diluted by a considerable quantity of burnt gas which has not had time to escape through the port causing a further loss of power. With the throttle nearly closed on a light load, the impact of the incoming mixture is so slight that the percentage of exhaust gas left in the cylinder is very high. This dilution is so great that with moderately low speeds (easily within the capacity of the four stroke cycle engine) it is either impossible to ignite the charge or it is impossible to ignite two in succession.

In marine service where the loads are constant, and the speeds fairly uniform, there is but little trouble from the last mentioned source, and as the fuel is usually a smaller item than the repair bill, the simplicity of the small two stroke engine with its freedom from mechanical troubles usually gives satisfactory results in the hands of the novice.

(27) Three Port—Two Stroke Cycle Engine.

The principal difference between the three port and two port types of the two stroke cycle engine is in the manner in which the charge is admitted to the crank case for the initial compression. In the two port motor, as previously described, the check valve "V" opens to admit the charge, and closes during its compression in order to prevent its escape through the opening by which it was admitted to the cylinder. With the three port type there is no check valve in the crank case, the admission and the retention of the charge being controlled by the movement of the piston in practically the same way that the piston controls the opening and closing of the exhaust and transfer ports in the cylinder.

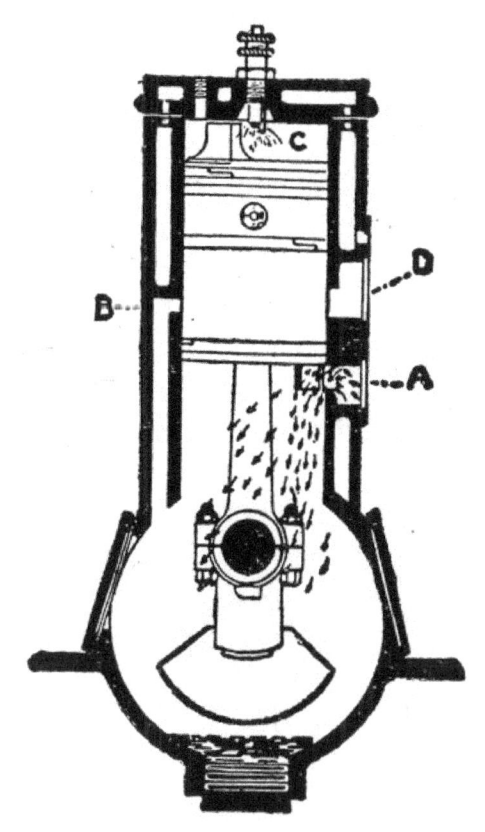

Fig. 6.

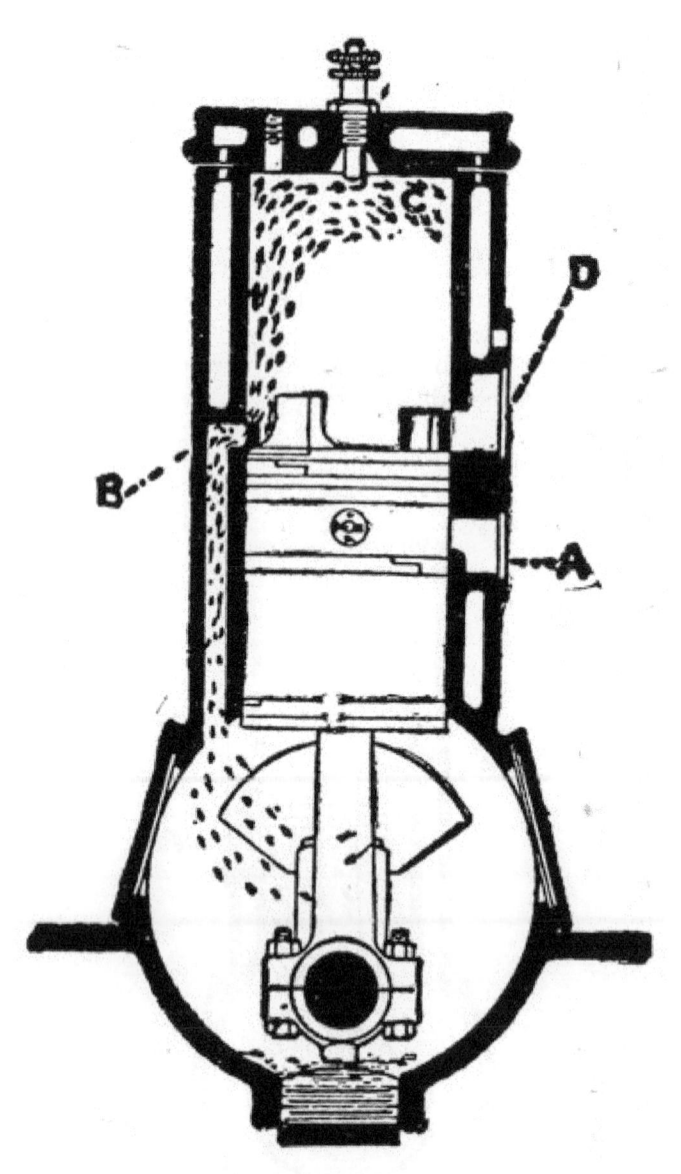

Fig. 7

Figs. 6–7. Diagram of Three Port—Two Stroke Cycle Engine in Two Positions.

By the piston control of the gases in the crank case, the valve is eliminated, which makes one less moving part to cause trouble and expense, and permits the use of the same type of carburetor that is used on the four stroke cycle engine. As the check valve opens and closes at a high speed, (twice that of the valves on a four stroke cycle engine), there is considerable wear on the valve seats due to the continuous banging, which results finally in a loss of the initial compression. When the initial compression is reduced in this way the engine loses power because of the reduction of the charge in the cylinder.

While the three port type is free from valve leakage troubles, it has a steady loss due to the high vacuum that exists in the crank chamber when the piston is on its upward stroke. This vacuum drags against the piston and absorbs a considerable amount of power until the piston reaches the upper end of the stroke. At this point the inlet port is opened and the vacuum is broken by the rush of the mixture through the inlet port. Besides the power loss, the vacuum has a bad effect on the lubrication of the main crank shaft bearing.

Elevation of Fairbanks-Morse Three-Port Two Stroke Marine Motor Showing Warming Device for Carburetor Air.

Described by strokes, the cycle of events in the three port, two stroke cycle engine is as follows:

STROKE 1. In Fig. 6, the piston is shown at the end of the compression stroke with ignition taking place in the combustion chamber C. The pressure due to the expansion drives the piston down on the working stroke at the same time causing the initial compression of the mixture in the crank case as shown by Fig. 7. The gas in the crank case cannot escape during compression as the inlet port A is covered by the piston.

(a) As the piston descends, its upper edge uncovers the exhaust port D, allowing the greater portion of the exhaust gases to escape and reduces the pressure in the cylinder to that of the atmosphere.

(b) Descending a little farther, the top of the piston uncovers the opening of the transfer port B, allowing the compressed gases in the crank case to enter the cylinder as shown by the arrows. These gases, guided by the deflector plate on the top of the piston are thrown upwardly, as shown by the arrows, and sweep the residual burnt gases before them through the exhaust port. The cylinder is now filled with the combustible mixture ready for compression.

STROKE 2. The piston now moves up on the compression stroke, compressing the charge in the cylinder and at the same time creates a vacuum in the crank-case. Just before the piston reaches the end of the exhaust stroke, the lower edge of the piston uncovers the inlet port A (See Fig. 7), which allows the mixture from the carburetor to flow into the partial vacuum and fill the crank case ready for the next initial compression. When the end of the stroke is reached, the charge in the combustion chamber C is fired and the cycle is repeated. It should be noted that the incoming gas and the initial compression are controlled entirely by the action of the lower edge of the piston on the inlet port A.

(28) Reversing Two Cycle Motors.

As the admission and exhaust in the two stroke cycle engine each occur once per revolution, and are controlled directly by the piston position at opposite ends of the stroke, it is evident that the direction of rotation is not affected by gas control or valve timing, as in the case of the four stroke cycle engine. The factor that does determine the direction of rotation in the two stroke engine is the time at which ignition occurs in regard to the angular position of the crank. By changing the relation between the crank position at the end of the compression stroke and the time at which the spark occurs, it is possible to reverse the engine even when it is running.

Should the engine be standing still in the position shown by Fig. 6, with the crank on the dead center, when ignition occurred, there would be no more tendency to turn the crank to the right than to the left, providing of course, that there was no effect from the momentum of a revolving fly wheel. If ignition occurred with the crank inclined ever so little toward the right, the pressure of the piston would force the crank downwards in a right handed direction. If the crank were inclined to the left, the tendency would be for left handed rotation.

If the ignition system were arranged so that the spark occurred when the crank was inclined towards the right every time that the piston came up on the compression stroke, we should have continuous rotation in a right hand direction. By shifting the sequence of the spark so that it occurred with the crank on the left we would cause the engine to stop and reverse to left handed rotation. This is exactly the method used in reversing two stroke motors in practice, the change in the ignition being accomplished by advancing or retarding the mechanism that dispatches the spark ("Timer" or "Commutator").

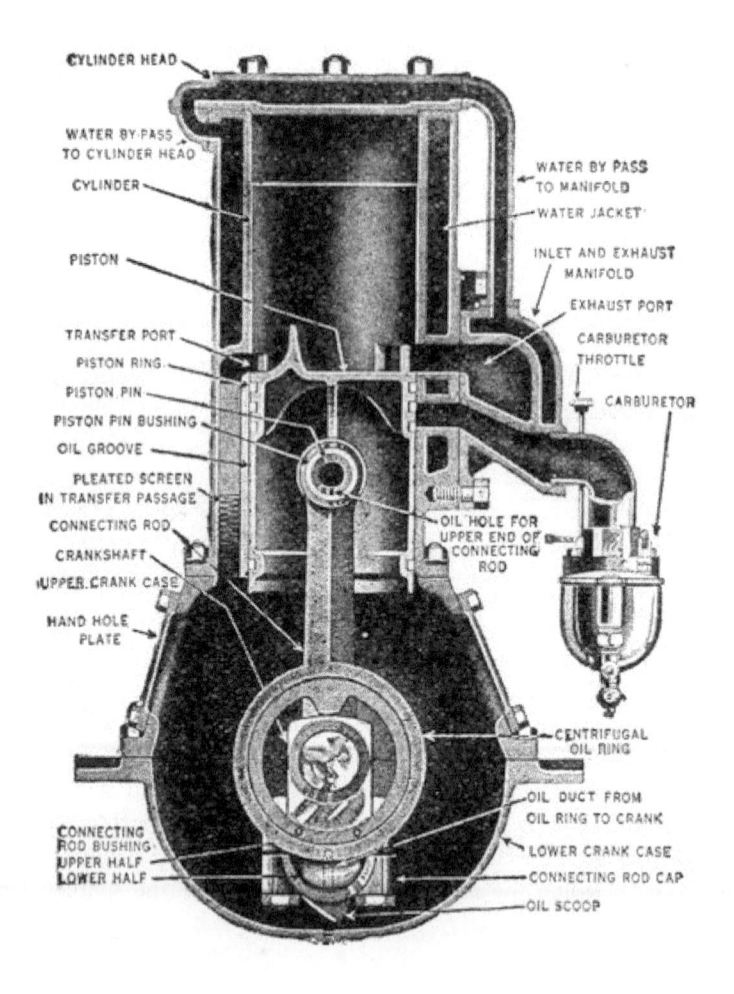

Fig. F-9. Cross Section of Fairbanks-Morse Three Port—Two Stroke Cycle Engine, with Parts Named.

This is an advantage not possessed by the four stroke cycle engine of the ordinary type, as the cams and valve mechanism require reversal as well as a reversal of the ignition system. This relation between the valve action and rotation in a four stroke cycle engine may be illustrated by the following example. Consider the piston at the end of the compression stroke in an engine designed for right hand rotation. After ignition, under the proper conditions, the piston would descend turning the crank to the right until it reached the bottom of the stroke, at which point the exhaust valve would open and relieve the pressure in the cylinder.

Let us now consider an attempt at reversing the engine by causing the spark to occur before the piston reached the end of the compression stroke with the crank still inclined toward the left. In this case the piston would force the crank down in a left hand direction until it reached the end of the stroke. The exhaust valve would not open to relieve the pressure, as the exhaust cam would be moving away from the valve rod instead of toward it. Should the crank swing a little past the dead center, because of its momentum, the inlet valve would be opened instead of the exhaust, and the contents of the cylinder would shoot through the intake pipe and carburetor. This would bring matters to a close as far as rotation was concerned.

The opening of the inlet valve on the reversed working stroke would occur as the inlet valve closes one stroke, or one-half revolution, before the end of the compression stroke. As the engine turned backward one-half revolution, the inlet cam would again be brought into contact with the inlet valve rod, opening the valve and allowing the burned gases to pass through the carburetor. Should the pressure be sufficiently reduced by inlet valve to allow the piston to reach the end of the second stroke, it would start on the third stroke by inhaling a "charge" of burnt gas through the exhaust valve which would now be open.

(29) Scavenging Engines.

As the piston does not sweep out all the cylinder volume because of the space left at the end of the cylinder for compression, more or less burned gas remains in the combustion chamber which dilutes the active mixture taken in on the suction stroke. Not only are the residual gases useless in generating heat but they also occupy a considerable space in the cylinder that might otherwise be filled with a heat producing mixture. Their diluting effect also prevents the complete combustion of a certain percent of the fuel actually taken into the cylinder for which the burnt gas is incapable of supporting combustion.

The amount of burnt gas remaining in the cylinder depends upon the cycle of the engine and also upon the valve timing and size of the exhaust piping. In the four stroke cycle engine the volume of residual gas is equal to the volume of the combustion chamber, in the two stroke cycle it varies from one-tenth to one-third of the entire cylinder volume, depending on the load and speed. With correct design and free exhaust passages, the gas held in the clearance space of a four stroke cycle engine is at a pressure considerably below that of the atmosphere, and consequently its actual volume is even less than the volume of the combustion chamber.

Many systems have been devised for the purpose of clearing the cylinder of burnt gas in order to minimize the loss of fuel in large engines, but owing to their complication have never been successfully applied to small engines of the automobile or marine types. In general, the "scavenging" is accomplished by pumping out the clearance space at the end of the scavenging stroke, while fresh air is admitted to the cylinder through the inlet valves, or by blowing out the clearance space by a blast of pure air furnished from an air pump attached to the engine.

There have been several systems proposed by which the gas in the cylinder is withdrawn by the inertia of the exhaust gas in specially designed ejectors, and by the compression of fresh air in the crank case of the engine. The former system known as "organ pipe ejection," is by far the simplest method of all as the ejector is simply a tube without moving parts, and it also possesses the additional advantage of reducing the back pressure on the piston. Unfortunately these advantages are obtained only at certain loads, and with certain velocities of the exhaust gases, which makes it impossible to obtain even approximately correct scavenging at other loads and speeds.

When air pumps are used for scavenging, a great percentage of the economy obtained is offset by the power required to operate the pumps. In addition to the frictional losses of the pumps, are the increased maintenance charges and repair bills.

CHAPTER IV
INDICATOR DIAGRAMS
(35) General Description.

A brief description of the indicator as a means of recording the pressures in the cylinder of a simple heat engine in relation to the piston position was given in paragraph (6), Chapter I, and as this instrument is so peculiarly adapted to locating the events taking place in the cylinder we will devote some space on its application to the practical gas engine cycles described in the preceding chapter. Since each event in the cycle is accompanied by a corresponding increase or reduction in pressure, the beginning or end of an event will be indicated on the diagram by a change in the vertical height of the curve above the atmospheric line, at some particular piston position. The piston position will be in the same relation to the total stroke as the pencil position will be to the horizontal length of the card.

If the event, for example, as indicated by a drop in pressure, be at the center of the card, it will show that the drop in pressure took place when the piston was in the center of the cylinder or at mid-stroke. Should the pressure change at a point one-quarter of the card length from the starting point of the pencil, it shows that the event took place in the cylinder when the piston had accomplished the first quarter of its stroke, and so on. It should be noted that horizontal distances on the indicator card denote piston positions, and the vertical distances, pressures.

As explained in a former paragraph the length of the vertical lines represents certain definite pressures, each inch of length representing so many pounds as per square inch, the exact amount per inch depending on the indicator spring strength or adjustment. To make this point clear, all of the indicator diagrams shown in this chapter will be provided with a scale of pressures at the left of the diagram by which the pressure at any point may be accurately measured off for practice. It should be noted that points on the curves which are above the atmospheric line represent positive pressures above the atmosphere, and that the points lying below the atmospheric line represent partial vacuums which may be expressed as being so many pounds per square inch below the atmosphere. The vacuum pressures indicate the extent of the "suction" created by the piston when drawing in a charge of air and gas.

Straight vertical lines show that the increase of pressure along that line has been practically instantaneous in regard to the piston velocity, for if the pressure increased at a slow rate this line would be inclined toward the direction in which the piston was moving, as the piston would have moved a considerable distance horizontally while the pencil was moving vertically. This inclination of the vertical line gives an idea of the rate at which the pressure increases in relation to the piston speed, the greater the inclination, the slower is the rate of pressure increase. Straight horizontal lines that lie parallel to the atmospheric line denote a constant pressure or vacuum.

The rate at which horizontal lines descend or incline to the atmospheric line represents the rate at which the pressure increases or decreases, in respect to the piston position (not piston velocity). A steep curve represents a rapid expansion or compression from one piston position to the next. A waving or rippling line indicates vibration due to valve chattering or explosion vibrations. A straight inclined line shows that the pressure is decreasing or increasing in direct proportion to the piston position.

(36) Diagram of Four Stroke Cycle Engine.

By referring to paragraph 25, Chapter III, it will be seen that the five events of suction, compression, ignition, expansion and exhaust are accomplished in four strokes, in the following order:

Stroke 1.　　Suction—(Mixture drawn into cylinder).

Stroke 2.　　Compression—(Mixture compressed).

Stroke 3. $\{$ Ignition.
Expansion (working stroke).

Stroke 4.　　Exhaust—(Scavenging stroke).

These events with the pressures incident to each drawn to some relative scale are shown graphically in Fig. 10 by four lines representing the four strokes of the piston. In order to show the relation between the diagram and the piston, a sketch of the cylinder with a stroke equal to the length of the diagram is shown directly beneath the curve. The vertical line IJ is the scale of pressures (somewhat exaggerated in order that the small vacuum and scavenging pressures shall be clearly shown). The line marked "atmosphere" represents atmospheric pressure and it is from this line that all measurements of pressure are taken.

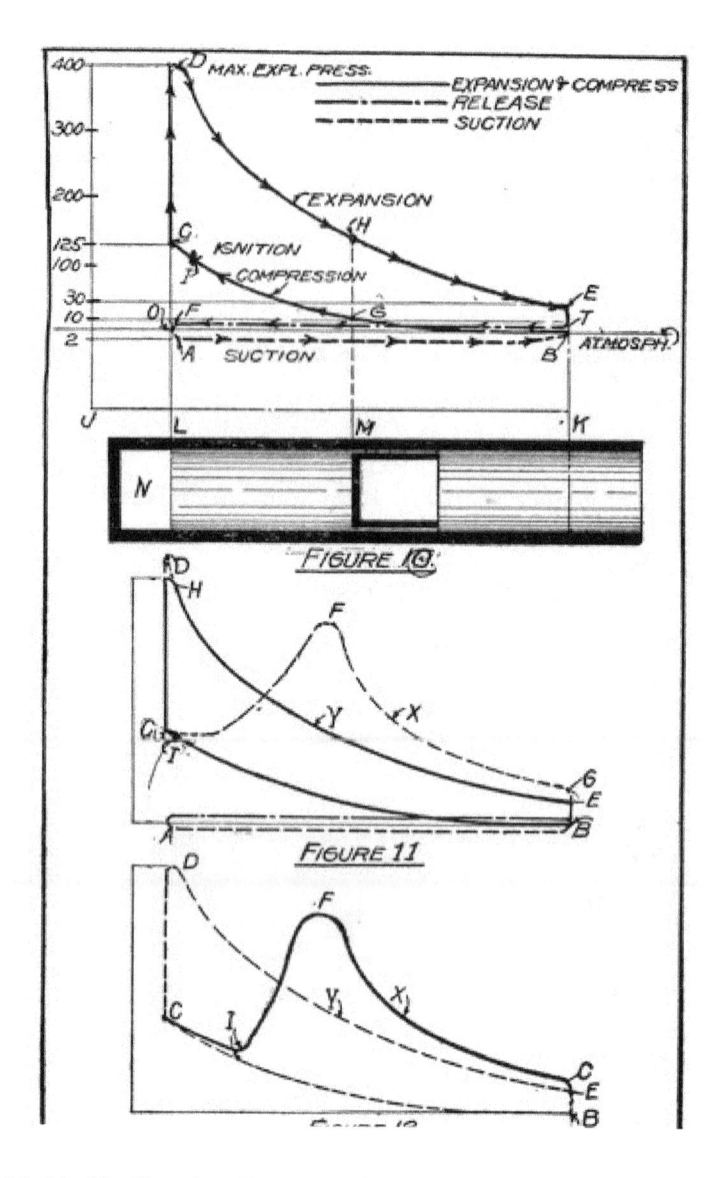

Figs. 10–11–12. Showing Respectively a Typical Four Stroke Diagram, Retarded Combustion and Retarded Spark.

Consider the piston starting on the suction stroke, the piston moving from the position L to K, or from left to right. The movement creates a partial vacuum in the combustion chamber N which is shown on the diagram as the distance OA, equal to 2 pounds below atmosphere according to the pressure scale. The suction line remains at this distance below the atmospheric line until within a short distance of the end of the stroke when it rises to meet the atmospheric line at B when the piston reaches the end of the stroke at K. This rise at the end of the stroke is due to the fact that the piston moves more slowly when approaching the end of the stroke while the velocity of the incoming gases remains nearly constant so that the piston exerts no pull nor suction. On the diagram the entire suction stroke is represented by AB.

The piston now returns on the compression stroke from K to J compressing the mixture in the combustion chamber N. On the diagram this stroke is shown beginning at B, with the pressure slowly rising until the pressure is a maximum at the point C at the end of the stroke. During the compression, the pressure has risen from that of the atmosphere at B to 125 pounds pressure at C as shown by the scale. At a point slightly before C is reached, ignition occurs, and the pressure rapidly rises from C to D, due to the expansion of the heated gas. In this case the combustion is practically instantaneous as shown by the straight, vertical combustion line CD.

At D the piston starts on the working stroke from left to right increasing the volume of the gas and at the same time diminishing the pressure because of the expansion until the maximum pressure of 400 pounds per square inch at D is reduced to 30 pounds per square inch at E, the line DE being called the expansion line. During this time the heated gas has been performing work on the piston. At E the exhaust valve opens and the pressure drops from E to T, a point still about 10 pounds above atmospheric pressure. Theoretically the pressure should drop instantly from E to atmosphere, or from 30 pounds per square inch to zero, but practically this is impossible because of the back pressure due the slow escape of the exhaust gases through the comparatively small valve openings and exhaust pipes. Since considerable pressure is exerted by the piston on the return stroke in forcing the gases out of the exhaust valve, the exhaust line TO on the diagram is nearly 10 pounds above the atmospheric pressure from T to O. At a point near O, the piston slows up on nearing the end of the stroke so the gases have more time to escape through the valves, and the pressure drops to the atmosphere, ready for the succeeding suction stroke.

It should be noted that the points A, B, E, and F represent periods of valve action. At A the inlet valve opens; at B the inlet closes; at E the exhaust opens; at F the exhaust closes, and at A the inlet again opens at the beginning of the suction stroke AB. That this is true is apparent from the fact the inlet must open at the beginning of the suction stroke, and both valves must be closed from the point B to the point E in order to prevent the escape of the compressed charge and expanded gases from the cylinder. At the end of the working stroke the exhaust valve must liberate the gases and remain open to the end of the scavenging stroke to eliminate the residual gas while the closed inlet valve prevents the burnt gases from being forced through the inlet pipe and carburetor.

As shown on the diagram, the exhaust valve closes at the same time that the inlet opens, as F, and O both occur on the same vertical line DL. This is true theoretically, but owing to the different conditions met in practice, the actual setting of the valves may vary slightly from that shown on the diagram. Some makers of high speed engines open the inlet slightly before the exhaust closes as it is claimed that the inertia of the exhaust gas passing through the exhaust pipe creates a slight vacuum that is an aid in filling the cylinder with a fresh charge. It should be borne in mind that this condition only exists when the piston has come to rest and exerts no pressure on the exhaust gas. The vacuum is due to the velocity inertia of the gas after it has been reduced to atmospheric pressure. Other makers close the exhaust valve a very little before the inlet opens, but no matter what the setting, the difference in the time of opening and closing is very small, and the results obtained probably differ by an almost negligible amount.

During the suction and scavenging strokes, the fly wheel of the engine is expending energy on the gas since it is moving a considerable volume at a fairly high pressure. In the case of the scavenging stroke, the piston is working against 10 pounds back pressure, which on a 10 inch piston would amount to a force of 785 pounds. With the 2 pound vacuum the drag on the piston would amount to 157 pounds, no small item when the velocity of the piston is considered. Of course the pressure of 10 pounds per square inch is rather high, but it is often attained with long and dirty exhaust pipes. It is items of this nature that cut into the efficiency of the engine, and increase the fuel bills, and it is only by the indicator that we can determine the extent of such "leaks" and remedy them.

Since the area of the indicator card represents the power of the engine, it is evident that we lose the power represented by the area included in the rectangle FEBO on the scavenging stroke plus the area BOA on the suction stroke. The area included in BCO represents the work taken from the engine in compressing the charge, but this is returned to us during the next stroke plus the benefits gained by compressing the mixture. The arrows show the direction in which the piston is moving during that event.

An actual engine does not follow the form of the diagram shown by Fig. 10 exactly because of certain conditions met with in practice such as imperfect mixtures, faulty valve and ignition timing, small valve areas or leakage. The combustion in the real engine is neither instantaneous nor complete but it approximates the "**IDEAL**" cycle just described more or less closely with a high compression and a fairly well proportioned mixture.

(37) Detecting Faults With the Indicator.

For the best results the gas must be completely ignited at the point of maximum compression, and the pressure must be established on the dead center, so that the indicator card will show a straight and vertical combustion line. As all gases require a certain length of time in which to burn, the ignition should have **LEAD**, that is, should be started before the end of the stroke so that combustion will be complete at dead center. The amount of ignition lead required depends on the fuel and the compression. In Fig. 10 the point of ignition (I) is shown as occurring before the end of the compression at (C), which insures a straight combustion line CD.

With a lean or slow burning gas, that is, a gas slower than used on the diagram, combustion would not be complete at the end of the stroke if the same point of ignition were used. This effect is shown by Fig. (11), in which the full line diagram BCDE represents the ideal diagram (Y), and BCFG represents the slow burning mixture with the same point of ignition (X). The compression curves of both diagrams are coincident as far as C, the ideal diagram shooting straight up at this point and the weak mixture diagram staying at the same level. When under the influence of the mixture (X) the piston starts from left to right and reaches the point F before the slow burning gas reaches its maximum pressure. During this part of the stroke there has been very little pressure on the piston and it will be noticed that the maximum pressure is far below that of the ideal diagram. This low maximum is due principally to the reduced compression under which the gas has been burning, from C to F.

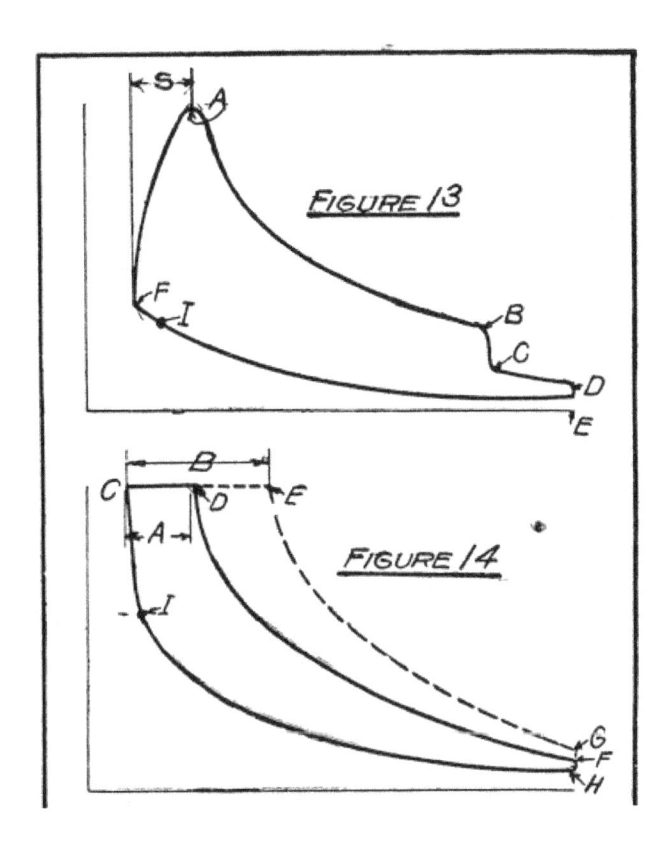

Figs. 13–14. The First Diagram (13) Shows a Two Port Two Stroke Diagram, the Second Shows a Typical Diesel Card.

As the gas has but a small part of the stroke left in which to expand, the pressure at the point of release is much higher than the release pressure of the ideal diagram, which means that a considerable amount of heat and pressure have been wasted through the exhaust pipe. Besides the heat loss, the high temperature of the escaping gas has a bad effect on the exhaust valve and passage. The great volume of gas passing through the exhaust valve also increases the back pressure on the scavenging stroke.

Delayed or retarded ignition will cause a low combustion pressure and slow combustion with any type of fuel or compression pressure as will be seen from Fig. 12. In this case the compression pressures of the ideal diagram Y and the diagram X showing the retarded spark are of course the same, the compression line extending from B to C in the direction of the arrows. At C the ignition occurs for curve Y, and the pressure immediately rises to D.

In the case of curve X, ignition does not occur until the point I is reached, the compression falling on the line CI with the forward movement of the piston as far as the point I. At this point the compression pressure is very low which results in the slow combustion indicated by the slant of the combustion line IF. The point of maximum pressure F is much below D of the ideal curve, and as there is no opportunity for complete expansion during the rest of the stroke, the release pressure is high causing a great heat loss. If running on a **LATE** or **RETARDED** spark is continued for any length of time the excessive heat that passes out of the exhaust will destroy the valves.

It is apparent that for the best results, the spark should occur slightly before ignition in order to gain the effects of the compression, and a high working pressure on the piston. It is also evident that the point of ignition should be varied for different mixtures that have different rates of burning. With engines that govern their speeds by throttling or by changing the quality of the mixture it is necessary for the best results, to vary the point of ignition with each quality of fuel that is admitted by the governor. The retard and advance of the ignition is very necessary on an automobile engine because of the throttling control and constant variation of the load and speed. All automobilists know of the heating troubles caused by running on a retarded spark.

(38) Two Stroke Cycle Diagram.

In the two stroke cycle diagram, the lines showing the suction and scavenging strokes are missing if the indicator is applied only to the working cylinder.

Starting at the beginning of the working stroke as at A in Fig. 13, the gas expands during the working stroke until the piston uncovers the exhaust port at B where the pressure drops to C. A slight travel uncovers the inlet port with the pressure still above atmosphere due to the pressure in the crank case filling the cylinder. The crank case pressure continues from C to D or to the end of the stroke, the pressure dropping slightly at the latter point.

The compression stroke now takes place with the piston moving from right to left, the compression pressure reaching a maximum at F. Ignition occurs slightly before the point of greatest compression, at I, and the expanded gas increases in pressure until the point A is reached. From this point the same cycle of events is repeated. Because of the dilution of the charge by the burnt gases of the preceding combustion, the mixture burns slowly as will be seen from the inclined combustion line FA. Due to this delayed combustion, the piston travels the distance S on the working stroke before the pressure reaches a maximum. This diagram is typical of the small marine type of two stroke cycle engine which has no further scavenging than that performed by the rush of the entering mixture. The diagram of the pressures and vacuums in the crank case are similar to those of suction and compression in the four stroke cycle type.

(39) Diagram of Diesel Engine.

A diagram of the Diesel engine is different in many particulars from that of an ordinary gas engine, as will be seen from the diagram in Fig. 14. The pressures rise in an even, gradual line from the end of the compression curve, and instead of having a sharp peak at the end of the combustion, as in a gas engine, the top of the curve is broad and greatly resembles the indicator diagram of a steam engine. The compression curve constitutes a greater proportion of the pressure line than that of a steam engine, the rise of pressure due to the ignition being very slight in comparison to the height of the compression curve. There is no explosion in the usual sense of the word, only a slight increase in pressure as distinguished from the rapid combustion in the gas engine.

Starting at the beginning of the compression stroke at H, the pressure of the pure air charge increases to about 500 pounds to the square inch at I, the point at which the fuel is injected. From I to C is the increase of pressure due to the combustion. The pressure stays at a constant height from C to D as the fuel supply is continued between these points, and is cut off when the piston reaches the position D. It will be seen that the admission of the fuel through the distance A covers a considerable proportion of the working stroke, and that the points of fuel injection and ignition are coincident.

From the point of fuel cut-off at D expansion begins and is continued in the usual manner to F, the point of release.

When the load is increased, the period of oil injection is also increased, the other events remaining constant. Should the light load require an oil injection period as shown by A, the greater load would require injection for the period B. In the latter case, the expansion line would be EG, which would produce a diagram having a greater area than the line DF, and there would be a great increase in the release pressure GH as well.

It will be seen from the diagram that the quantity of air taken into the cylinder and the compression pressure remain constant with any load, and that for this reason it is possible to have a constant point of ignition, or rather point of fuel injection. As there is no mixture compressed, there are no difficulties encountered at light loads due to attenuated mixtures. An excess of air over that required to burn the fuel is also present at every load within the range of the engine. For the sake of simplicity, the suction and scavenging lines on the Diesel engine have been omitted, but they are the same in all respects as the corresponding lines shown in the diagram, Fig. 14.

(40) Gas Turbine Development.

In the attempt to gain mechanical simplicity, small weight, and diminutive size of the steam turbine, many able experimenters have endeavored to obtain an internal combustion motor in which the energy of the expanding gas is converted into mechanical power by its reaction on a bladed wheel, but so far the problem is far from being solved. In 1906 two experimental turbines were built by René Armengand and M. Lemale, of the constant pressure type, one of which developed 30 Brake horse-power and the other 300 horse-power.

A 25 horse-power De Laval steam turbine was altered by Armengand says Dugald Clerk so that it operated with compressed air instead of steam. The compressed air was passed into a combustion chamber together with measured quantities of gasoline vapor, and the mixture was ignited by an incandescent platinum wire as it entered the chamber, thus maintaining a constant pressure with continuous combustion. Around the carborundum lined combustion chamber was imbedded a coil in which steam was generated by the heat of the burning gas, the steam being used to reduce the temperature of the gas from 1800°C to about 400° as it issued from the orifice and came into contact with the running wheel. The working medium was therefore composed of two elements, the products of combustion and the steam at the comparatively low temperature of 400°C.

The constant pressure maintained in the combustion chamber was about 10 atmospheres, and the hot gases were allowed to expand through a conical Lava jet in which the expansion produced a high velocity, and reduced the temperature of the fluid. At this reduced temperature and high velocity the gases impinged upon the Laval wheel, and rotated the wheel in the same way as steam would have done. The experiments showed that under these conditions the total power obtained from the turbine separate from the compressor was double that necessary to drive the compressor.

In the large 300 H. P. turbine the first part of the combustion chamber was lined with carborundum, backed by sand, but the second part was surrounded by a coil through which water was circulated. The water kept the temperature of the combustion chamber within safe limits, and after absorbing heat, it passed also around the jet nozzle, and was discharged into the passage leading to the jet, and there converted into steam by the hot gases. A mixture of products of combustion and steam thus impinged upon the turbine wheel. The expanding jet was arranged to convert the whole of the energy into motion before the fluid struck the wheel; the temperature was thus reduced to a minimum before the gases touched the blades. Notwithstanding this, the wheel itself had passages through which cooling water flowed, and each blade was supplied with a hollow into which water found its way. In the large turbine the compressor was mounted on the turbine spindle; it was of the Rateau type, and consisted of an inverted turbine of four stages, which delivered the compressed air finally to the combustion chamber at a pressure of 112 lb. per sq. in. absolute. The efficiency of this turbine compressor was found to be about 65 per cent. The total efficiency of the combined turbine and compressor was low, as the fuel consumption amounted to nearly 3.9 lb. of gasoline per B. H. P. hour. An ordinary gasoline engine with a moderate compression can readily give its power at the rate of 0.5 lb. of gasoline per B. H. P. hour. The combined turbine and compressor was stated to have run at 4,000 R. P. M. and to have developed 300 H. P. over and above the negative work absorbed by the compressor.

A gas turbine in which there was no compression was built in the following year by M. Karovodine which gave 1.6 horsepower at a speed of about 10,000 revolutions per minute.

It contained four explosion chambers having four jets actuating a single turbine wheel, which wheel was of the Laval type, about 6 inches diameter, having a speed of 10,000 R. P. M. The explosion chambers were vertical, and had a water jacket surrounding the lower end. The upper portion contained the igniting plug on one side, and the discharge pipe connecting with the expanding jet on the other. In the lower water-jacketed part there was provided a circular cover, held in place by a screwed cap. This circular plate was perforated with many holes, and it carried a light steel plate valve of the flap or hinging type, which pulled down by a spring contained within the admission passage. This spring could be adjusted, and the lift of the valve was regulated by means of a set screw passing diagonally through the water jacket. Air was admitted at one side by a pipe leading into the valve inlet chamber and a corresponding passage or pipe admitted gasoline and air or gas to mix with the air before reaching the thin plate valve. Adjusting contrivances were supplied in both air and fuel ducts. To start the apparatus, an air blast was forced through the valve, carrying with it sufficient gasoline vapor to make the mixture explosive. The electrical igniter was started, and the spark kept passing continuously. Whenever the inflammable mixture reached the upper part of the combustion chamber ignition took place, and the pressure rose in the ordinary way, due to gaseous explosion. The gases were then discharged through the pipe and nozzle on the Laval wheel. The cooling of the flame after explosion and the momentum of the moving gas column reduced the pressure within the explosion chamber to about 2 lb. per sq. in. below atmosphere. Air and gasoline vapor then flowed in to fill up the chamber, and as soon as the mixture reached the igniter, explosion again occurred. In this way a series of explosions was automatically obtained, and a series of gaseous discharges was made upon the turbine wheel. Diagrams taken from the explosion chamber showed a fall in pressure during suction of 2 lb. per sq. in.; ignition occurred while the pressure was low, and the pressure rapidly rose to about 1 1-3 atmospheres absolute. The pressure propelling the gas column and jet was thus only 5 lb. per sq. in. above atmosphere. The pressure rapidly fell, and the whole process was repeated again. According to the diagrams taken, a complete oscillation required about 0.026 second, so that about 40 explosions per second were obtained.

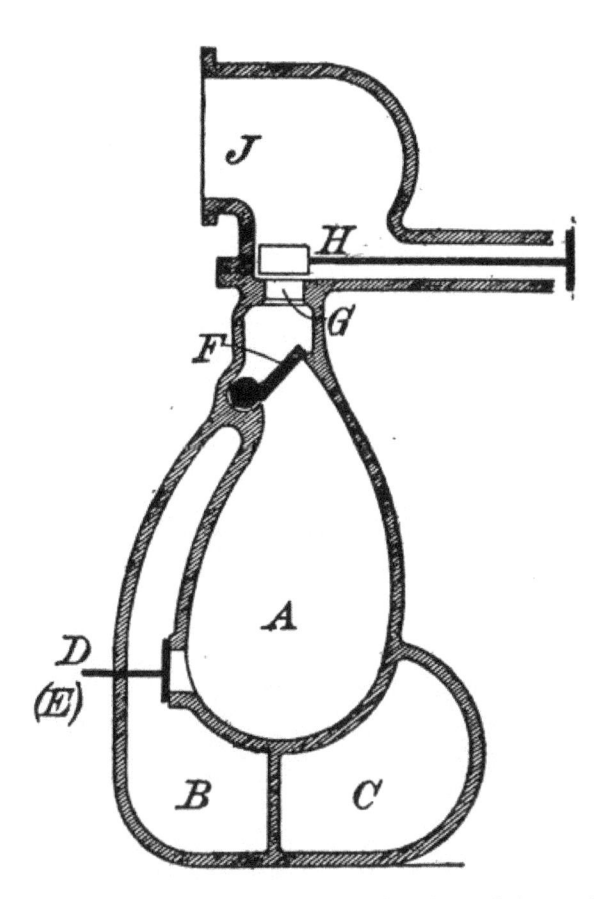

Fig. 15. Cross-Section of the Combustion Chamber of the Holzwarth Gas Turbine. From the Scientific American.

The most promising type of turbine that has been built to date is that designed by Hans Holzwarth, an engineer of some prominence in the steam turbine field. A 1000 horse-power machine has been built at this writing and as experimental machines go has made most remarkable performance.

The turbine in general arrangement outwardly resembles the Curtis steam turbine, in that the turbine wheel rotates in a horizontal plane, the spindle or shaft is vertical and a dynamo is mounted on this spindle above the turbine. In the Holzwarth turbine ten combustion chambers are provided, each of a pear or bag shape. They are arranged in a circle around the wheel, and are cast so as to form the base of the machine. The wheel is of the Curtis type, with two rows of moving and one row of stationary blades.

In this turbine the energy of the fuel is liberated intermittently by successive explosions, instead of by continuous combustion, and in much the same way that the explosions occur in a reciprocating engine. Tests made on the new machine have shown that it is in no way inferior in efficiency to the ordinary type of motor, and that at full load, the weight per horse-power is only about one-quarter of that of the reciprocating engine. The weight factor, as is well known, is of the utmost importance in marine service and should prove of value to the marine engineer, if this alone were its only characteristic.

Any of the ordinary power gases may be used with success, as well as vaporized liquid fuels, and the lower grade oils such as crude and kerosene have given much better results in the turbine, than in reciprocating engines, even at this early stage of its development. As the heat losses are much smaller than met with in ordinary practice, the temperature is higher, which, of course, greatly facilitates the vaporization of the lower grade liquids.

Mr. Holzwarth does not give the dimensions of his turbine wheel, but from the drawings and some of the velocities given by him it appears to be about 1 m. in external diameter. The lower part of each combustion chamber carries gas and air inlet valves, and the upper part carries a nozzle arranged to cause the gases to impinge upon the first row of moving blades. This nozzle is connected to and disconnected from the combustion chamber by means of an ingeniously operated valve. The explosion chambers are charged with a mixture of gas and air, which appears to attain a pressure of about two atmospheres within the chamber before explosion. The air and gas are supplied under sufficient pressure from turbine compressors, actuated by steam raised from the waste heat of the explosion and the gases of combustion, so that whatever work is done in compression is obtained by this regenerative action, and does not put any negative work upon the turbine itself. The combustion chambers are fired in series, by means of high-tension jump spark ignition.

Referring to the cut, the explosion chamber A is filled intermittently with the explosive mixture at a low pressure (about 8 to 12 pounds per square inch). When ignition has occurred, the pressure of explosion opens the nozzle valve F, allowing the compressed gases to flow through the nozzle G to the bladed turbine H, on which the energy is to be expended. The expansion of the heated gases in the nozzle reduces the pressure to that of the exhaust, with the resulting increase in the velocity of the gas. By means of fresh air, the nozzle valve F is kept open throughout the expansion and scavenging periods.

After the expansion has been completed, the air that is forced through the valve D, at a low pressure, thoroughly scavenges or removes the residual burned gases left in the combustion chamber and nozzle, forcing it into the exhaust. When the scavenging has been completed, the nozzle valve and the air valve D are closed. The combustion chamber A is now filled with pure cold air, which not only enables a fresh charge of gas to be introduced into the chamber but which also aids in keeping the chamber cool.

Pure fuel gas, or atomized oil, is now injected through the fuel valve E, forming an explosive mixture ready for the ensuing cycle of events. A number of these chambers are arranged around the turbine wheel in order to have a uniform application of power, by having the several chambers working intermittently. This is in effect, the same proposition as increasing the number of cylinders on a reciprocating engine.

CHAPTER V
TYPICAL FOUR STROKE CYCLE ENGINES
(41) Essential Parts of the Gas Engine.

On all gas engines of accepted type are found certain devices necessary for the performance of the events or cycles outlined in the preceding section.

For the sake of simplicity these devices are treated as a part complete in itself. The details of construction, and the refinements found necessary in the actual construction will be described in the succeeding chapters.

The names and purpose of these essential components, and their relation to the operation of the engine as a whole, will be found in the following outline:

1. The **CARBURETOR** is a device whose purpose is to vaporize the liquid fuel, and mix the vapor thoroughly and in correct proportions with the air required for the combustion, in the engine cylinder.

The combustible mixture thus formed is drawn into the cylinder of the four stroke cycle engine or into the crank chamber of the two stroke cycle engine.

GENERATOR VALVES or **MIXING VALVES** are similar to the carburetor in principle but are slightly different in detail.

2. The **CYLINDER** is the containing vessel in which the combustion and expansion of the gas takes place.

The cylinder as its name would suggest has a circular opening or bore extending from end to end, the bore being smoothly finished to receive the reciprocating piston.

3. The **PISTON** is a plunger or movable plug fitting the bore closely enough to prevent the escape of gas, but at the same time is capable of sliding freely to and fro.

When pressure is established in the cylinder from the combustion, pressure is also exerted on the end of the piston tending to force it out of the cylinder. The extent of this force is governed by the area of the end of the piston and also by the pressure of the gas.

Thus the purpose of the piston is to convert the pressure of the expanding gas into direct mechanical force, and also to transform the increasing volume of gas into motion. Another, and no less important function of the piston is to compress the combustible gas in the upper end of the cylinder for ignition.

Piston and Connecting Rod of the Sturtevant Aero Motor, Showing Three Piston Rings.

4. The **CONNECTING ROD** (Sometimes called the Pitman) transmits the pressure on the piston to the crank, the connecting rod being the means through which the to and fro motion of the piston is transmitted into the rotary motion of the crank; its action being similar to that of the human arm turning the crank of a pump or windlass.

5. The **CRANK** receives the pressure and motion of the piston from the connecting rod, changing the reciprocating motion of the piston into the rotary motion required by the machinery which the engine drives.

In the majority of cases the crank revolves, while the cylinder stands still, but in some of the recently developed aeronautic motors this is reversed, the cylinders revolving with the crank stationary. The relative motion, however, is the same in both cases.

(6.) The **CRANK SHAFT**, of which the crank is an integral part, transmits the rotary motion of the crank to the driving pulley.

(7.) The admission and release of the gases to and from the cylinder are controlled by the **INLET VALVE** and **EXHAUST VALVE**, respectively, in a four stroke cycle engine.

The valves are merely gates, allowing the gas to flow, or stopping it, at the proper intervals, depending on the event taking place at that time in the cylinder.

In the two stroke cycle engine there are no valves, the admission and release of the gas being controlled by the position of the piston, and the openings cut in the cylinder walls.

6. **IGNITION** or the firing of the combustible charge is accomplished by the **IGNITION SYSTEM**. In most modern engines the mixture is ignited when it is under the greatest pressure or at the end of the stroke.

For maximum efficiency the mixture should be ignited when it is under the greatest pressure or compression. The time at which ignition occurs is also controlled by the ignition system.

7. The **GOVERNOR** regulates the speed of the engine; either by changing the richness of the mixture, by changing the number of working strokes in a given time or by altering the quantity of gas admitted to the cylinder, or sometimes by a combination of these methods.

8. The **BELT WHEELS** or **PULLEYS** are the means of transmitting the power of the engine to the work to be performed. The engine is generally connected to the driven machinery by a belt connecting the engine pulley with the pulley of the driven machine.

9. The **FLY WHEELS** by reason of their mass and their momentum, store up a portion of the energy expended during the working stroke, and return it to the engine in order to carry it through the idle strokes of compression, admission and expulsion. In some engines the fly wheels serve in double the capacity as pulleys.

10. The **BASE** or **FRAME** of the engine acts as a foundation for the various working parts, holding them in their proper positions.

(42) Application of the Four Stroke Principle.

While the five events of every commercial four stroke cycle engine are accomplished in exactly the same order, or routine as explained in paragraph (8), Chapter 3, the actual design and method of applying the cycle varies greatly in different makes of engines. This great difference in the details of construction often makes it difficult for the novice to identify the cycle of operations in that particular engine. The different forms of valve gears that are used to perform the same functions in the cycle are good examples of the variation in design, some makers using the poppet or disc type, some the sliding sleeve, and others the rotary type.

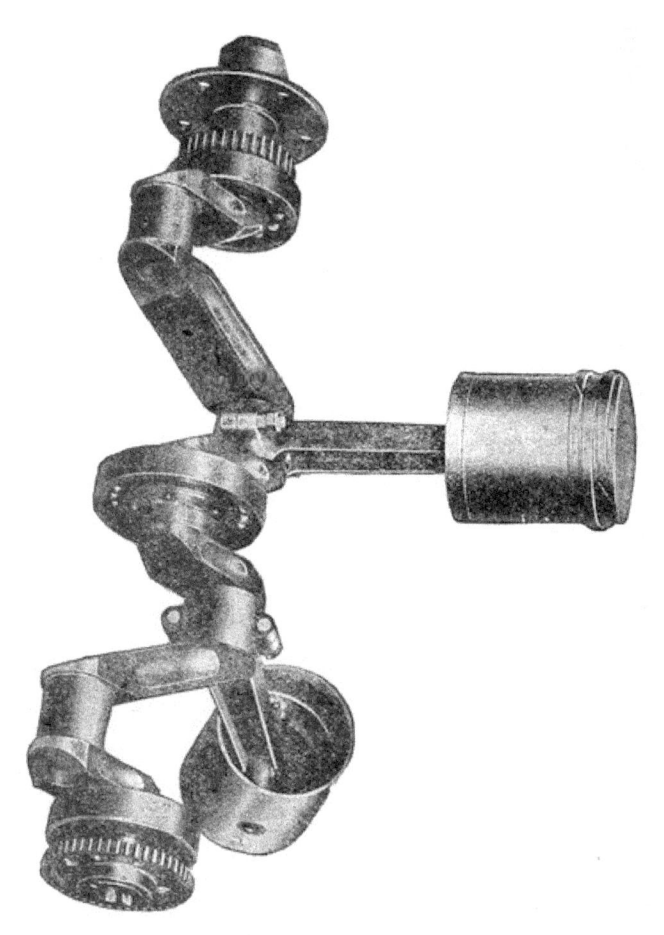

Fig. 16. Ball Bearing Crank Shaft, Pistons and Connecting Rods of the "Maximotor," in Their Relative Positions.

Multiple cylinder engines vary in the cylinder grouping or arrangement, the arrangement and number of cylinders depending on the service for which the engine is intended, the amount of vibration permissible, or the weight. The question of speed also introduces modifications in the design, but no matter what valve arrangement is adopted or what grouping of cylinders is used, a four stroke cycle engine performs the five events of suction, compression, ignition, expansion and exhaust in four strokes, in each and every cylinder. With the exception of fuel injection (which in reality corresponds to the ignition event) in the four stroke Diesel engine, the indicator cards of all four stroke cycle engines passes the same characteristics as the diagram shown in Fig. 10.

In this chapter, the engine will be described without regard to the fuel used, or to the means adopted in vaporizing it, for the vaporizing appliances are considered as being external to the engine proper, except in some of the heavy oil engines, and as the fuel is gasified before entering the cylinder the question of fuel does not affect the general construction of the engine. The majority of engines are readily converted from gasoline to gars, or in some cases kerosene, by changes in the vaporizing device, and with the exception of changing the compression pressure, little further alteration is needed. Since the vaporization and admission of the heavier oils, such as crude oil and kerosene has a more intimate relation to the engine than the use of gasoline or gas, the heavy oil engines will be described in a separate chapter in order that the process of oil burning may be more fully explained. It should not be understood that the cycle, or principle of the oil engine differs from that of any other engine, but that the vaporizer forms such a close connection with the engine proper that they must be described as one unit.

(43) Horizontal Single Cylinder Engine.

An example of a modern single cylinder engine operating on the four stroke cycle principle is the "Muenzel" engine shown in Section by Fig. 17. It is of the single acting type, that is, the pressure of the gases acts only on the left end of the piston which reciprocates in a horizontal direction. Surrounding the cylinder in which the piston slides, is the water jacket (shown by the short horizontal dashes) which keeps the cylinder walls from becoming overheated by the successive explosions of the mixture. The cooling water is pumped into the jacket through the pipe shown over the cylinder, and flows out of the jacket through an outlet near the bottom of the cylinder.

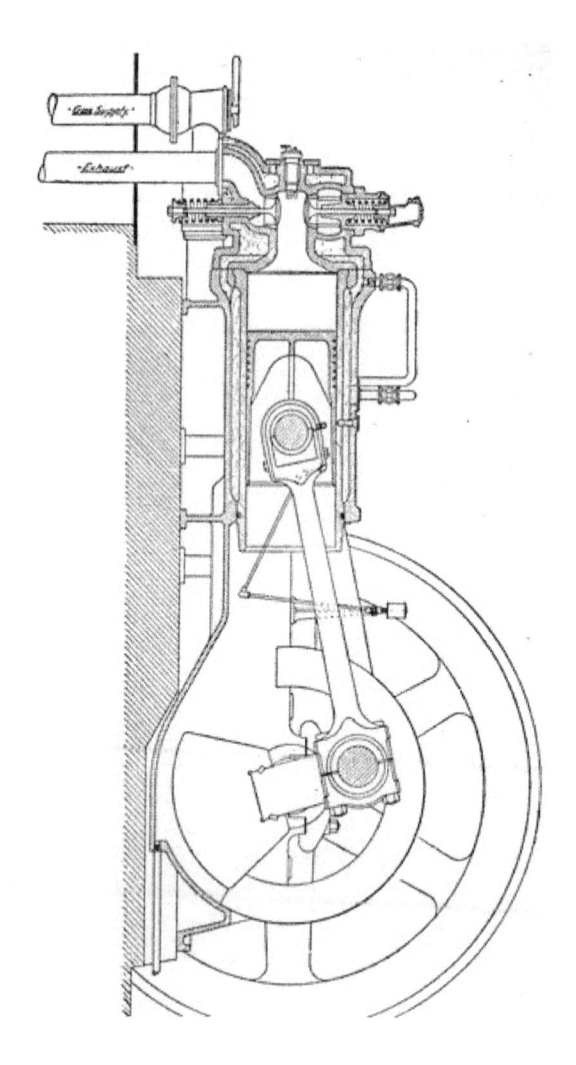

Fig. 17. Longitudinal Section Through the Muenzel Horizontal Engine.

Both the inlet and exhaust valves are situated in an extended portion of the combustion chamber to the left of the piston, the upper valve being the inlet and the lower valve, the exhaust. The valves are held on their seats by means of coil springs that act on the upper ends of the valve springs. Admission of the explosive mixture is controlled by the upper valve, and the release of the burnt gases by the lower. Pipes at the bottom of the cylinder marked "Gas Supply" and "Exhaust" convey the gases to and from the inlet and exhaust valves respectively.

Fig. 18. Elevation of Muenzel Engine Showing Lay Shaft and Valve Connections.

The inlet valve, and the inlet valve spring are held in one unit by a removable metal housing known as a "Valve Cage", which is arranged so that the cage, valve, and spring may be removed as one piece from the cylinder casting when the valves need attention by removing a few bolts. As the cage is directly over the exhaust valve, and is considerably larger in diameter, it is possible to remove the exhaust valve through the opening left by the removal of the inlet valve cage. Both valves are surrounded by a water jacket, as are the passages that lead to them.

Both the inlet and exhaust valves are opened and closed at the proper moments in the stroke by means of cams mounted on the horizontal cam shaft shown by Fig. 18 through a system of levers. The cam shaft is the shaft running parallel to the engine bed from the crank-shaft to the end of the cylinder and turns at one-half the speed of the crank-shaft. At a point directly below the inlet valve in Fig. 18, will be seen an enlargement on the shaft on which rests the rod running from the inlet valve to the cam shaft. This is the cam.

A cylindrical casing shown above the cylinder contains the governor which maintains a constant speed at all loads by operating a valve in the intake pipe which varies the quantity of mixture entering the cylinder in proportion to the load. The governor is driven from the cam-shaft by spiral gears. The igniter which furnishes the spark for igniting the gas is located between the two valves at the extreme left of the combustion chamber (Fig. 17).

It should be noted that the cylinder head which closes the left end of the cylinder, and which carries the valves is separate from the main body of the Cylinder. By unscrewing the bolts that hold it to the cylinder, the head may be removed when it becomes necessary to remove dirt and carbonized oil from the combustion chamber, or when it becomes necessary to remove the piston. The cylinder barrel in which the piston works may also be removed through the opening left by the piston head when it becomes worn, and another barrel or liner may be substituted, thus practically renewing the engine at a small fraction of the cost of a new cylinder. The liner is fastened firmly to the outer cylinder casting at the left but is free to slide back and forth in the casting at the right hand end, this end being provided with a packed joint. This play given to the liner allows it to expand and contract freely with the different changes of temperature without causing strains either in the cylinder or in the liner.

(44) Multiple Cylinder Engines.

Since the power exerted by a single cylinder four stroke cycle engine is intermittent, the explosive force exerted on each power stroke is much heavier than would be the case if the power application were continuous, as the explosions must be heavier to compensate for the idle periods. To reduce the strain on the engine and the vibration as well and to obtain an even turning moment it has been customary to provide more than one cylinder on engine of over 10 horse-power capacity. In this way the total power is divided among a number of cylinders, and as no two cylinders are under ignition at any one time the turning moment is more even, the vibration is less, and the strain on the engine is considerably reduced.

Dividing the power in this way makes it possible to reduce the weight of the engine as less material is required to resist the strains and a small flywheel may be used because of the even engine torque. In order to gain the full benefit of this reduction in weight, the builders of aeronautic motors have carried the multiplication of cylinders to an extreme, the Antoinette for example having sixteen cylinders. Engines having more than six cylinders exert a continuous pull as the impulses "overlap," that is, ignition occurs in one cylinder before another cylinder in the series ends its working stroke. The greater the number of cylinders, the more continuous will be the torque or turning moment. The multiple cylinder engine may be considered as a group of single cylinder engines connected together, and receiving their fuel from a common source, the only difference between the single and multiple being in the inlet and exhaust piping and the ignition system.

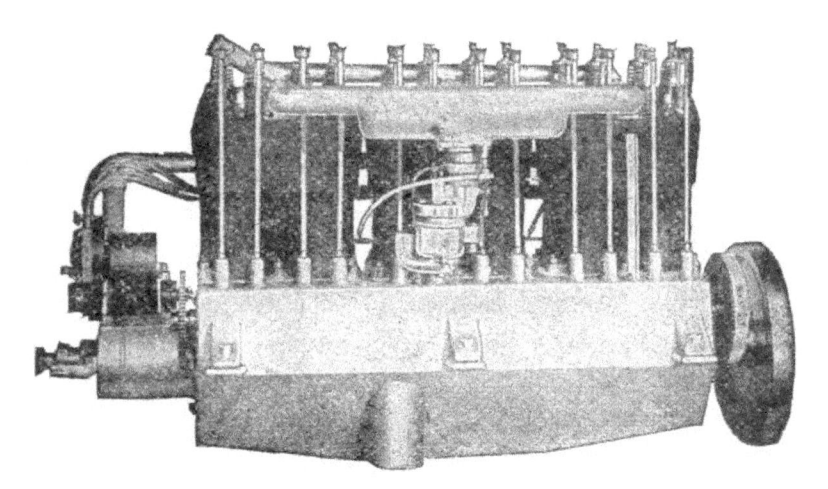

Fig. F-12. Six Cylinder Maximotor.

Fig. F-13. Four Cylinder Buffalo Motor for Marine Service.

As a single cylinder four stroke cycle engine has one working impulse in every two revolutions, a two cylinder engine will have an impulse for every revolution as there are twice as many impulses in the same time. It should be remembered that the number of impulses given per revolution by a four stroke cycle engine is equal to the number of cylinders divided by two. Thus, a six cylinder engine has $6 \div 2 = 3$ impulses per revolution, and an eight cylinder, $8 \div 2 = 4$ impulses, providing of course, that the engine is single acting.

Arrangement of the cylinders varies with the service for which the engine is intended and the perfection of balance that is required, the principal arrangements being the "V," the "upright," the opposed, the "radial," "tandem," and "twin." The upright engine has the cylinders all on one side of the crank-shaft in a straight line, as in the four cylinder automobile engine. In this form, each cylinder has an individual crank throw the number of throws being equal to the number of cylinders. This engine is fairly well balanced in the four, six and eight cylinder types, as one-half of the connecting rods and throws are up, while the other half are down, but as the connecting rods do not all make equal angles with the center line of the cylinder at the same time there is a slight unbalance in the four and six cylinder types. Because of the ignition sequence, two cylinder vertical motors are in no better balance than the single cylinder type since both crank throws and connecting rods are on the same side of the shaft at the same time. For this reason the two cylinder engine is most commonly built in the opposed type which gives perfect balance.

In "V" type arrangement, one-half of the cylinders are set at an angle of about 90° with the rest of the cylinders, or in the two cylinder "V" the cylinders are set in the same plane, perpendicular to the shaft, at angle varying from 57½° to 90°. The "V" type arrangement is adopted where light weight and compactness are the principal requirements, as the weight and length are both reduced by putting the cylinders opposite to one another by pairs, the "V" being practically one-half the length of an upright having the same number of cylinders. This arrangement permits the use of one-half the number of crank throws used in the vertical type as each crank throw acts for two cylinders. For the reason that both the cylinders of a two cylinder "V" act on a common crank throw, the two cylinder "V" is in no better balance than a single cylinder engine.

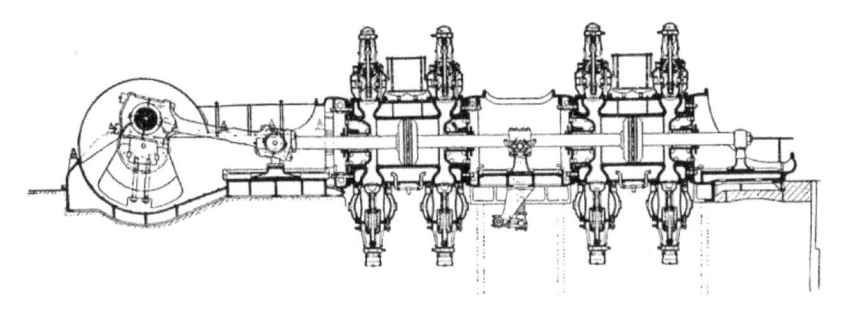

Fig. 18-a. G. H. H. Double Acting Tandem Cylinder Engine (German). It will be Noted that an Inlet and Exhaust Valve Are Placed at Both Ends of Each Cylinder. The Exhaust Valves Are Below and the Inlets Above the Cylinders. As this Engine is of the Four Stroke Cycle Type, Each Cylinder Gives One Impulse per Revolution, or Two Impulses per Revolution for Both Cylinders. The Piston and Piston Rod Are Both Cooled by Water, and Are Supported by the Cross Heads so that Their Weight is Taken Off the Cylinder Bore.

An "opposed" type engine is in the most perfect mechanical balance of any engine as the crank shafts and connecting rods are not only on opposite sides of the crank-shaft, but make equal angles with the center line of the cylinders as well, at all points in the revolution. The explosive impulses occur at equal angles in the revolution as in the four and six cylinder vertical type. An opposed engine may be considered as a "V" having a cylinder angle of 180°. In the opposed type, one crank throw is provided for each cylinder, the pistons of the opposite cylinders traveling in opposite directions at the same time.

A "radial" or "Fan" type motor, as the name would suggest has the cylinders arranged in one or two rows around the crank case, each cylinder being on a radial line passing through the center of the cylinder with one crank throw for each row. The Gnome engine illustrated elsewhere in the book is an example of this type, the seven equally spaced cylinders acting on a common crank throw. When more than seven cylinders are used on this engine, as in the fourteen cylinder engine, two cranks are provided, each crank serving seven cylinders. This arrangement cuts down the weight of a motor enormously because of the short crank shaft and case. With the ignition properly timed and the cylinders correctly spaced the firing impulses occur at equal angles.

"Tandem" cylinders are employed only on stationary engines, the cylinders being placed on the same center line, one in front of the other, and when this arrangement is adopted it is the usual practice to make the cylinders double acting. The two pistons are connected by a rod known as the "piston rod" which extends from the rear end of one cylinder into the front of the following cylinder. Tandem cylinders require too much room for use on automobiles or motor boats, and for this reason are seldom seen in this service.

The "twin" engine is a modification of the vertical cylinder arrangement, both cylinders being on the same side of the shaft and in line with one another. It is the type most generally used on very large stationary engines that have more than one cylinder, and instead of being vertical as in their prototype are generally laid horizontally. Since the twin engine is generally double acting, the crank throws are placed on opposite sides of the shaft.

(45) Four Cylinder Vertical Auto Motor.

A common type of four cylinder vertical motor is shown by Fig. 19, which is of the type commonly used on automobiles. In order to show the general construction of the cylinder, each cylinder is cut through at a different point. The cylinder at the extreme left is shown in elevation, or as we would see it from the outside. In the second cylinder from the left, the section is taken through the valve chamber, which projects from the side of the cylinder. A section through the center of the cylinder is shown on the third cylinder, and the fourth cylinder is in elevation.

On cylinder No. 1, (left) is seen the exhaust pipe (32) and the inlet pipe (31) entering to valve chamber and connected to the exhaust valve and inlet valve respectively. The pipes are held in place by the clamp or "crab" (33). The exhaust pipe connects with the exhaust valve of each cylinder, and terminates at the fourth cylinder as shown by (32). Screwed into the top of the valve chamber on cylinder No. 1 are the two spark plugs (34) and the relief cock (35).

Referring to cylinder No. 2, the inlet valve (42) is shown at the left of the chamber and the exhaust valve also shown by (42) is shown at the right. Above the valves are the spark plugs (34) which project into the space above the valves. Pressing against the lower ends of the valve stems and holding the valves tight on their seats are the springs (44) which fit into the washers (45) fastened to the stems. The valve stems terminate in a nut at (48). The valve stem guides (43) form a support for the valves and at the same time form an air tight connection for the stems to slide in.

Immediately beneath the stems are the push rods (46) which are provided with an adjustment (48) at the upper end, and a roller (49) at the lower end. The rollers (49) rest directly on the cams mounted on the cam shaft (27), and as the irregular cams revolve, the push rods are moved up and down which in turn act on the valve stems and raise the valves at the proper moment. The cams raise the valves and the springs close them. The two cams (exhaust and inlet) appear as two rectangular enlargements on the shaft (27). The bearings (53), support the cam shaft, one being supplied for each cylinder.

At the extreme left of the crank shaft is shown the half time gear (20) which meshes with the gear on the crank-shaft and drives the cams. Next to this gear is the large cam shaft bearing 26. It should be noted that the section through the valve chamber taken on cylinder No. 2 is at a point considerably back from the center line of the cylinders and not in the same plane as the section shown on cylinder No. 3, which is taken through the center line of the cylinders.

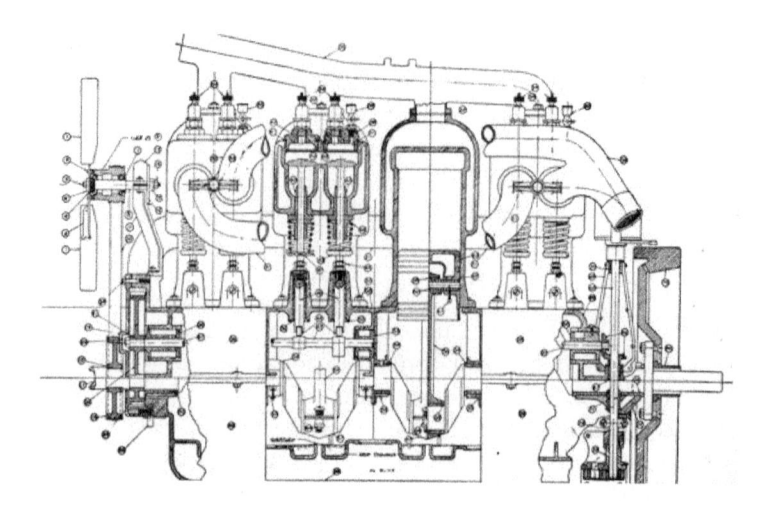

Fig. 19. Cross-Section Through Typical Four Cylinder Automobile Engine. Courtesy of the Chicago Technical College.

In the section of cylinder No. 3, we see the water space surrounding the upper portion of the cylinder with the opening (37) connected to the water manifold (36), through which the water leaves the cylinder and passes to the radiator. At the lower end of the stroke is the piston, one-half of which is shown in section and one-half in elevation so that internal and external appearance may be readily seen. The piston pin (60) is located approximately in the center of the piston to which it is secured by means of the set-screw (61).

By means of the connecting rod (56), the motion of the piston is transmitted to the crank-shaft throw (54), both ends of which are provided with bronze bushings (59) and (58), fitting on the piston pin and crank-pin respectively. Between each crank throw are the main crank shaft bearings (55) which are provided with the bronze bushings (54). Below the connecting rod ends is the small drip trough containing oil into which the pipes on the rod ends dip when passing around the lower end of the stroke. When the pipes enter the oil puddle a small amount of lubricating oil is driven into the crank-pin bearing because of the force of impact, this force also causing oil to splash about in the crank case for the lubrication of the main crank shaft bearings and cam shaft. In order to maintain a constant level of oil in the puddle so that the bearings shall receive a constant supply of oil, a small overflow opening is placed in the center of the puddle which allows an excess of oil to overflow into the return oil sump below.

This excess of oil drains by gravity back to the oil circulating pump (73), at the right which again forces the oil to the various bearings. In this way, the same oil is used over and over again until it becomes unfit for lubricating purposes because of dirt or decomposition. The oil pump is driven from the cam-shaft through the level gears (66) and the vertical shaft (72). To the right of the oil pump is the fly-wheel (75) which furnishes the power for the idle strokes of the engine.

At the upper end of the vertical shaft that drives the oil pump is an extension (68) which passes through the bearing (70) and drives the ignition timer shown at the top of the housing (69). The timer controls the period of ignition in the cylinders in regard to the piston position so that the spark occurs at the end of the compression stroke. At the extreme left of the engine is the radiator fan (1) which is driven from the crank-shaft pulley (16), the belt (10), and the fan pulley (1122). This fan increases the amount of cold air that is drawn through the radiator, (mounted to the left of the engine) and increases its capacity for cooling the jacket water of the engine. The water circulating pump is located on the opposite side of the motor.

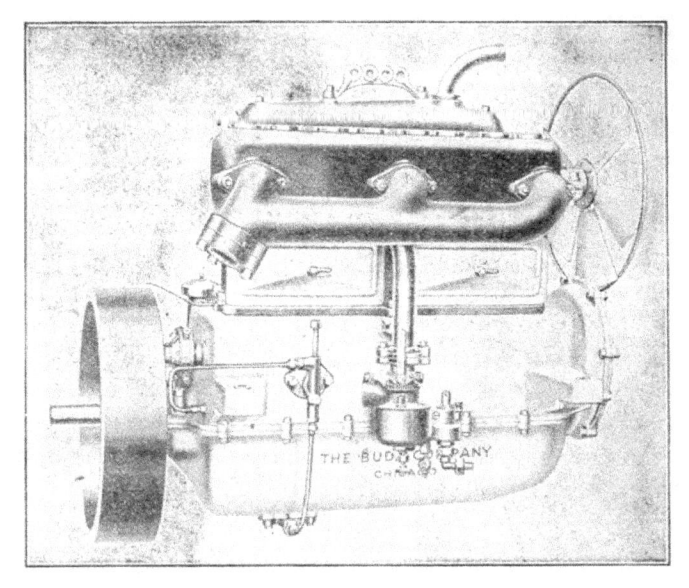

Fig. 19-a. Buda Four Cylinder Automobile Motor. Carburetor Side.

Fig. 19-b. Buda Motor, Pump Side, Cylinders "En Bloc."

In this motor both the inlet and exhaust valves are located on the same side of the cylinder which arrangement classifies the engine as an "L" type, the extended valve pockets forming an "L" with the center line of the cylinder. In the motor shown by Figs. F-14-F-15, the inlet and exhaust valves are on opposite sides of the cylinder as shown in the cross-section, which classifies the motor as a "T" type, as the valve chambers together with the cylinder forms a "T." The latter type of motor has several advantages over the "L" type, but as it requires two cam shafts, one for the inlet and one for the exhaust valves, it is not adopted by the builders of the cheaper grades of automobiles. Since the exhaust valves are on the opposite side of the cylinder, in the "T" type, the inlet air is not expanded nor the output diminished by the heat of the exhaust passages. The piping is less complicated which permits of a more effective arrangement of the carburetor and magneto. Since the piping in the latter type can be arranged to better advantage, less back pressure is the result.

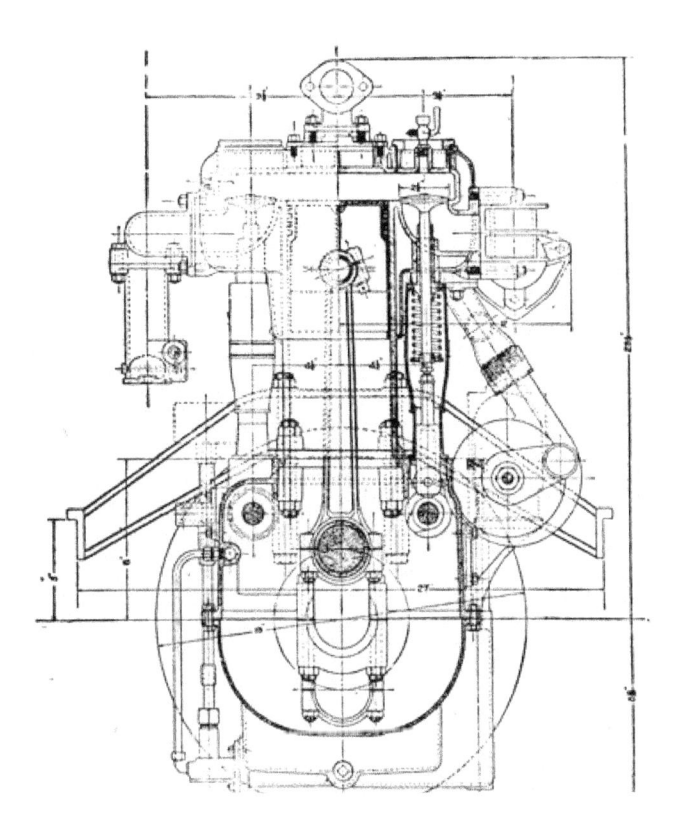

Fig. F-14. Cross-Section Through Wisconsin Truck Motor. "T" Type.

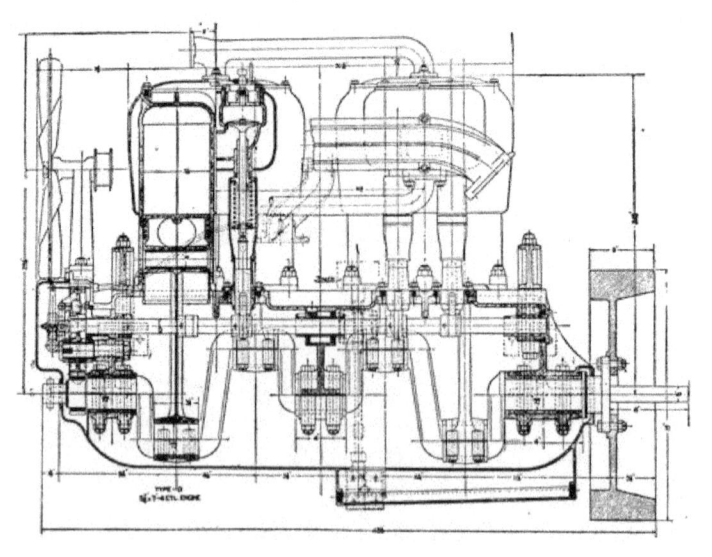

Fig. F-15. Longitudinal Through Wisconsin Truck Motor.

As in the previous case, the valves are acted on directly by the cams and push rods, one cam shaft being provided on each side of the cylinders. In order to reduce the noise made by the push rods and springs, all of the springs are enclosed by sheet metal housings or tubes. The circulating pump is shown at the left nearly on a line with the left hand cam shaft, the pump outlet being inclined toward the cylinder so that it enters the water jacket under the exhaust valves. Water leaves the jacket by the pipe shown on the cylinder tops.

From the longitudinal section it will be seen that the cylinders are cast in pairs, two cylinders to the pair, instead of singly as in the previous case. The large pipe crossing at about the center of the cylinders is the exhaust pipe (shown in front of the left pair), and the pipe shown under the exhaust is the water inlet pipe from the circulating pump. It will be seen from the longitudinal section that the main crank-shaft bearings are fastened to the upper half of the crank case, and are entirely independent of the lower half which acts simply as an oil shield. This construction allows the oil shield (lower half) to be removed without disturbing the adjustment of the bearings, when it becomes necessary to inspect the internal mechanism.

Six Cylinder Rutenber Automobile Motor, with Cylinders Cast in Pairs.

Large removable plates cover the top of the water jackets so that it is a simple matter to clean out the water space in case that it becomes coated with deposits from the water. This is an important feature as a great many of the heating troubles may be overcome by having access to the interior of the water jacket. The water outlet pipes connect with the jacket covers. Both cam shafts are driven by the gears at the right which connect with the crank shaft pinion. Fan is belt driven from an extension to the cam shaft.

All bearings are supplied with oil by a high pressure force feed pump, the crank pins receiving their supply through channels drilled in the crank shaft and pin, which in turn are connected to the oil supply of the main bearings, no dependence being placed on a splash system. After leaving the bearings, the oil drops into the crank case and drains into the sump shown at the left of the longitudinal section. From the sump, the oil returns to the oil pump from which point it is returned to the circulating system under high pressure.

(46) Stationary Four Cylinder Engine.

An English stationary engine, the Browett-Lindly, similar in many respects to the automobile engines just described, is shown in longitudinal and cross-section by Figs. 20 and 21. This is of the "L" type of valve arrangement, but instead of having the valves side by side as in the preceding case, the inlet valve is placed over the exhaust as will be seen from the cross-section view.

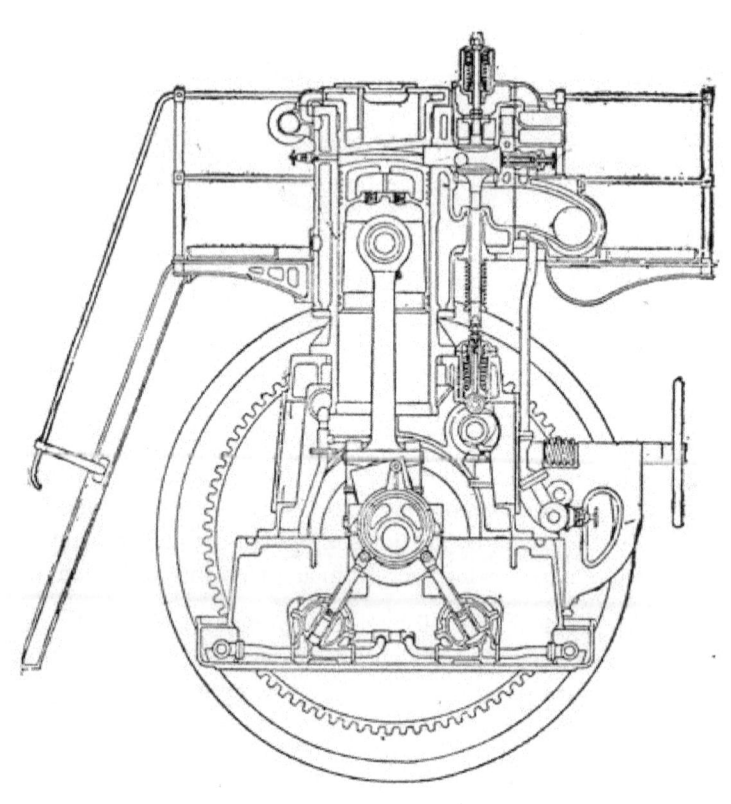

Fig. 21. Cross-Section Through Browett-Lindly Engine.

The exhaust valve is operated directly from the cam shaft by the push rod as in the auto engines, but the inlet valve receives its motion through a long vertical rod and horizontal lever, the latter being located on the cylinder head as shown by the longitudinal section. A supplementary valve is mounted loosely on the stem of the inlet valve, and this valve is held against the seat of the gas inlet port by a short spring.

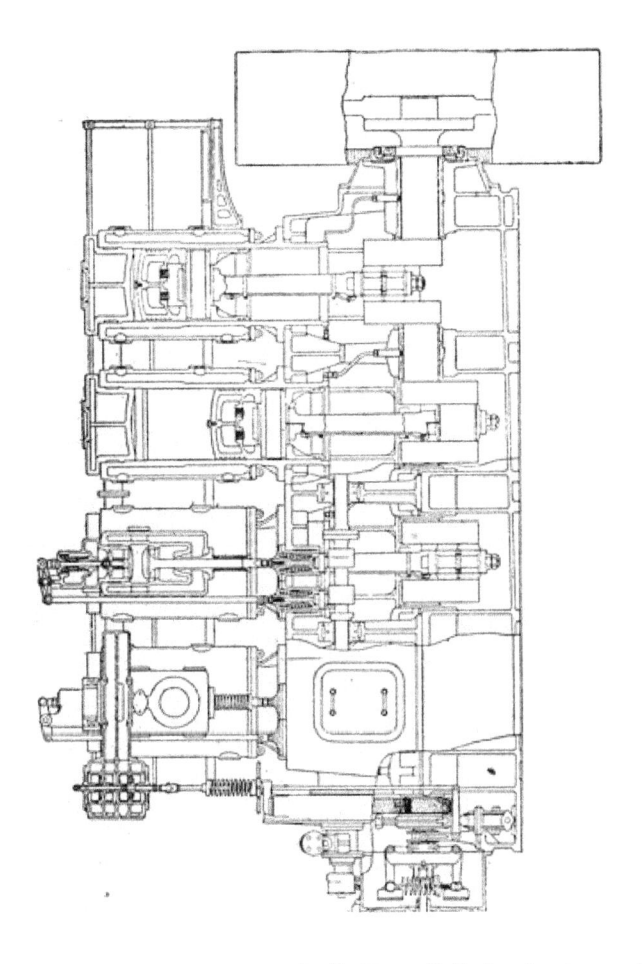

Fig. 20. Section Through Browett-Lindly Four Cylinder Stationary Engine.

A collar on the main valve spindle opens this gas valve, and, by adjusting the position, a certain amount of lag can be given, so that air first enters the cylinder and then, by further travel of the main valve, the gas valve opens and the combined charge is taken in. This prevents any "back fires" as the gas and air are entirely separated until they enter the cylinder.

Starting is effected by means of compressed air, and is entirely automatic. No compression release is provided, as this is unnecessary under the system adopted. By opening the main compressed air valve compressed air is admitted to two valve boxes placed underneath the cam shaft, and the pressure of air raises the valves against their levers and cams. Should the swell on the cam be opposite a lever as it will be in the correct starting position, the valve cannot close, and the compressed air then passes to the cylinder through a check valve on the face of the cylinder, and the engine starts. The automatic check allows the cylinders to take in a charge of mixture on the second stroke and firing takes place immediately. When the explosion pressure is greater than the air pressure the check remains closed and no more starting air enters the cylinder.

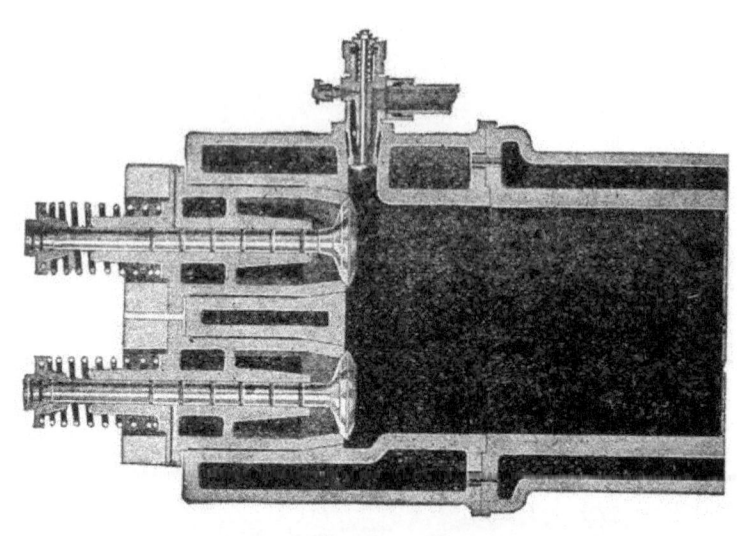

Fig. 21-a. Section Through Cylinder of Fairbanks-Morse Type "R E" Engine, with Valves in the Head.

Governing is effected by varying both the quantity and quality of the mixture.

The main valve, plunger, and rod springs, and all springs on the valves and valve motion, are arranged to be in compression. The exhaust valves are of cast-iron, and are fitted with renewable seats in the cylinders. The admission valves are of nickel steel, and are arranged in boxes, which, when removed from the cylinders, provide the ports which give access to and space for the removal of the exhaust valves which are withdrawn vertically.

Forced lubrication is fitted throughout all bearings, valves, plunger guides, governor, cam shaft, etc., the oil under pressure being supplied by two valveless pumps, either of which is sufficient to maintain the working pressure of oil.

The normal output of the engine is 400 brake horse-power, with an allowable overload of 40 horse-power for ½ hour. The exhaust pipe is water jacketed, each section being supplied from the small pump shown at the end of the cross section.

Double ignition is provided for an emergency, by two high tension magnetos, each of which is connected to a separate set of plugs. When starting the engine, an ordinary spark coil and storage battery are used until the engine gets up to speed, when the coil is cut out and the magneto is thrown in.

(47) The "V" Type Motor.

An example of the "V" type motor is shown by Fig. 22, which is a front elevation of the Frontier aeronautic motor, a type that occupies a minimum of space with a minimum of weight.

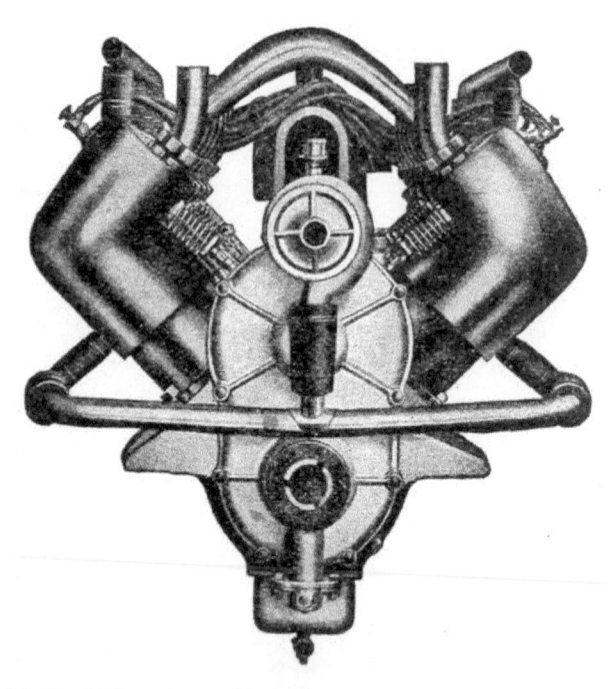

Fig. 22. End Elevation of Frontier 8 Cylinder "V" Type Motor.

The cylinders are cast separately and are furnished either with iron or copper water jackets, the copper jackets being deposited over the cylinder barrels by an electrolytic process in much the same way as that of the celebrated French Antoinette. Bolts passing through flanges on the bottom of the cylinder fasten them to the base. A special aluminum alloy is used for the base which is cast in a single piece with webs to receive the bearings. A unit crank-case insures perfect alignment, prevents a greater part of the oil leakage, and forms a much stronger construction than the usual split pattern. A chamber is provided for the cam shaft at the apex of the case through which issue the push-rods. Shafts and piston pins are hollow. All push rods are adjustable for wear and have steel balls running on the cams which eliminate the possibility of mis-timing through wear.

Lubrication is by a bronze pump geared from the crank-shaft and is connected to an oil tank located in the base from which the oil is forced through the crank-shaft up through the hollow connecting rods to the piston pins, thence to the cylinder walls, the surplus returning to the tank in which the strainer is located.

The circulating pump is driven from the cam shaft as shown in the cut and supplies the cylinders and radiator with water through the copper water manifolds which are designed to give an equal supply to each cylinder. Exhaust manifolds are of seamless steel tubing.

The cylinders are $4\frac{1}{8}$ bore × $4\frac{3}{8}$ stroke, and develop 60 to 70 horse-power at 1,100 revolutions per minute, which speed has been attained with an 8-foot 6-inch propeller having a pitch of 5 feet. Without radiator or propeller, the iron jacketed motor weighs 312 pounds, and copper jacketed weighs 290 pounds, the latter making a difference of 22 pounds in the weight.

A high tension Bosch magneto is used which is mounted on a pad cast on the top of the crank-case and is driven from a gear meshing with the cam shaft gear. Connection is made from the magneto to plugs placed over the inlet valves in the valve caps.

A 100 horse-power aero engine of the "V" type is shown by Figs. 23–24–25, which is built by the All British Engine Company for the aeronautical branch of the English War Department. It has eight cylinders of 5 inch bore, by $4\frac{1}{4}$ inch stroke, and develops its rated horse-power at 1,200 revolutions per minute. *Data from "Aero," London.*

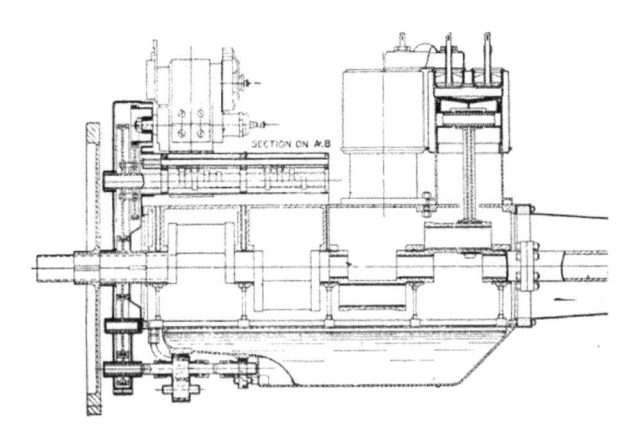

Fig. 23. Longitudinal Section Through A. B. C. 100 Horse-Power "V" Motor.

The crankshaft, which is of three per cent nickel chrome steel, having an ultimate tensile strength of 157,000 lbs. per sq. in., is of distinctly large diameter, and is carried in plain bearings lined with white metal. It is provided with four throws, each crank pin being arranged to take the big end bearings of two connecting rods from cylinders on opposite sides of the crank case. There is a bearing between each throw, and in order to reduce the overall length of the engine the cylinders are staggered on the crank case. The H section connecting rods are stamped out of steel having a tensile strength of 90,000 lbs. per sq. in., and for the purpose of lubrication a hole is drilled from end to end down the center of the web. As mentioned before, the cylinders are staggered, and there is no overhanging of the big end bearings at the point of attachment to the connecting rod. The bearings themselves are lined with white metal. The small end bearings are provided with phosphor bronze bushes, and the piston pin is of steel bored out hollow and hardened.

Fig. 24. Valves and Valve Motion of A. B. C. Motor. ("*Aero,*" London.)

A very interesting detail of the engine is the combination of the water outlet pipe from the top of the cylinder with the bearings for the rocking arms (which are steel stampings) actuating the valves. This is shown in Fig. 25. A hollow steel column is bolted to the top of the cylinder and protrudes from the water jacket, which is fastened to it with the usual shrunk ring. To this column is attached a hollow T shaped pipe of phosphor bronze, the column of the T piece forming the outlet for the water. On one arm of the T piece the exhaust rocker takes its bearing and on the other the inlet rocker. Each T piece arm is connected to its fellow on the next cylinder by means of rubber pip.

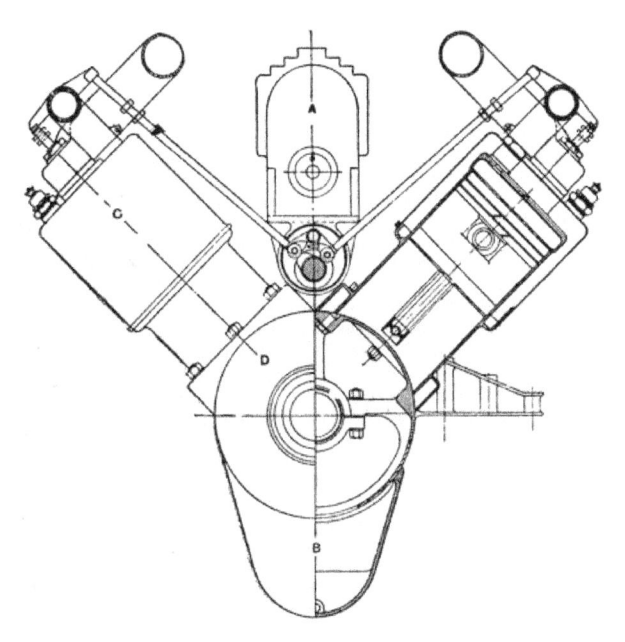

Fig. 25. End Elevation of A. B. C. Motor.

A small bracket projecting from the T piece forms a saddle on which the valve spring rests. This is a plain semi-elliptical leaf spring which works both valves. It is slotted at each end and slightly turned up so as to engage with a cotter pin passed through a slot in the end of the valve stem.

The crank case is of rather unusual design, being absolutely circular in section and machined all over. It is practically a tube with flanged portions bolted on to form the ends. Having no horizontal joints, it is strong and easily kept oil tight. Three radial arms, with slight webs and reinforced with steel columns down the center, support each bearing. The crank case is carried by four feet, which are arranged to accommodate three different widths of engine bearer. To the fore end of the crank case is bolted a long conical aluminum nose carrying at its extremity a compound push and pull ball bearing 6 in. in diameter, which supports an extension shaft bolted to the crankshaft by means of a flanged coupling.

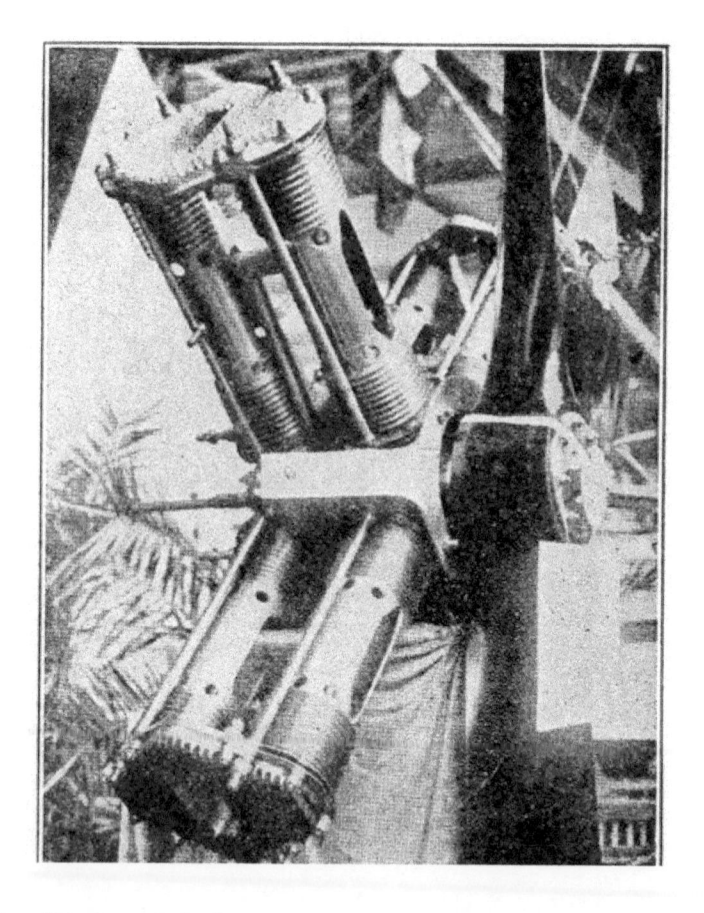

Fig. 24-a. "Sixteen" Cylinder Favata Radial Type Aero Motor, Consisting of Four Groups of Two Cylinders Per Group. Cylinders are of the Double Acting Type and are Stationary.

At the outer end of this extension is a flange to which the propeller is bolted, but the arrangement is specially devised to make quick detachment possible. The boss of the propeller has a hollow hub and is plate bolted permanently to it by twelve bolts.

The direct nose is interchangeable with a speed reduction gear so that the propeller can be driven at a lower speed than the engine. Fitting this gear nose raises the center line of the propeller-shaft some $5\frac{1}{4}$ in. The gears are carried on substantial ball bearings, plain bearings being used also in such a way that they take up the load if the ball bearings through any cause should fail. The reduction is by means of silent chains. The arrangement of the gear wheels is plain from the drawing, and it will be noticed that there is no intermediate wheel between the crankshaft pinion and the camshaft wheel, which are of steel and phosphor bronze respectively. A separate gear wheel is provided on the camshaft for driving the magneto. The water and oil pumps are carried low down outside the crank case, and are driven by intermediate wheels at double the engine speed. The shafts are joined together through Oldham couplings, so that it is possible to remove the pumps separately. Both these pumps are of the gear type.

The camshaft is made in one piece with the cams, and is hardened, being drilled out for lightness. It is enclosed in a casing of steel tube, which is practically separate from the crank case, being attached thereto at one end by the timing gear case and at the other by a saddle. The camshaft is carried in six bearings. An interesting point is the fact that the gear wheels are bolted to flanges on the shafts instead of being attached by keys. Carried in the tube directly above the camshaft is a second shaft forming the fulcrum of the rocking arms for the cam rollers. A very interesting point is the provision of an arrangement for lifting the exhaust valves. The little rocking arms carrying the rollers which bear upon the cams are provided with webs, parallel with the camshaft and between it and the shaft carrying the rockers is a third shaft, the sides of which normally just clear the webs of the rocking arms on either side. This shaft is provided with wedge shape pieces along it, so that by sliding it along the wedges lift the rocking arms clear of the cams, and thus, through the tappet rods and rockers, the valves themselves are opened.

Fig. 26. Mesta Engines on Test Floor.

Not the least interesting particular of this engine is the thorough way in which the lubrication is carried out. Four of the bolts which attach the caps of the main bearings are prolonged through the bottom of the crank case, and serve to carry a detachable oil sump which holds sufficient oil for a run of six hours. As already mentioned, the oil pump is driven at twice the engine speed, and maintains a pressure of something like 110 pounds per square inch. It delivers directly into a straight steel tube placed along the bottom of the crank case, and from this tube a vertical tubular connection is taken to each of the caps of the main bearings. The crankshaft and crank pins are hollow, and, as in the previous engine, in the hollow portions tubes of a slightly smaller diameter are placed, the tubes being expanded over at the ends, so that closed annular spaces are formed which are used as lubrication leads. The lubricating oil passes through the main bearings into these annular spaces in the shafts, from them to the annular spaces in the crank pins, and so to the big-end bearings. From the big-end bearings it travels up the connecting rods to the gudgeon pins. It is interesting to note at this point that the connecting rods work in slots in the crank case which just allow sufficient clearance for their travel, in order to prevent the flooding of oil into the cylinders. A steel-lined oil lead is taken up to the saddle which supports the tubular camshaft casing at the propeller end of the crank case. The bearings carrying the camshaft are cut away at their lower edges clear of the tube so that the oil can flow along the full length of the casing, the level being sufficient to allow the cams to dip. Precautions are taken to keep oil from flowing out of the bearings, and the casing over the gears is specially arranged to prevent the oil from flooding below.

(48) Mesta Gas Engines.

The Mesta four stroke cycle, double acting gas engine, built by the Mesta Machine Co., Pittsburgh, is an excellent example of American big engine practice. Mesta engines are built in sizes from 400 horse-power up to the largest used, and is built either in tandem or twin tandem units. While the engine does not differ widely in either principle or construction from engines of the same size it has several features worthy of note that are not found on other engines.

Up to the medium sizes, the cylinders are cast in one piece, the largest cylinders being made in two parts of cast steel with air furnace iron bushings. The central part of the cylinder is open as will be seen from the cuts, and is covered with a cast iron split band bolted at the center line. The valve chambers are located directly opposite one another on a vertical center line, the inlet valve being at the top and the exhaust valve at the bottom. This arrangement gives a better distribution of the mixture, increases the output with given size of cylinder and equalizes the stresses occasioned by the explosions. As the engine is double acting in all cases there is one inlet and one exhaust at each end of the cylinder.

Both the inlet valve and the corresponding exhaust valve on each end of the cylinder are operated by a single eccentric on the horizontal lay-shaft shown running below and parallel to the cylinders. The regulating valves which are controlled by the action of the governor are perfectly balanced against the pressure in the cylinder which results in a very small resistance to the governor action, therefore no oil relay nor similar complications are required. Any of these valves are easily removed for clearing, a point of great importance when running on a gas that is laden with tar or other impurities.

Fig. 27. End View of Mesta Engine.

The chrome-vanadium piston rod carries the pistons floating free from the cylinder walls reducing the wear on the bore, while the piston rings maintain a gas tight contact with the cylinder walls. Each piston rod is made in two halves, the joint between the sections being made between the cylinders at which point the rods are supported by an intermediate crosshead and guide. Both parts of the rod are interchangeable. The pistons are made in one casting. As will be seen from the accompanying cuts the front end of the piston rod is carried by a cross-head which relieves the pressure on the piston and packing glands.

Speed regulation is performed by the governor by controlling both the quantity and the quality of the mixture. Independent valves in the gas and air passages are actuated by the governor according to changes in the load. This method of control combines all of the good features of quantity and quality regulation.

Make and break ignition is used, with the igniter trip gear so designed as to allow all of the igniters to be timed from one lever, or adjusted independently as the case may require. Each combustion chamber is supplied with two igniters, one at the top and one at the bottom, which insures regular and rapid combustion and therefore gives a maximum of efficiency and reliability.

Compressed air is introduced into the cylinders for starting at a period corresponding to the power stroke in normal operation. This is accomplished by cam operated poppet valves located in the air main and check valves in the cylinders. By this system the engine can be started and put on full load in less than one minute.

(49) Knight Sliding Sleeve Motor.

The Knight motor was the first four stroke cycle automobile motor to employ an annular slide valve in place of the usual poppet valve. Its success has led to the development of several other motors of a similar type which follow the construction of the original engine more or less closely. Being free from the slap bang of eight to twelve cam actuated poppet valves which hammer on their seats at the rate of a thousand blows per minute, the Knight motor is free from noise and vibration. Instead of the jumping of a number of small parts, there is only the slow sliding of the sleeves over well lubricated surfaces. They make no noise because they strike nothing and can cause no vibration because they are a perfect sliding fit in their respective cylinders.

Besides insuring noiseless operation, the valves increase the output, efficiency and flexibility of the motor for they are positively driven and are not affected in timing by fluctuations in the speed. The wear of the reciprocating increases the efficiency of the sleeve instead of destroying it. With poppet valves at high speeds, the valves do not seat properly in relation to the crank position owing to the inertia of the valves and to the gradual weakening of the valve springs which delays the closing of the valves. Carbon also gets on the seats of the poppet valves and prevents proper closure. These faults cannot exist with sliding sleeves when they are once set right as they are positively driven through a crank and connecting rod.

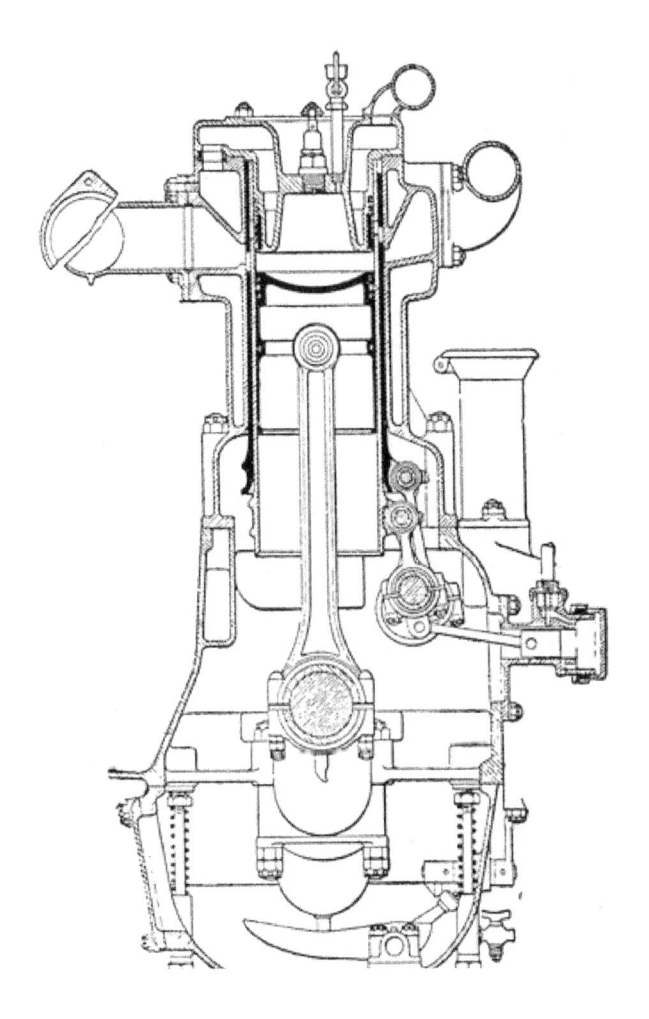

Fig. 28. Section Through Knight Motor Showing the Sleeves, Eccentrics, and Automatic Adjustment for Lubrication. Inlet is at the Right, Exhaust at the Left.

At high engine speeds the velocity of the exhaust and inlet gases is very high in the poppet valve type due to the many restrictions and turns in the passages which causes back pressure and a considerable loss of power. With the sliding sleeve type an ideal form of combustion chamber is possible and the passages to and from the chamber are short and direct. Very large port areas with a low gas velocity are also possible. The sleeves are more effectively cooled than the poppet type, being in direct contact with the water cooled walls for their entire length. Because of the large port areas, the cylinders receive a full charge of mixture, and as a result the engine accelerates and gets under way with remarkable ease.

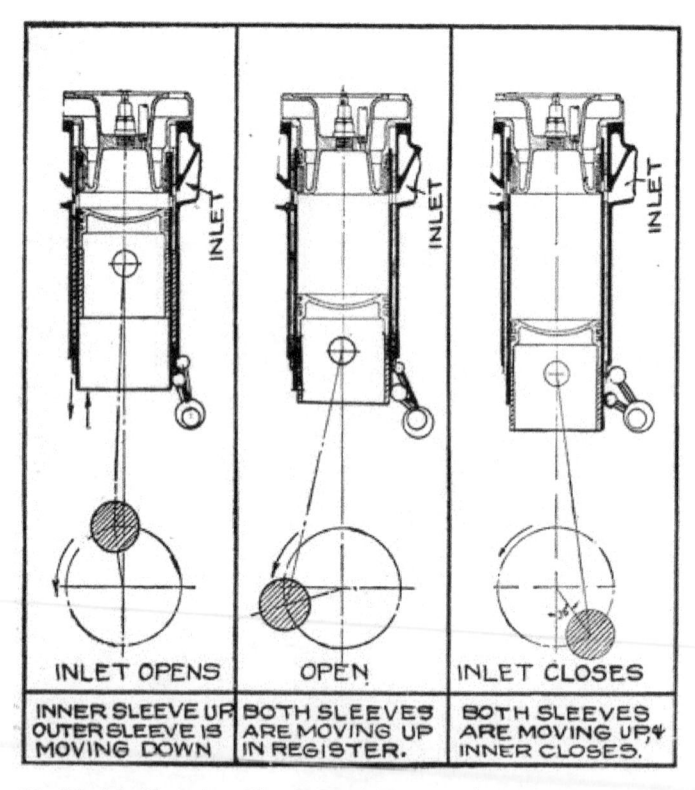

Figs. 28–29–30. Showing Sleeve Positions on the Inlet Stroke. (Knight Motor.)

The arrangement of the slide valves, or sleeves, is shown by Fig. 28, which also gives an idea of the cylinder form, and the location of the piston. Fitting the engine cylinder closely, one within the other, are the two sliding valve sleeves, and within the inner sleeve slides the power piston.

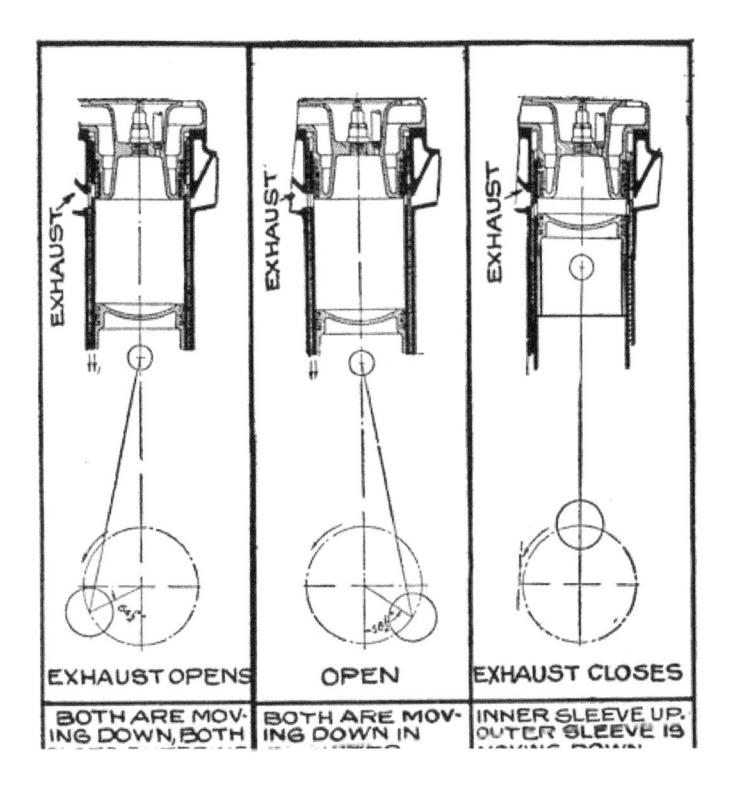

Figs. 31–32–33. Showing Sleeve Positions on the Exhaust Stroke.

Each sleeve has two slots cut in it, one on each side, which form an outlet and inlet for the exhaust and inlet gases respectively. When the slots on the intake side of both the outer and the inner sleeves register, or come opposite to one another, and also opposite to the intake pipe, a charge of gas is drawn into the cylinder. After the explosion has taken place, the sliding motion of the sleeves brings the other two openings, on the exhaust side, opposite to one another, and opposite the exhaust pipe, which allows the burnt gas to escape to the atmosphere through the exhaust manifold.

The sleeves are driven from cranks on the half-time shaft shown at the side of each cut, through the small connecting rods, which gives them a reciprocating motion. Like the cam shaft on a poppet valve motor, the lay shaft runs at half the crank shaft speed, since the engine is of the four-stroke cycle type. The lower ends of the sleeves, to which the connecting rods are fastened, are made thicker than the portion within the cylinder, and are heavily ribbed for strength in the overhang.

The sleeves are of the same composition of cast iron as the cylinder and are provided with oil grooves cut in their outer surfaces for gas packing, and the distribution of oil. Leakage between the inner sleeve, and the cylinder head is prevented by a packing ring, or "junk" ring that is fastened to the bottom of the inwardly projecting cylinder head. The junk ring not only prevents the leakage of gas during the explosion, but it also serves another purpose.

The exhaust ports or slots in the inner sleeve are above the junk ring during the explosion, in which position they are protected from contact with the burning gas. The life of valves is greatly increased by this protection. It will be noted that the entire surface of the sleeves is in contact with water jacketed surfaces, making perfect lubrication and smooth working possible. The two spark plugs for the dual ignition system are shown in the depressed cylinder head.

Complete water jacketing encircles the cylinders, cylinder heads, the circulation area enclosing the plugs and the gas passages so that a uniform heat is maintained the entire length of the piston travel.

The half-time shaft, the magneto, and the water pump are driven by a silent chain from the crank case; this drive being found superior to the gears commonly used for this class of work. The cranks on the half-time shaft are made in one integral piece with the shaft.

Although the piston on the Stoddard-Dayton Knight motor has a stroke of $5\frac{1}{2}$ inches, it is scarcely as much as this considered as friction producing travel, because the inner sleeve in which it rests moves down in the same direction $1\frac{1}{8}$ inches.

This distribution of the working stroke to two surfaces reduces the wear on the side of the sleeve caused by the angularity or thrust of the main connecting rod. On the compression stroke, both outer and inner sleeves go up in the same direction as the piston, the inner sleeve moving the faster. On the exhaust stroke and suction stroke the sleeves move in a direction opposite to the direction of the piston, but on these strokes there is very little work performed by the piston and consequently little thrust is produced on the sleeves and walls of the cylinder.

It is a valuable feature to have the sleeves descend with the piston on the working stroke because this is the stroke in which the piston has the greatest amount of side thrust.

The up and down movement of the sleeves is very little compared with that of the piston. A stroke of $5\frac{1}{2}$ inches gives a piston speed of 916 feet per minute at a speed of 1,000 revolutions per minute. The stroke of the sleeves is $1\frac{1}{8}$ inches and its speed is but 93.7 feet per minute, or a little more than one-tenth that of the piston. This fact makes the problem of lubrication a feasible one, the slow-movement of the sleeves distributing the oil thoroughly between them as well as between the outer sleeves and the cylinder walls.

The action of the valves, and their position at different points in the cycle, is shown in diagrammatic form by Figs. 28–29–30–31–32–33, the particular event to which each diagram refers being marked at the foot of the cuts. The direction of the sleeve movement is indicated by the arrows at the bottom of the sleeves. Particular attention should be paid to the position of the slots in the sleeves.

The first three diagrams show the position of the inlet slots that govern the admission of the combustible gas from the carburetor. Fig. 28 shows the slots coming together to form an opening in the inlet port as the lower edge of the outer sleeve separates from the upper edge of the inner sleeve. The outer sleeve is now moving rapidly downward while the inner sleeve is slowly rising, and as their motion is opposite the opening is quickly formed. Fig. 29 shows the full opening with the slots in register.

When closing (Fig. 30) the outer sleeve is nearly stationary while the inner sleeve is rising rapidly. When the inner sleeve port is covered by the lower edge of the junk ring, the valve opening is closed, the slot in the outer sleeve remaining opposite the inlet opening.

The exhaust port opens (Fig. 31) when the lower edge of the slot in the inner sleeve leaves the junk ring in the cylinder head, the sleeve moving rapidly downward at the moment of opening. To obtain a rapid opening of the exhaust, the ports are arranged so that the inner sleeve is just about to reach its maximum speed at the time of opening.

The outer sleeve closes the port (Fig. 33), closure starting when the upper edge of the outer sleeve coincides with the lower edge of the cylinder wall port. At this time the outer sleeve is traveling downward at maximum speed, so that the closing of the exhaust is as rapid as the opening.

The lubrication of the Knight motor is accomplished by what is known as the movable dam system, which overcomes the tendency of the motor to over-lubricate. A movable trough is placed under each connecting rod, in the crank case, that is connected to the carburetor throttle lever in such a way that the opening and closing of the throttle raises and lowers the troughs.

When the throttle is opened, raising the troughs, the points on the ends of the connecting rods dip deeper into the oil which creates a splashing of oil on the lower ends of the sliding sleeves. In this way the oil is fed to the engine in direct proportion to the load and the heat produced in the cylinder. When the motor is throttled down, the points barely dip into the oil.

An excess of oil is fed to the troughs by an oil pump, which keeps them constantly overflowing. The overflow is caught in the pumps located in the crank case, and returned to the circulation so that it is used over and over again.

Claims of great efficiency are made for this system, there having been many tests made showing 750 miles per gallon of oil, while even as high as 1,200 miles per gallon has been made under favorable conditions.

The oil pump is contained in the crank case, and is of the gear type, insuring positive action. The pump also acts as a distributer, a slot being cut in one of the gears which register successively with each of the six oil leads. In this way it is possible to obtain the full pump pressure in each lead should they become obstructed in any way.

In the upper half of the crank case are cored passageways through which the air passes before reaching the carburetor. These passages not only eliminate the rushing sound of the intake air, but also form an efficient method of warming the air supplied to the carburetor and cooling the crank-case. It is possible to furnish warm air after the engine has been idle for several hours, as the oil in the crank case remains warm longer than any other part of the engine.

(50) Reeves Slide Sleeve Valve.

A simple and compact form of slide sleeve valve gear has been developed in England that is of more than passing interest. It permits of a maximum area for both the inlet and exhaust gases which of course keeps the velocity and back pressure at a minimum for a given valve lift. The small lift also insures noiseless operation and a small amount of wear. The sleeve is balanced at the end of the working stroke. The combustion chamber is nearly hemispherical in shape which reduces the heat loss to the walls.

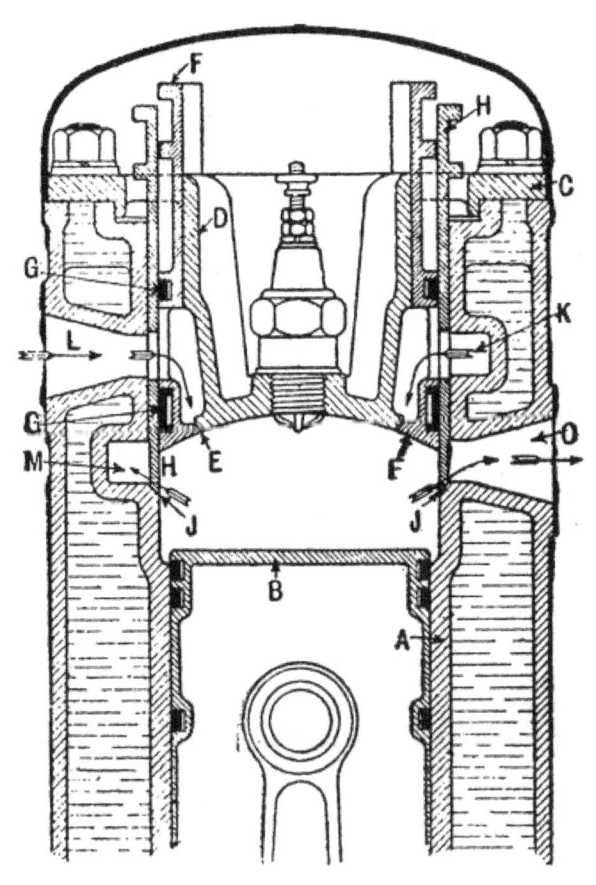

Fig. 34. Reeves Slide Valve Gear.

Referring to the section of the end of cylinder given in the diagram, (34) A is an open-ended water-jacketed cylinder in which the piston B works. At the upper end of the cylinder is attached a ring C forming an extension of the stationary cylindrical head D carrying the sparking plug. At the lower

- 139 -

end of the head D is provided a seating E for the sliding cylindrical inlet valve F, which takes its bearing around the circular head. This inlet valve is provided with expanding rings G to keep it gas-tight. Surrounding the inlet valve F is a second cylindrical exhaust valve H, which is provided with an angular seating at J. The outer circumference of the cylindrical exhaust valve H bears against the walls of the cylinder.

Cast in the cylinder is an annular space K communicating with a passage L for the admission of the inlet gases. These pass through suitable ports cut in the sides of the exhaust valve H and the inlet valve F, so that they are free to pass through the space made when the inlet valve F is lowered from its seat. A similar type of annular space M is cast in the cylinder in connection with an opening O for the passage of the exhaust gas when the cylindrical valve H is raised from its seating at J.

The cylinder head is not water jacketed as the builder states that the continual passage of the intake gases keeps it reasonably cool. The exhaust passages are thoroughly water cooled.

(51) Argyll Single Sleeve Motor.

The Argyll sliding sleeve automobile motor is unique in the fact that only one sleeve is used to control both the inlet and exhaust gases instead of the two sleeves commonly used on the Knight motor. This sleeve, instead of having either a purely vertical or horizontal motion, has a peculiar combination of the two, that is to say, it moves a certain amount in rotation within the cylinder, and an equal amount vertically, the combined motion constituting an ellipse. The external appearance of the engine is shown by Fig. 35, which will give an idea of the general arrangement of the cylinders, ports and piping.

In Fig. 36, is shown the successive movements and events determined by the sleeve, and the method of opening and closing the inlet and exhaust ports by the elliptical movement of the sleeve. The shaded ports are one of the inlet and one of the outlet ports, respectively, which are cast in the cylinder wall, and are afterwards machined true. The dotted port, which changes its position in each diagram, is one of the ports in the moving sleeve, its position in each of the figures is marked by the event that is occurring in the cylinder at that time.

In diagram 1, the shaded port to the right is the exhaust port, and the shaded port to the left, the inlet, this relative arrangement being true, of course, in each of the succeeding diagrams. It will be noted, that in the position shown, in the exhaust stroke (beginning of stroke), the sleeve port has just started on its downward stroke, moving also a trifle to the right as it progresses. Its progress to the right may be more clearly seen by consulting diagram 2, for the movement.

By consulting the other five figures it will be seen that the dotted port, in its relation to the shaded ports, first moves out to the right, and then reverses, moving to the left, and this combined with the up and down movement constitutes an elliptical path. In diagram 6 the exhaust is closed, and the inlet port has just begun to open, the dotted port now starting to move out to the left, and to rise.

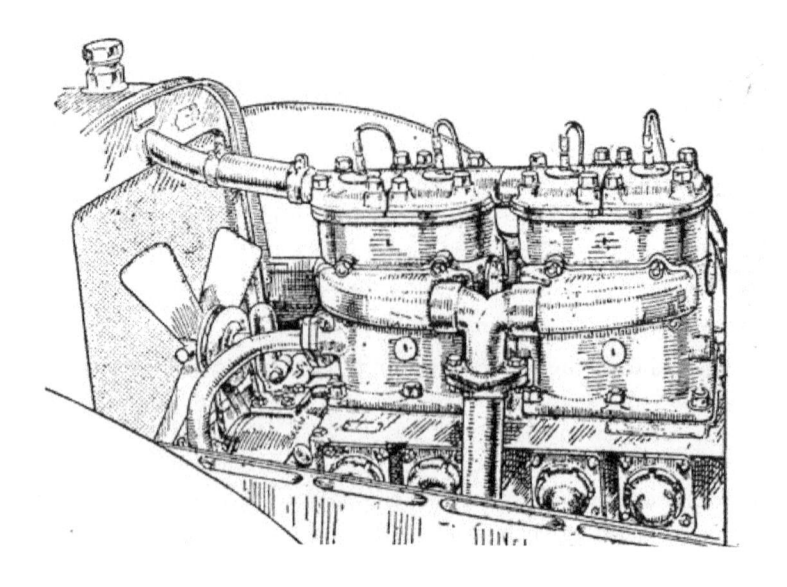

Fig. 35. Elevation of Argyll Single Sleeve Motor from The Motor, London.

In diagram 10, the inlet is nearly closed, the sleeve port passing away from the cylinder ports to the water jacketed portion of the cylinder above.

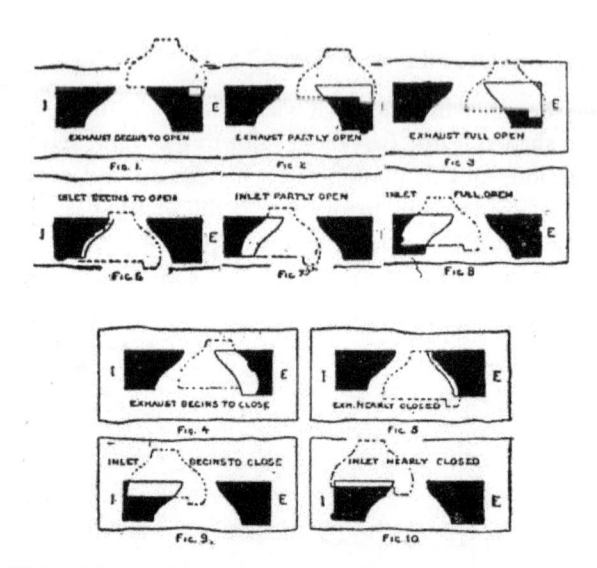

Fig. 36. Valve Motion Diagram of Argyll Motor Showing the Valve Positions at Different Parts of the Working Stroke.

This series of diagrams shows the operation of the duplicated port of the sleeve (which port is the one shown dotted) in relation with one of the inlet ports and one of the exhaust ports in the cylinder wall, the latter ports being marked respectively I and E. The elliptical movement referred to in the text can be traced by following the different positions of the dotted port in the sleeve. In the top row of diagrams it is seen to come downwards and also to move over to the left, whilst in the lower set it rises—bearing still to the left—until, after Fig. 10, it goes higher up for the compression and explosion strokes, during which it bears over to the right and comes down again ready to commence once more the cycle, as in Fig. 1. The other ports in the cylinder wall are the same as those shown, and the other ports in the sleeve are akin in shape to half of the dotted port, but they are without the little tongue cut in the base of this double purpose port. This little tongue in the duplicated port is designed to give as much lead to the exhaust opening as possible, without interfering with the correct timing of the inlet port. The way in which it just misses interfering with the closing of the inlet port is seen in Fig. 10. We are indebted to "The Motor" for these cuts.

(53) Sturtevant Aeronautical Motor.

The cylinders of the Sturtevant aeronautical motor are of the "L" type and are cast separately with the cylinder barrel and water jacket in one integral casting. A special iron is used for these castings that has an ultimate tensile strength of 40,000 pounds per square inch. The valves which are easily accessible through valve covers, are operated directly from the cam shaft without valve rockers. A hollow cam shaft is used with integral cams to insure a maximum of strength with a minimum of weight, and bearings are placed between each set of cams. A bronze gear fitted on the cam shaft meshes with a gear on the crank shaft without intermediate idlers.

Fig. 41. Six Cylinder Sturtevant Aero Motor.

Like the cam shaft, the crank is bored out from end to end with a propeller flange applied on a taper at one end of the shaft. A bearing is provided between each throw with an additional thrust bearing at the forward end of the shaft which may be arranged to take either the thrust or the pull of the propeller. Lubricating oil is applied to all the bearings under a pressure of twenty pounds per square inch, this pressure being maintained by a gear pump attached directly to the end of the cam shaft. The oil is transferred from the pump to the bearings through passages cast in the base, no piping being used. Oil enters the hollow crank shaft at the main bearings and is conducted through the arms to the connecting rod bearings. The oil flying from the crank shaft falls into the oil sump at the bottom of the case where it is cooled before being used again. A second gear pump in tandem with the first takes the oil from the sump and forces it through a filter into the tank.

Fig. 43. Crank Shaft of Sturtevant Motor.

Fig. 42. Crank Case of Sturtevant Motor.

This system enables the use of a more efficient filter than with the suction type and eliminates any danger of its becoming clogged and stopping the oil supply, since, in the event of such an occurrence the pump would furnish sufficient pressure to burst the filter. However, the filter is particularly accessible and may be instantly removed for cleaning without disturbing the oil. The tank regularly fitted to the motor holds sufficient oil for three hours' use. If the engine is required to operate for a longer time without opportunity for replenishing the oil supply, a larger tank can be used. As no oil is allowed to accumulate in the base with this system of lubrication, the motor can be operated continuously at an angle.

Water circulation is maintained by a centrifugal pump of large capacity, the impeller of which is mounted directly on an extension of the crank shaft, eliminating the usual bearings and its grease cup.

The ignition is provided by a high-tension Mea magneto, its special construction permitting the motor to be started under a retarded spark avoiding the danger of back kick from the propeller.

The cylinder and all exposed parts are rendered absolutely weather-proof by means of a heavy coat of nickel plating.

(54) The Rotating Cylinder Motor.

While it is the common belief that the rotary cylinder gasoline motor is of French origin it may safely be said that this type of motor was in actual use in America for several years before it even reached the experimental stage in Europe. The Adams-Farwell Company of Dubuque, Iowa, were driving automobiles successfully with a rotary cylinder motor before Orville Wright flew at Fort Meyer, Va. Although the original Farwell motor more than proved its right to existence by faithful service under the most exacting conditions, the motor never received the consideration that it deserved, probably because of its great divergence from what is known as "accepted practice."

In Europe no such prejudice existed, and consequently the type made rapid strides, although, to the writer's belief, the European model is inferior in many ways to the original American type. The fact that this type of motor holds practically all of the world's aviation records speaks for its practicability in spite of its unusual construction.

With the rotary motor, the cylinders and crank case revolve about a stationary crank shaft, the latter part not only serving as a point of reaction of the cylinders but as a support and intake pipe as well. Since the crank throw remains stationary, the cylinders and pistons revolve about two different centers, the cylinders revolving about the crank case and the pistons and connecting rods about the crank pin. Since the pistons, cylinders, and connecting rods must necessarily revolve together, as one unit, there is absolutely no reciprocating motion in regard to the crank shaft except for a very slight movement due to the difference in angularity of the connecting rods. The motion of all the parts is strictly rotary in every sense, except for the relation of the pistons to the cylinders, and the motion is as continuous as in a turbine. This insures freedom from vibration. As the cylinders and crank case have considerable inertia there is no need of the added weight of a fly-wheel. The movement of the piston in the cylinder bore is brought about by the difference in the centers about which these parts revolve. This gives cylinder displacement without the reversal of stresses or shock or jar.

Because of the revolving cylinders, the mixture is supplied to the crank case through a hollow shaft, the gas being drawn into the cylinder on the suction stroke through an inlet valve placed in the head of the piston. As a rule, the exhaust is direct to the air through the exhaust valves and without manifolds or mufflers. The motion of the cylinders through the air multiplies the efficiency of the radiating Fins.

(55) The Gyro Rotary Motor.

In the Gyro motor, made by the Gyro Motor Company of Washington, D. C., are embodied all of the principles of the typical revolving motor, but with extensive improvements in the design and in the details. It weighs 3¼ pounds per horse-power, complete. This light weight is due to the design of the motor and to the use of alloy steels, and is attained without sacrificing strength or durability.

Each cylinder is machined out of a heavy 3½ per cent tubular nickle steel forging that weighs nearly 40 pounds. After the metal is removed and the cylinder worked down to size, the shell weighs but 6½ pounds. The radiating ribs on the outside of the cylinder are machined out of the solid bar, and are arranged in helicoid or screw-like formation around the cylinder barrel. This adds to the strength of the cylinder and also aids in the circulation of the air. The comparative thickness of the cylinder wall may be seen from Fig. 44. The stiffening effect of the radiating ribs will also be noted. The crank case to which the cylinders are fastened is of vanadium steel, and is divided into two parts. In addition to supporting the cylinders, the crank case also serves as a mixing chamber for the gasoline and air. By removing the bolts seen between each cylinder, the entire working mechanism can be laid bare for inspection. The exterior of the case carries the exhaust valve mechanism and the ignition distributer. The crank shaft is a nickel steel forging with an elastic limit of 110,000 pounds. It is bored hollow throughout its length and serves as an intake manifold by conveying the mixture from the carbureter, attached to its outer end, to the crank case.

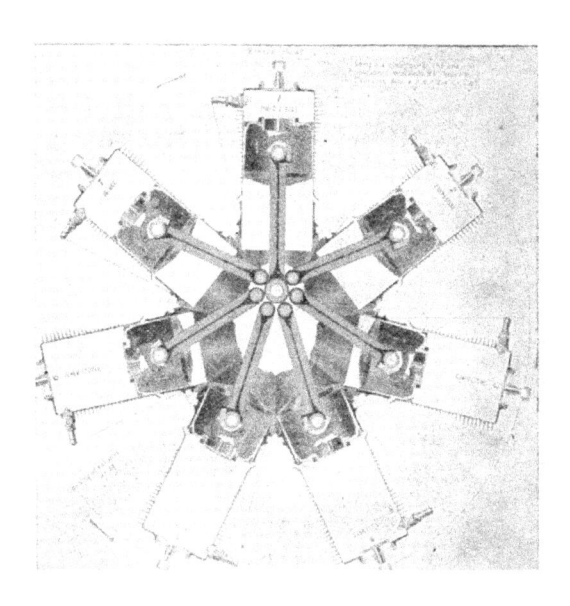

Fig. 45. Section Through Rotary Gyro Motor.

The intake valves in the heads of the piston are mechanically operated by a specially patented movement which consists of two parts, a counter-balancing member, and an operating member. The counter balance balances the valve against the disturbing influence of the centrifugal force, while the operating member, which is fastened to the connecting rod, controls the opening or closing of the valve by the angular position of the connecting rod. This valve action insures a full opening of the valve and a full charge during practically all of the suction stroke.

There are two separate paths provided for the exhaust gases, one being through the auxiliary exhaust ports at the end of the stroke, and the other path through the exhaust valve located in the cylinder head. The auxiliary ports may be seen in the cross-section directly below the piston head in cylinders 4 and 5. The auxiliary ports are uncovered by the piston at the inner end of the working stroke, and it is at this point that the greater percentage of the exhaust leaves the cylinder. These ports or holes are formed on a projecting annular ring in which enough material is provided to make up for the strength lost by boring the ports. As these ports are, in the majority of cases, bored at an acute angle with the center line of the cylinder, it is impossible for the cylinder oil to escape.

All exhaust valves are operated by levers and push rods connected to a cam mechanism on the outside of the crank case. A single cam ring operates all of the valves except where a step-by-step compression is desired. The exhaust mechanism is provided with a simple device by which the closing of the exhaust valve may be delayed through any portion of the exhaust stroke, thus reducing the compression and adding to the facility of cranking. The motor is started with the compression entirely released in which condition it can be spun about its shaft with ease.

After giving the motor its initial spin, the compression and spark are thrown in and the engine begins its normal operation. The compression release lever may be used for starting or slow running and in cutting off the power regardless of the ignition advance or retard.

One connecting rod, called the "master" rod, is an integral part of the spider that contains the ball bearings of the crank pin, thus controlling the angular relation between the connecting rods and cylinders. The remaining six rods are, of course, articulated on the spider by pins so that the rods may move in regard to the spider when in different parts of the stroke. The shell of the pistons is of a fine grade of iron, very thin and elastic, so that it may conform readily to the outline of the cylinder bore. The head of the piston consists principally of the intake valve cage, the cage carrying the piston pin as well as the valve.

Oil is supplied by a positive pump that measures the lubricant in exact proportion to the load on the engine. Both the oil and the gasoline mixture enter the crank case through the hollow crank shaft and mingle in the form of a vapor. This oil mist reaches every moving part and results in perfect lubrication. The pistons are provided with oil shields which carry the oil directly to the cylinder walls and prevent the loss of oil through the exhaust valve.

Ignition is performed by a high tension magneto through a distributer which directs the current to the proper cylinder. As in all rotary engines, the Gyro has an uneven number of cylinders (3, 5, and 7) in order that the cylinders receive firing impulses through equal angles of rotation. An even distribution of firing is impossible with an even number of cylinders, as two adjacent cylinders out of six alternately fire together and then 180° apart. This produces a very jerky turning movement, and is productive of much vibration. In the seven cylinder motor the magneto is driven by gears having a ratio of 4 to 7, and the high tension current is distributed to the cylinders by 7 brushes, the leads from the brushes being taken direct to the spark plugs.

(56) Gnome Rotary Motor.

The Gnome was the first rotary aviation motor built in Europe and is still one of the most capable flight motors abroad as its many victories and records prove. It is built in four sizes, 50, 70, 100, and 140 horse-power, the 50 and 70 horse-power motors having 7 cylinders, and the 100 and 140 horsepower, having 14 cylinders, which consist of two rows of 7 cylinders per row. The cylinders of all sizes rotate about a stationary crank shaft while the pistons rotate in a circle, the center of which is the crank pin. Vibration is practically eliminated at high speed as the pistons do not reciprocate in the ordinary sense of the word, but simply revolve in a circle, the reciprocating relation between the cylinders and pistons being obtained by the difference in the centers of the two revolving systems. The cooling effect of the radiating ribs is greatly increased by the air circulation set up by the rotation of the cylinders. This method of cooling introduces a great loss of power due to the blower action of the cooling ribs, this loss often amounting to 15 per cent of the output of the engine.

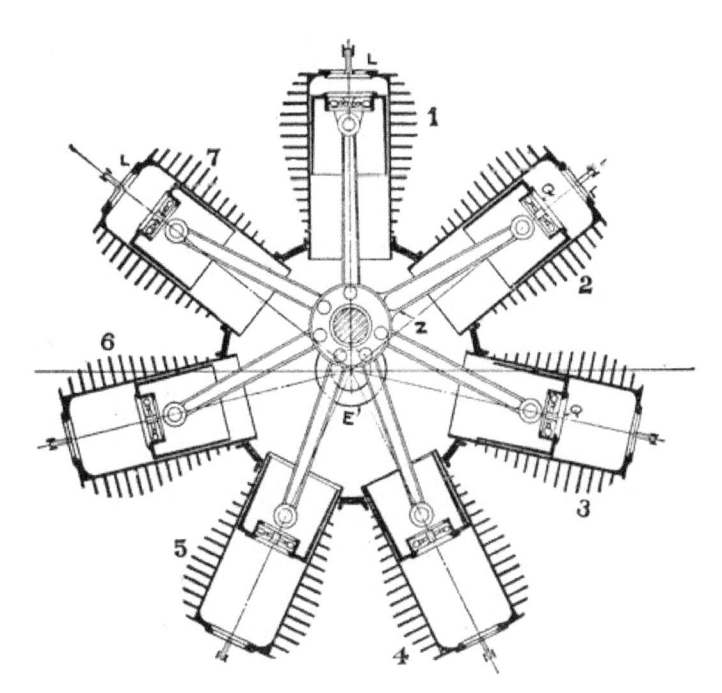

Fig. 50. Cross-Section Through the Seven Cylinder Rotary Gnome Motor, Showing the Crank Shaft Arrangement and Valves.

The crank shaft is stationary and acts as a support for the engine, one end being fastened into a supporting spider which forms a part of the aeroplane frame. The crank shaft is hollow and also serves to conduct the mixture from the carburetor fastened at its outer end to the crank-case of the motor. Only one crank throw is provided on the seven cylinder engine as the cylinders are all arranged in one plane which passes through the center of the crank throw. In the fourteen cylinder engine where the cylinders are in two rows, there are two crank throws, one for each row of cylinders.

The seven cylinders are arranged radially, as will be seen in Fig. 50, each being spaced at an equal distance from the crank shaft and at equal angles with one another, the arrangement in general being similar to that of the "Gyro" motor shown in the preceding section. All cylinders are turned out of solid forged steel bars, the cylinder walls being only 1.2 millimeters thick after the machining operation. This results in the strongest and lightest cylinder possible to build, as all superfluous material is removed and the chances of defects in the material are reduced to a minimum as the character of the metal is revealed by the extended machining operations.

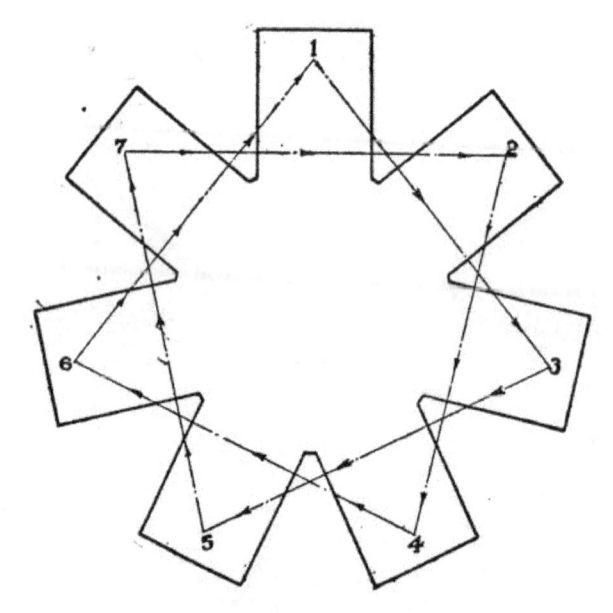

Fig. 51. Firing Diagram of Seven Cylinder Rotary Motor. On Starting at Cylinder No. 1, and Following the Zig-Zag Line in the Direction of the Arrows, it Will be Seen that Ignition Occurs at Every other Cylinder at even Intervals Through Two Revolutions, Ending at Cylinder No. 1.

As the motor operates on the four stroke cycle system, an odd number of cylinders is chosen in order that the firing may be carried out through equal angles in the revolution to obtain a uniform turning movement. Since a four stroke motor must complete two revolutions before all of the cylinders have fired, or completed their routine of events, it is evident that the number of cylinders must be odd in order to bring the last cylinder into firing position in the last revolution. When seven cylinders are used, the cylinder are fired alternately as they pass a given fixed point, that is, one cylinder is fired, the next skipped, the third fired, and the fourth skipped, and so on around the circle, so that the firing order in terms of the cylinder numbers is 1, 3, 5, 7, 2, 4, 6. The cylinders fired in the first revolution in order are 1, 3, 5, 7, and in the second revolution, 7, 2, 4, 6, the cylinder 7 being common to both revolutions. The cylinders are numbered according to their position on the engine, and **NOT** according to the firing sequence. See Fig. 51.

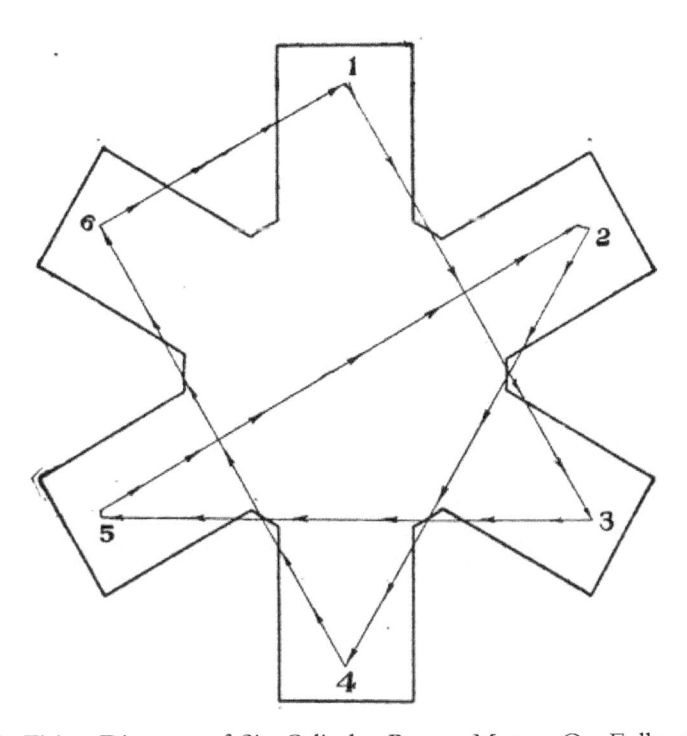

Fig. 52. Firing Diagram of Six Cylinder Rotary Motor. On Following the Zig-Zag Line it Will be Seen that All of the Cylinders Are Not Fired at Equal Intervals. In Some Cases Two Adjacent Cylinders Fire in Sequence, and in Others Two or Three Spaces are Jumped.

With a six cylinder engine it is possible to fire the cylinders in two ways, the first being in direct rotation; 1, 2, 3, 4, 5, 6 thus obtaining, six impulses in the first revolution, and none in the second. The second method is to fire them alternately, 1, 3, 5, 2, 4, 6, in which case the engine will have turned through equal angles between impulses 1 and 3, and 3 and 5, but through a greater angle between 5 and 2, and even again between 2 and 4, and 4 and 6. See Fig. 52.

Mixture is drawn into the cylinder by the suction of the piston through an inlet valve in the piston head, in practically the same way as in the "Gyro" motor, but unlike the latter motor, the valve is lifted by the suction (automatic valve) and not by the mechanical actuation of the connecting rod. The inlet valve is balanced against the effects of centrifugal force by a small counter-weight in the piston head, and the valve is held normally on its seat by a flat spring acting on the valve stem. The gases are brought into the crank case from the carburetor through the hollow crank-shaft as described elsewhere. See Fig. 53.

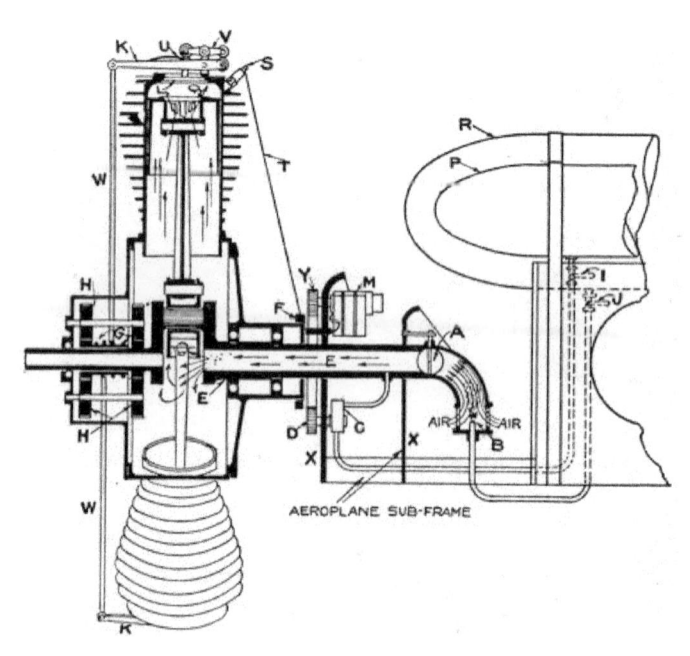

Fig. 53. Longitudinal Section Through Gnome Rotary Motor.

All exhaust valves are located in the cylinder head and are actuated by long push rods that are moved by individual cams in an extension of the crank case. The exhaust valves are counter-balanced against centrifugal force and are retained on their seats by a flat spring. The counter weights do not entirely overcome the effects of the centrifugal force but allow a slight excess to exist which will permit the engine to run with a broken spring. All of the exhaust gases escape directly to the atmosphere without piping or mufflers.

Owing to the fact that the advancing or leading face of the cylinder is cooler than the trailing face, the cylinder bore is thrown out of line by the difference in expansion between the two sides. Because of this distortion of the bore, a special form of piston ring is used, which, by its flexibility, adapts itself to variations in the bore. These rings are of brass and are shaped like the pump leathers of a water pump so that the pressure of the explosion acting on the inside of the ring tends to force the thin shell against the cylinder. In spite of this precaution, the compression pressure is very low at the best, in the most of cases not over 45 pounds per square inch. The exhaust valve screws into the end of the cylinder and may be removed, complete with its seat, for the frequent regrinding necessary to efficient operation. After the cylinders are ground with the greatest care and accuracy, the finishing is carried still further by wearing-in the cylinder with an actual piston carrying an "obturateur" or piston ring.

Fig. 54. Gnome Motor on Testing Stand. From Scientific American.

The bushing into which the spark plug screws is not integral with the cylinder as in a cast construction, but is welded into the side of the cylinder head by means of the autogenous process. It is also evident that this construction enables the inlet valves to be easily removed, since these screw into the piston head. Both inlet and exhaust valves in the Gnome engine are removed with the greatest ease, special socket wrenches being supplied for the purpose. The castor oil, which is used as a lubricant, and the gasoline, are fed by a positive acting piston pump to the hollow crank shaft. The lubricant and fuel then pass through the automatic inlet valve in the head of the cylinder.

Fig. 55. Gnome Motor Running On Test Stand. From Scientific American.

The spark produced by the high tension magneto is led to the proper cylinder through a brush that presses on a revolving ring of insulating material in which is imbedded 7 metallic segments, one of the segments being connected to a corresponding cylinder. As the distributor ring revolves the segments come into contact with the brush in the proper order. The magneto is stationary and is supported by a bracket in an inverted position. A pinion on the magneto shaft meshes with a large gear mounted on the revolving crank case so that the armature of the magneto always bears a positive relation to the piston position. As the engine requires seven sparks for every two revolutions, or 3½ sparks per revolution it is evident that the magneto must turn 1.75 times as fast as the engine, if the magneto is of the ordinary type that generates two sparks per revolution. In other words the magneto speed is to the engine speed as 7 is to 4.

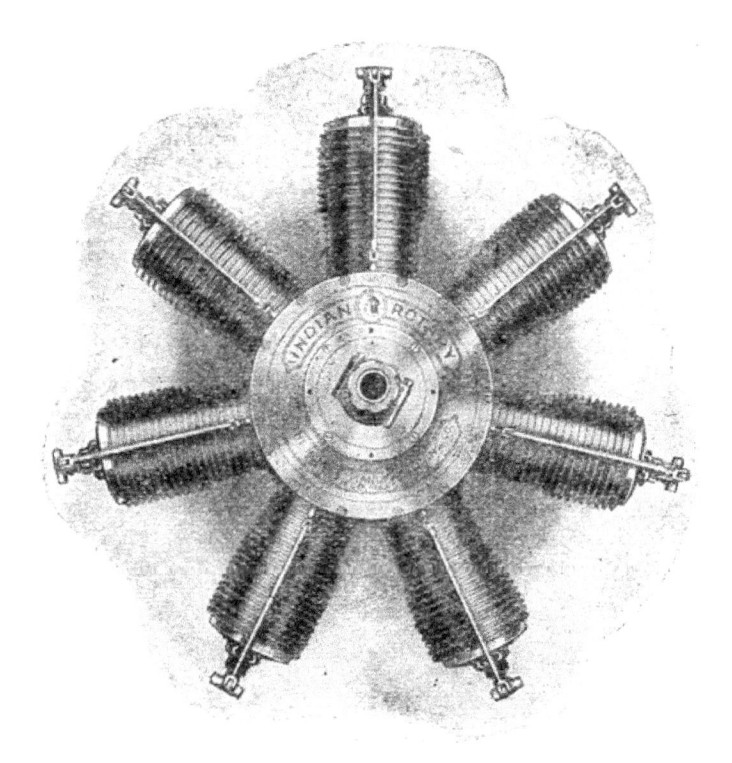

The "Indian" Rotary Aero Motor.

The arrangement of connecting rods is interesting, the big end of one rod being formed into a cage for the reception of the crank-pin ball race. The outer circumference of the cage carries the pins to which the other six connecting rods are fastened. It is necessary that one rod be integral with the cage to prevent its rotation in regard to the cylinders. Annular ball bearings are used on both the main bearings, for the thrust bearing to take the thrust of the propeller, and on the large end of the master connecting rod. The large ends of the auxiliary connecting rods and the small ends of all the rods have plain bearings.

CHAPTER VI
TWO STROKE CYCLE ENGINES
(30) The Junker Two Stroke Cycle Engine.

The Junker two stroke cycle engine stands unique among the large stationary units not only in the principle of its working cycle but in its construction as well, and while it may be considered freakish when compared to standard practice it has proved its value in many European installations. The combustion occurs in the center of an open ended cylinder between two pistons that are forced in opposite directions by the expansion of the gas, and as there is a single acting piston in each end of the cylinder at the end of the stroke, there is no need of stuffing boxes, cylinder heads or valves.

It is apparent that by moving the pistons in opposite directions, the effective piston velocity is twice that of the actual velocity of either of the pistons, and that it is therefore possible to gain a high heat efficiency at high piston velocities with a low rate of rotation. The double pistons increase the scavenging effects, reduce the losses to the cooling water and increase the efficiency at light loads. A marked reduction in weight over the four stroke cycle engine is made possible because of the absence of valves and valve gear.

This engine is of the injected fuel type that is the fuel is sprayed into the combustion chamber after the completion of the compression stroke in a manner similar to the Diesel engine. By prolonging the injection of fuel after the piston has started on the outward working stroke it is possible to maintain the maximum pressure due to the combustion for a considerable period. This gives an indicator card that is very similar to that of a steam engine as the flat top of the Junker's card due to the continued combustion and pressure corresponds to the admission line of the steam engine. As ignition is caused by the high temperature of the compression, almost any low grade oil may be used even down asphaltum oils and coal tar.

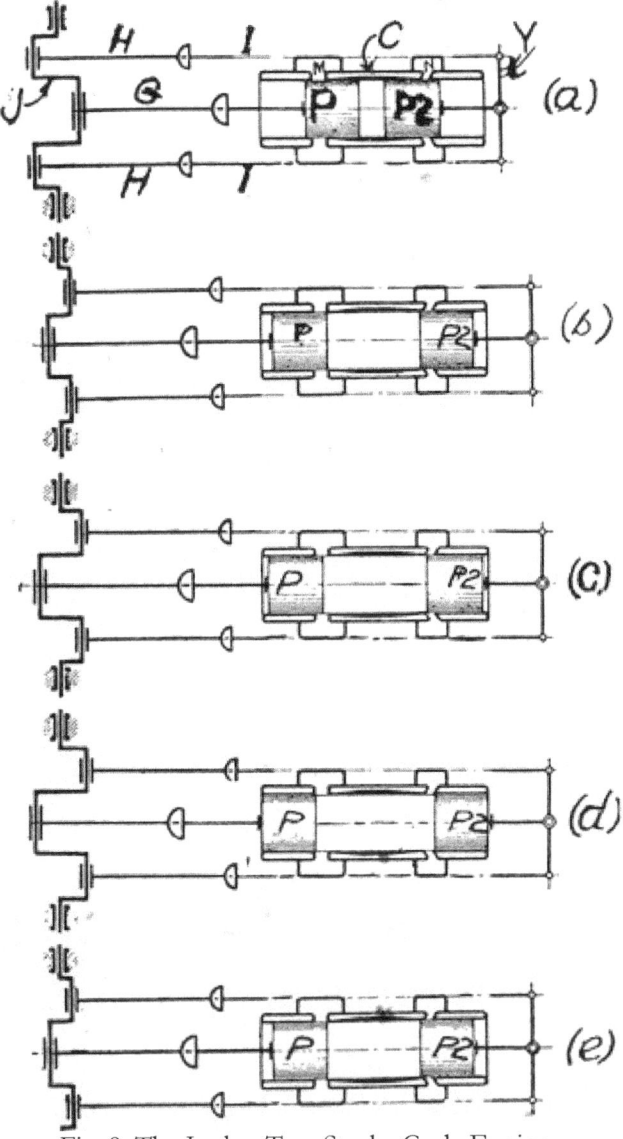

Fig. 8. The Junker Two Stroke Cycle Engine.

In Fig. 8 five piston positions corresponding to five events are shown by the diagrams a, b, c, d, e. From the diagrams one may also get an idea of the arrangement of the principal parts of the engine and their relation to one another. P and P2 are the two pistons, C the open ended cylinder, G the connecting rod of the inner piston P, H-H the two connecting rods of the piston P2, I-I the side rods of the piston P2, and V is the three throw crank shaft which is acted on by the three connecting rods H-H-G. The piston P2 is connected to the side rods through the yoke Y. It will be noted that the crank throws controlling the piston P2 are 180° from the crank connected to piston P, which causes the pistons to move in opposite directions.

With the pistons together at the inner dead center, the space between them is filled with highly compressed air from the previous combustion stroke. At this point the fuel is injected into the highly heated air, and the expansion of the charge begins, the combustion proceeding under constant pressure during the first part of the stroke, or during that part of the stroke in which the fuel is admitted to the cylinder. When the supply of fuel is cut off the working stroke continues by the increase of volume, or expansion of the gas, the gases being reduced to nearly atmospheric pressure at the end of the stroke with the pistons at the position shown by diagram (b). At this point the piston P is just opening the edge of the exhaust port M, allowing the products of combustion to escape to the atmosphere through the annular exhaust passage that surrounds the port M.

As the pistons continue to move outwards the gases continue to issue from the exhaust port at practically atmospheric pressure until the position shown by diagram (c) is reached by piston P2. At this point P2 is just opening the inlet port N allowing fresh air to enter the cylinder for the purpose of scavenging the engine. The passage of the air through the intake port N and out through the exhaust port M continues until the pistons pass the outer dead center, shown by diagram (d), and begin to come back on the return stroke. In diagram (e) the pistons have traveled far enough to close both ports, and as the space between them is filled with pure air from that furnished by the port N, the pistons will continue to move toward one another on the compression stroke. When they have reached the end of their travel as shown by diagram A, the fuel is injected into the cylinder and combustion occurs due to the temperature of the high compression temperature.

This is the complete cycle of events made in two strokes, and it will be noted that the cycle has been accomplished without the use of valves. The compressed air for scavenging the cylinder is provided by air pumps that are driven from the connecting rods by a link motion. One low pressure pump for the scavenging and one high pressure pump for spraying the fuel into the cylinder against compression are provided. As the inside of the piston is always exposed to the atmosphere through the open ends of the cylinder and is never exposed to the heat of combustion, perfect cooling is secured, and as a matter of course, perfect lubrication.

In the two cylinder engine in which four pistons are used, the cylinders are arranged in tandem with the two adjacent pistons, and the two outer pistons connected respectively. In fact the second cylinder pistons are duplicates of those just shown and are connected to the linkage in such a manner as to have the corresponding pistons in one cylinder act with the corresponding pistons in the second.

(34) Koerting Two Stroke Cycle Engine.

One of the most prominent of the two stroke cycle scavenging engines built for heavy stationary service is the Koerting engine. Because of its peculiar scavenging arrangement, and as it is of the double acting type, it will serve to illustrate the cycle of that class of engine equipped with independent air pumps. Several of these engines are in use in Europe that have an output of over 4,000 horse-power, the general arrangement of which is the same as shown in the accompanying diagram Fig. F-11.

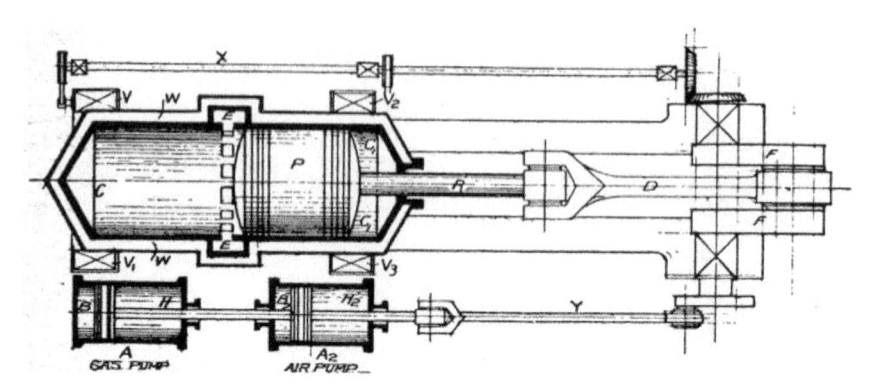

Fig. F-11. Koerting Two Stroke Cycle Engine with Scavenging and Charging Cylinders.

Since the engine is double acting, two similar combustion chambers are provided at each end of the piston as shown by C and C_1, and as each of the chambers gives one impulse per revolution because of the two stroke cycle, the single cylinder shown in the figure delivers two impulses per revolution to the crank-shaft. In order to have one exhaust port serve for both combustion chambers, the annular port E is placed in the center of the cylinder so that it is alternately opened to C and then C_1 as the piston travels to and fro, the port being covered by the piston at intermediate points in its travel. As the piston must cover the port for a considerable portion of the stroke, it is made very long, nearly as long as the stroke. The piston rod R that connects the piston with the crank passes through the cylinder head of chamber C_1, surrounded by a gas tight packing that prevents the leakage of the charge from C_1.

Unlike the ordinary type of two stroke cycle engine, the two combustion chambers are provided with mechanically operated inlet valves, V-V_1-V_2-V_3 that are opened at definite points in the stroke by the lay shaft X which is driven from the crank shaft. As the exhaust port E serves all of the functions of an exhaust valve, there are no valves provided at this point. Exhaust pipes connected to E carry the burnt gases to the atmosphere.

Two auxiliary air pumps of the double acting type are provided, shown at A and A_2, one pumping gas and the other air. They are driven from the crankshaft through the connecting rod Y, and are proportioned so that together they force a mixture of the correct proportion for complete combustion into the working cylinder at a pressure of about ten pounds per square inch. Air and gas are compressed on one side of each pump piston in the spaces B and B_2, and the air and gas are drawn in on the other side as at H and H_2. The connections from the compressor cylinders to the working cylinder are arranged so that the two crank ends of the compressor cylinders discharge into the crank end of the working cylinder, and the front ends of the compressors discharge into the front end of the working cylinder, the exact moment of discharge being controlled by the inlet valves V-V_1-V_2-V_3. The pumps are arranged so that only pure air is admitted at first in order to force the products of combustion through the exhaust port so that they will not contaminate the following mixture of air and gas. The inlet valve opens immediately after the piston of the working cylinder uncovers the port E and reduces the pressure of the burnt gases to that of the atmosphere.

By the action of the admission control, the scavenging air first admitted, is prevented from mixing with the residual gas from the previous explosion, and in the same way the device prevents the loss of fuel through the exhaust ports, thus overcoming the principal objections of the simple two stroke types described earlier in this chapter. The compressor cylinders provide only enough air and mixture for one stroke and no reservoir is provided for a surplus of air or mixture.

As the piston moves forward, on the compression stroke and covers the exhaust port, the inlet valves also close, and the compressor pistons arrive at the end of their stroke so that no more air or mixture is delivered to the inlet valves. At the end of the compression stroke ignition occurs and the expansion or working stroke begins. The piston again moves to the right on the working stroke until the front edge uncovers the port E where the exhaust gases escape to the atmosphere.

The valve gear on the gas compressing cylinder is arranged so that no gas is delivered to the inlet valves of the working cylinder until the air cylinder has provided sufficient air to insure perfect scavenging of the products of combustion, this preventing the fuel from becoming contaminated with the burnt gas. Speed regulation for varying loads is effected by shifting the valve gear of the gas pump so that the gas is delivered at an earlier or later period in the stroke of the working piston, thus causing a variation in the quantity of gas delivered to the working cylinder. This is controlled by the governor directly on the valve gear of the pump or upon a by-pass in the pump cylinder or both. The by-pass, when open returns all of the gas in the passage leading to the inlet valve, that is beyond a certain pressure to the cylinder, so that the gas is delivered to the cylinder at a constant pressure, and therefore in proportion to the load and point of cut off.

This method of governing produces a mixture that varies in richness with the different loads that are carried by the engine, but as the air enters the cylinder first and is prevented from mixing to any extent with the gas by the shape of the cylinder heads, the igniting value of the mixture is not disturbed particularly as the rich gas remains in the cylinder heads and in contact with the igniters.

Like all large engines, the Koerting is started by compressed air taken from a reservoir. A special starting valve is provided for each end of the cylinder which is operated from the cam shaft by means of an eccentric. The air valves may be thrown in or out of gear by a clutch.

(57) Two Stroke Cycle Rail Motor Cars.

A unique application of the two stroke cycle motor will be seen in Fig. 56 which shows a Fairbanks-Morse two stroke cycle motor direct connected to the driving wheel of a railway motor car. The three cylinders are mounted between the driving wheel with the ends of the axle terminating in the crank cases of the motors. Access to the bearings is had through a cover on the crank-case. The simplicity of this motor and its freedom from valves, cams, springs, gears, and other trouble causing parts makes it particularly adapted for the service that it performs in the hands of unskilled track laborers. As there is no water to freeze or leak, and as the lubricant is mixed with gasoline, the car needs very little more attention than the old type hand car.

The car is started by opening the gasoline supply cock, closing the ignition switch, and pushing the car along the track until the first explosion occurs. The speed is controlled in the usual manner by means of the spark advance and throttle. As the motor is of the two stroke cycle type, it may be reversed by simply changing the position of the timer without the use of the gears. The speed is the same in either direction. By the use of three cylinders, three impulses are obtained per revolution which gives a distribution of power equal to that of the ordinary six cylinder, four stroke cycle automobile motor.

Fig. 56. Two Stroke Cycle Fairbanks Motor for Driving Railway Section Cars.

For larger cars built for carrying large gangs of men, a three cylinder motor is used which drives through a clutch and gears, similar to that used on automobiles. It is located near the center of the axle and is supported on a frame that is independent of the car proper. This motor unit is easily removed from the car for inspection with all of the parts intact. A universal coupling is provided on the motor shaft to prevent strains due to changes in the alignment from being thrown into the motor. The motor of this car is started with a crank, and may be left standing with the motor running. As with the two cylinder car, the engine is reversible, and is lubricated by mixing the lubricating oil with the gasoline.

(58) Rotating Cylinder Two Stroke Cycle Motor.

An unusual type of two stroke cycle engine is that designed by M. Farcot for aeronautic work. It is of the rotating cylinder type in which the cylinders rotate about a stationary crankshaft, and unlike all previous two stroke motors, whether of the revolving or stationary cylinder type, no initial compression is performed either in the crank-case or otherwise.

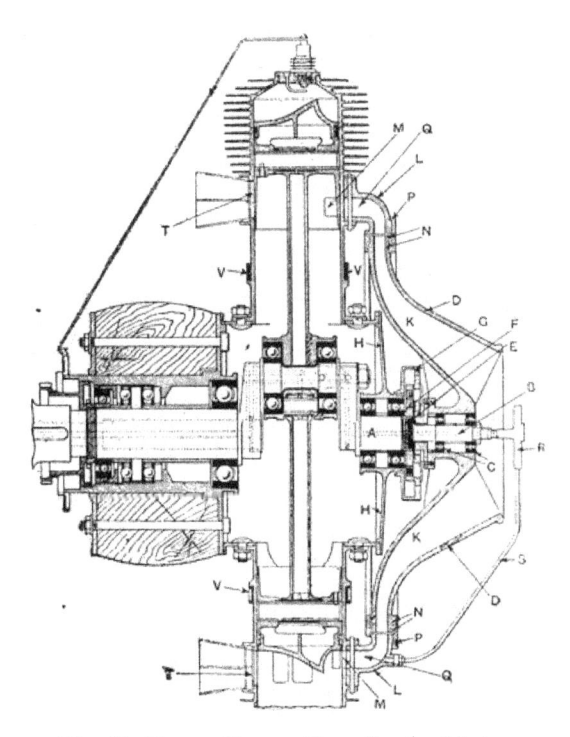

Fig. 63. Farcot Rotary Two Stroke Motor.

Undoubtedly the two-cycle rotating multi-cylinder engine has a future when some of the particularly difficult designing problems involved in its production have been successfully tackled. Crank case compression has had its devotees, but so far it has entailed the use of a low compression, owing largely to the difficulties involved in lubricating the bearings and maintaining gas-tight joints, besides other defects. Some of these barriers appear to have been surmounted in this design.

Fig. 63 of the accompanying drawings is a sectional side elevation of the engine, which, it will be seen, is similar in general disposition to the usual arrangement of the rotating cylinder type. In this particular case, however, the short end A of the stationary crankshaft is reduced in diameter at B, and on this part are mounted ball bearings C carrying the circular casing of a rotating centrifugal blower D. To the inner end of the hub of this blower is attached a gear wheel E, the teeth of which mesh with small intermediate pinions carried on a spider F attached to the crankshaft. These pinions are in turn driven by an internally toothed ring G attached to the hub of the crank case H. Thus the blower D is driven in the opposite direction to the crank-case and at a higher speed. In the interior of the blower casing radial blades K are provided.

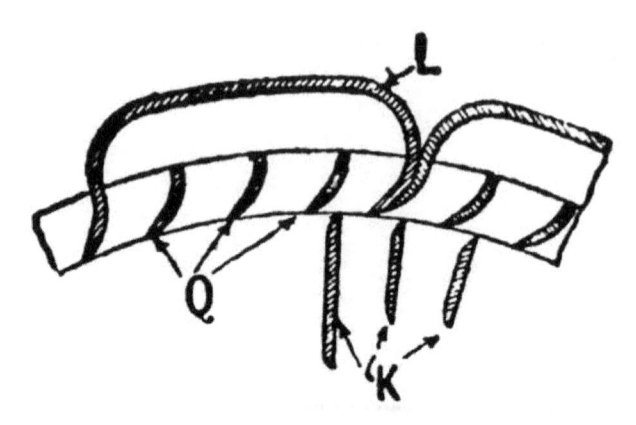

Fig. 64. Farcot Fan Plates.

A hollow annular casing L is bolted to the cylinders, and communicates with their interiors by means of inlet ports M covered and uncovered by the pistons.

The blower casing D has on either side circumferentially flanged rings N, which are a running fit in circular register slots provided in the annular casing L and its cover plate P, in order to provide a gas-tight joint between the opposite revolving casings D and L. Fan blades Q are also provided in the casing L to accelerate still further the incoming gas. The arrangement of the two sets of blades is made clear in the sectional sketch (Fig. 64). It will be realized that by means of this compound blower device a considerable pressure can be attained.

The crankshaft is drilled to provide a feed for the gasoline, which is atomized by a device R in the large central opening of the blower casing D by means of pressure fed from the annular casing L through suitable leads S.

As each piston nears the bottom of its stroke, exhaust ports T, provided with expansion cones for the purpose of increasing the velocity of the exhaust gases, are opened. The inlet port M is then uncovered, and the compressed charge rushes into the combustion chamber.

The general design of the engine is made plain by Fig. 63, but there is one other point to which reference should be made, and that is the provision of rings V, one on either side of the cylinders, to enhance the strength of the construction.

Although the difficulty of compression appears to have been cleverly tackled in this invention, the possibility of the compressed mixture in the inlet casing and blower becoming ignited at the moment of admission by a residue of exhaust gas in the combustion chamber still exists. However, the effect of such a backfire should not prove quite so serious as in some designs. Apart from other considerations, owing to the large area of the blower intake, such an occurrence should merely have a more or less elastic braking effect.

(60) Gnome Radial Two Stroke Motor.

The builders of the famous Gnome four stroke cycle rotary motor, Sequin Frères, have recently developed a radial two stroke cycle motor that bids fair to supplant their original type. Referring to the diagrammatic cross-sections which show only a single cylinder unit, a very long tubular piston will be seen that is divided into two independent chambers, A and B. Both chambers are placed in communication with the outside space, C and D.

The upper end of the piston is continued above the top division head of the chamber A, and the extension is provided with the slot F. Near the center of the piston, the walls of the piston are run out into a flat circular plate or trunk piston E, which is the actual piston head that receives the force of the explosion. The piston E reciprocates in the large cylinder H, which is reduced at its upper end to the diameter of the main piston barrel, for which it affords a sliding support, or guide, and also serves to aid the exhaust port closure. The lower end of the cylinder H is enlarged in diameter as shown by K so that a clear annular space is left between the cylinder walls and the piston head E, when the latter is at the bottom of the stroke. The cylinder diameter is then reduced to the diameter of the main piston barrel.

The motor operates as follows:

Suppose the piston to be ascending (Fig. 1), compressing the mixture above the piston head in the cylinder E, and at the same time the volume of the space M, below E, is being increased until the piston reaches the position shown in Fig. 2.

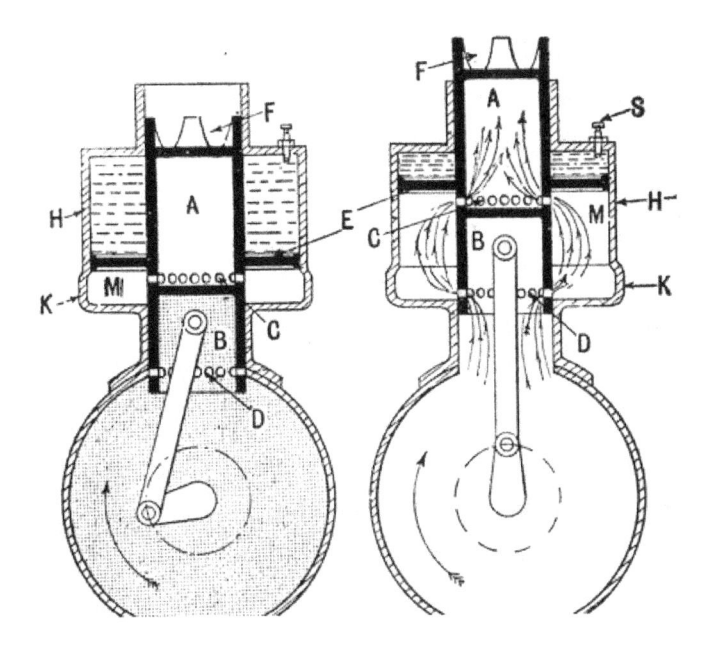

Fig. 65. Gnome Rotary Two Stroke Motor Diagram. Diagrams 1 and 2.

Referring to Fig. 1; the interior chamber A of the piston is in direct communication through the holes C with the space M, consequently as the piston goes up, a partial vacuum will be formed in these two chambers. When the piston reaches the top of its stroke as shown in Fig. 2, the holes D in the lower end B of the piston are uncovered as they rise into the increased diameter of the cylinder, and therefore the mixture is sucked in from the crank case until the chambers A and M are filled to atmospheric pressure.

The spark now occurs at the plug S, and the explosion takes place, driving the piston downwards as shown by Fig. 3, just before the exhaust takes place. The volume of the chamber M has now been decreased with the result that the mixture will have been compressed into the chamber A.

In Fig. 4, the piston has now reached the bottom of the stroke, and the ports F have opened as the slots carry below the upper end of the cylinder where the bore is increased. At the same time, as the piston plate E passes the bottom of the cylinder H into the enlarged diameter K, the compressed mixture in A and M rushes through the annular space opened around E into the combustion chamber and drives out the residual burned gases which still remain after the explosion. On starting the second revolution the piston rises and the cycle repeats as shown by Fig. 1.

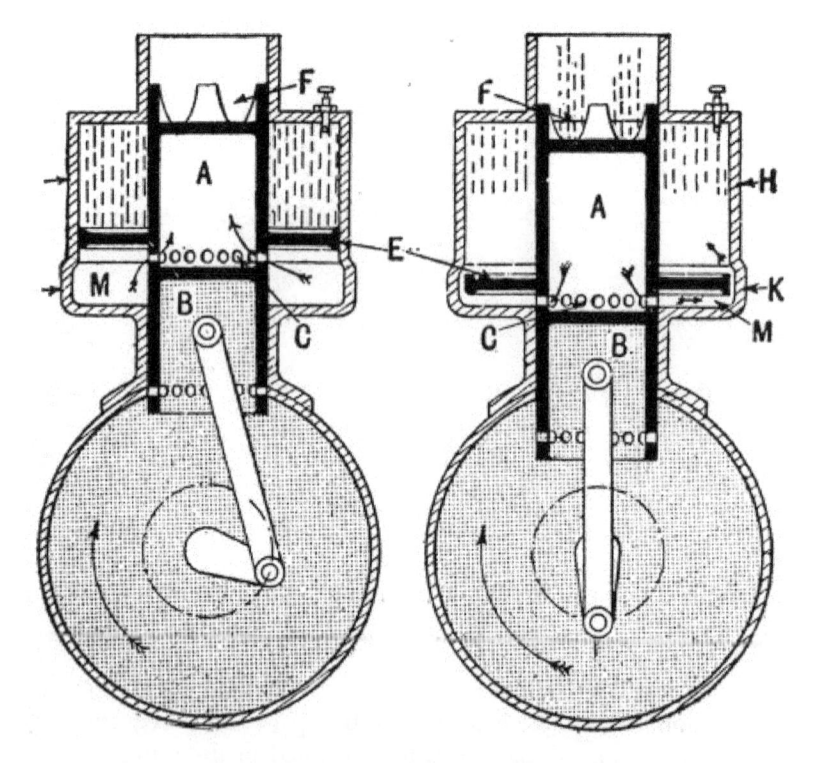

Gnome Rotary, Diagrams 3 and 4.

This engine may be built with any number of the cylinder units described, preferably with an uneven number, as in the case of the Gnome radial four stroke cycle, and with twice the number of impulses of the four stroke type a very uniform turning movement should be had.

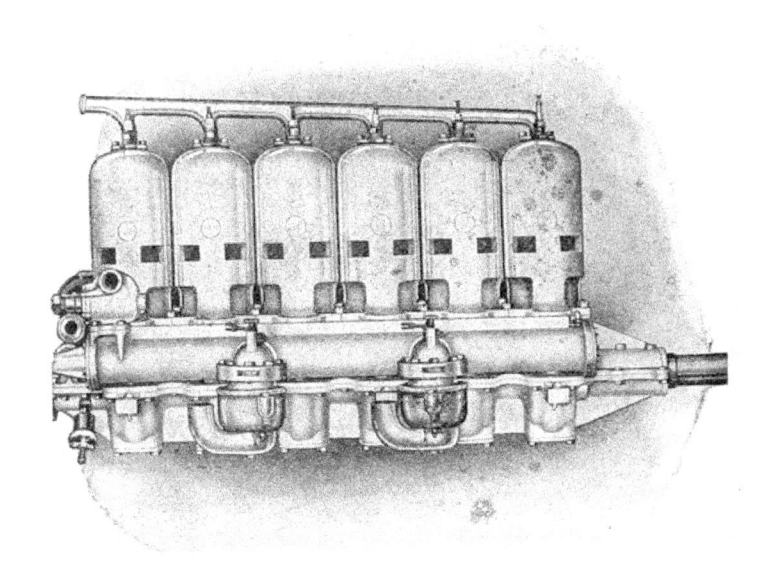

Fig. 64-b. Roberts Two Stroke Aero Motor Using a Rotating Tubular Valve that Controls the Mixture from the Carburetor so that it Enters Only One Crank Case at a Time. This Gives Each Cylinder an Equal Charge of Gas.'=

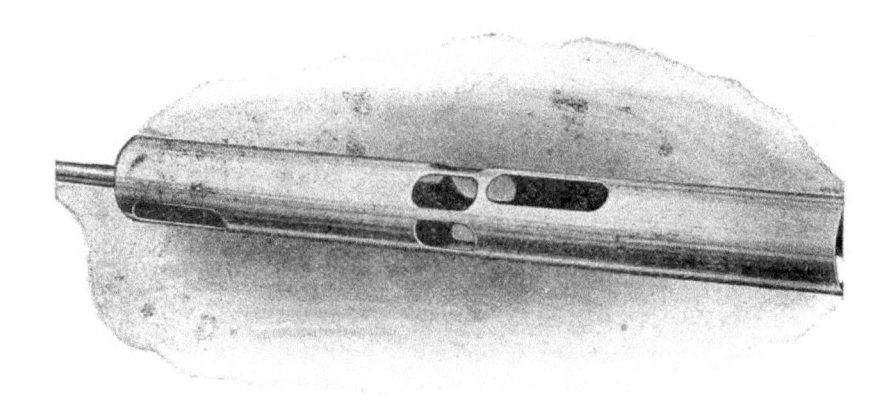

Fig. 64-c. Roberts Distributor Valve. The Ports Are Cut in the Valve so that Only One Crank Case is in Communication with the Carburetor at Any One Time. The Central Hole Connects with the Carburetor.

Since the valves are the parts that give the most trouble in the four-stroke cycle Gnome, this motor should be better adapted for aviation than the original type of Gnome.

(62) Variable Speed Two Stroke Motor.

A variable speed two stroke cycle motor is described by C. Francis Jenkins in the *Scientific American* that seems to solve many of the problems encountered in designing a two stroke cycle motor for automobile purposes. As is well known, the present design of the crank-case compression type is wasteful of fuel, and ignites irregularly at low speeds and light running, and as nearly all automobiles are well throttled for a greater portion of the time it means that this type of motor is working under the greatest disadvantage.

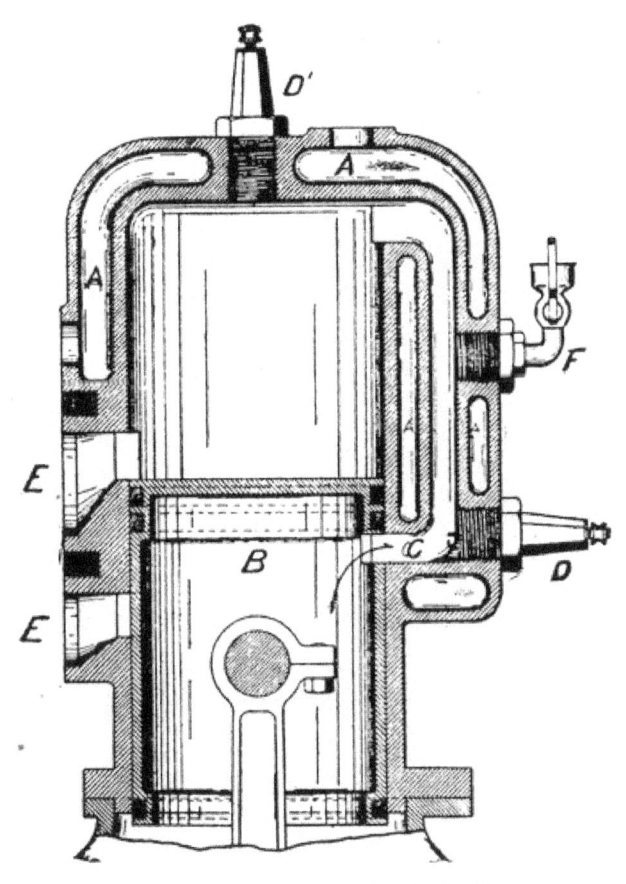

Fig. 66. Jenkins Two Stroke Cycle Motor.

Since the greater part of the trouble is due to the dilution of charge by the residual gases, and as the spark plug of the motor is situated in the most diluted portion of the gas, it would seem that a change of spark plug location, or a change in the circulation of the fresh mixture in the cylinder would be a great aid in remedying the difficulty. With the spark continually in contact with fresh undiluted mixture it would be possible to run it as low speeds as with the four stroke motor, with a corresponding increase in the efficiency, and opportunity to run with a constant advance of the point of ignition. This is accomplished by any or all of the following conditions:

(1.) By keeping good gas separate from bad.

(2.) By placing the spark near the intake port.

(3.) By leaving the plug in its present position and deflecting the fresh gas to meet it.

(4.) By changing the location of the inlet port.

Fig. 58-a. Two Cylinder Marine Engine, of the Two Stroke Type. Built by Fairbanks-Morse and Company.

In the motor invented and described by Mr. Jenkins, the method given by (4) is adopted as shown by Fig. 66, in which the spark plug is placed at the point of admission of the gas and in a confined passage. The operation of the motor is as follows:

Carbureted gas is drawn into crank-case from the carburetor (not shown) in the usual manner, i. e., by the upward movement of the piston; and by its downward movement is forced through the rectangular port in the wall of the piston into the combustion passage within the water-jacket when the port in the piston wall registers with the lower end of this combustion passage, and drives ahead of it the bad gas remaining after the previous explosion. If the throttle is wide open the combustion space above the piston will be completely filled, and on the ignition of the charge the maximum pressure will be exerted on the piston. If, however, the throttle is but slightly open, the combustion passage only may be filled and none overflow into the combustion space above the piston. This small charge will be just as efficient in proportion to its volume as was the large charge, for it was compressed to practically the same extent and none was mixed with the bad gas of the previous explosion. It will, therefore, be obvious that the spark-plug is always swept by the fresh charge, be it large or small, and the ignition will be just as certain in one case as in the other, although the charge and consequent impulse may be only just sufficient to keep the engine turning over, and without missing a single explosion.

Fig. 64-d. Roberts Cylinder Showing Cellular Screen in the Intake Port. This Screen Prevents Crank Case Fires by Chilling the Cylinder Flame Before it Enters the Crank Case.

In the motor built to test and demonstrate this design, provision was made for a second spark-plug to be located in the top of the cylinder for speed work, if this was found necessary. No opportunity has yet been had for making track tests, though without regret, as this two-cycle motor will run idle without missing or "stuttering," which was the thing heretofore impossible.

CHAPTER VII
OIL ENGINES
(31) Diesel Oil Engine.

The Diesel engine marks the greatest progress in the internal combustion field made in the last few years. It marks a distinct advance in both thermal efficiency, and in the character of the fuel that it has made a commercial possibility. By the use of cheap fuel heretofore unavailable for any type of prime mover, such as the asphaltum residual oils, coal tar, etc., it has lowered the cost of power production to a point where it is unapproached by any type of heat engine. Besides its thermal efficiency, the engine is free from the annoyances due to delicacy of the auxiliary appliances such as the carburetor, and ignition system which are indispensable with the ordinary type of gasoline engine.

This engine belongs to that type of engine in which combustion takes place at constant pressure (Brayton Cycle), that is the combustion pressure is maintained at a constant value for a considerable distance on the working stroke of the piston. This method differs from the Otto cycle in which the combustion proceeds at a constant volume, or the type in which combustion is completed before the piston moves forward on the working stroke.

In the Diesel cycle the first stroke of the piston draws pure air into the cylinder; the piston then moves forward on the compression stroke, compressing the air to 500 or 600 pounds per square inch and raising the temperature of the air to about 1,000 degrees C, the exact temperature and pressure depending on the character of the fuel used in the engine. The high pressure is obtained by using a small clearance space in the end of the cylinder. At the end of the compression stroke a spray of oil is injected into the cylinder which is instantly ignited by the high temperature of the compressed air.

The oil continues to burn as long as it is sprayed into the cylinder, this period being from one-quarter to one-third of the working stroke. After the oil is cut off, the hot gas is expanded to the end of the stroke at which point the pressure is very considerably reduced due to the mechanical work performed. It should be noted that the type of engine just described performs the complete cycle in four strokes, the fourth stroke being the scavenging stroke as in the ordinary four stroke cycle engine. While the four stroke cycle type of Diesel engine is by far the most common type, it is also built as a two stroke cycle that is similar to the two stroke cycle gas engine previously described except that pure air is received and compressed in the air compressor in place of the combustible mixture.

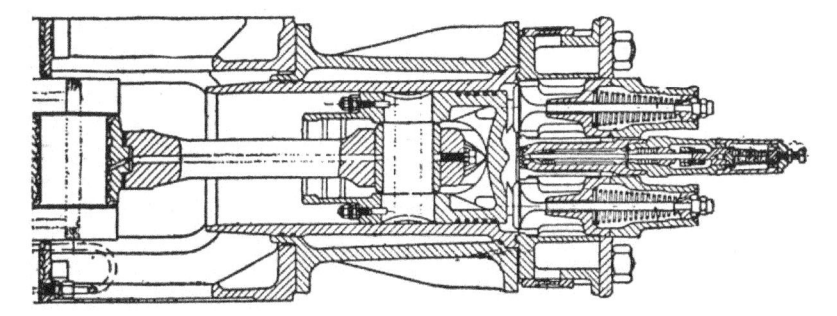

Fig. 9. Cross Section of Four Stroke Cycle Diesel Engine. The Center Valve is the Fuel Admission Valve.

It will be noted, that as there is no fuel in the cylinder during the compression stroke that there is no danger from preignition from an over heated charge, nor is there trouble from decomposed fuels due to a gradually increasing temperature so often met with in oil engines that compress the entire mixture. As the clearance space is exceptionally small there is a minimum of residual gas held in the cylinder after the explosion with the result that the fuel is completely consumed, and that a full charge is taken into the cylinder.

The speed and output are regulated by controlling the point in the working stroke at which the oil spray is cut off, and as this has no effect on the maximum pressure developed in the cylinder, as in the case of the ordinary gas engine control, the pressure charge under varying loads is not so severe. Because of the high compression, and the continued combustion, there is a very gradual increase of pressure. Since the amount of pure air admitted to the cylinder is the same at no load as at full load there is always sufficient air for the complete combustion of the fuel, and as there is a constant compression pressure there is a constant ignition temperature and constant quantity of the working medium. Because of the high compression obtained by the Diesel type, it has an efficiency that is far beyond that of any other form of internal combustion motor.

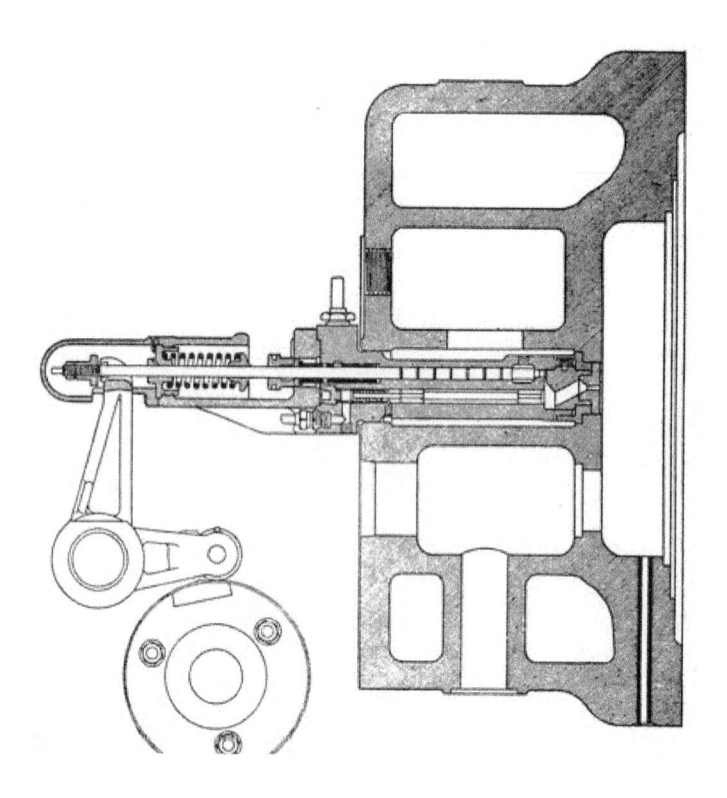

Fuel Nozzle of the Koerting Diesel Engine Showing Operating Cam and Lever, and Compressed Air Connection.

Since the fuel is introduced gradually into the combustion chamber the combustion pressure rises very slowly so that it is not an explosive engine in any sense of the word, the combustion pressure rising steadily from the compression pressure to the maximum in proportion to the supply of fuel. In the ordinary type of gas engine with a compression pressure of from 60 to 70 pounds per square inch the pressure rises abruptly to about three and one-half times the compression pressure, with a correspondingly rapid drop in the pressure on the expansion stroke. In the Diesel engine the drop of pressure in expansion is much more gradual, the indicator diagram expansion curve being nearly horizontal. The uniform pressures thus obtained result in smooth action and even driving power, obtained with no other type of engine.

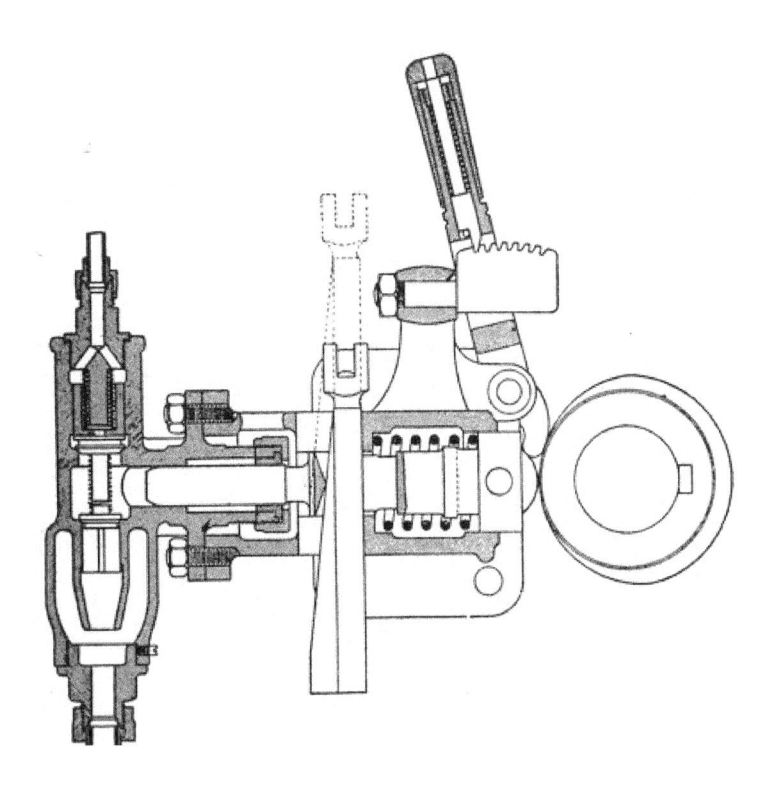

Fuel Pump of Koerting Diesel Engine with Operating Cam.

As the fuels used vary from the lightest hydrocarbons to the heaviest crude oils, there are many types of oil injection valves in use, the valves being in general divided into two classes, those in which the oil is vaporized mechanically by the pressure of a force pump, and those in which the fuel is vaporized by the atomizing effect of compressed air. Atomization by compressed air is however, the most common method since less trouble is experienced with the air pumps than with the liquid force pumps. The compressed air is supplied by pumps that are either operated by the main engine or by an independent compressor engine.

The fuel valve is a plug screwed into the cylinder containing an inwardly opening check valve in the inward end. The hole in the center of the plug receives the oil charge under a few pounds pressure from the tanks, during the compression stroke of the engine, and at the end of the compression stroke, a blast of air at a pressure of about 250 pounds above the compression pressure blows it into the cylinder in the form of a fine spray. Injection valves of the forced feed type consist of a plug with a small passage and a needle valve for regulating the spray. Fuel is pumped into the valve at about 250 pounds above the compression pressure of the engine by a small single acting pump which is built so that the length of the stroke may be adjusted to meet the load. In practice the length of stroke is regulated by the governor, so that the full contents of the pump are delivered at full load, and a reduced amount with a short stroke at small loads. On issuing from the fuel nozzle, the liquid strikes a gauze screen by which it is broken up into very fine spray.

Fluidity is practically the only factor that governs the quality of fuel that may be used with the engine, since exceptionally heavy oils and tars cannot be successfully sprayed. In Fig. 9 is shown a cross-section of a Diesel engine cylinder in which the center valve in the cylinder head is the fuel valve, and the valves to the right and left are the air inlet and exhaust valves respectively. The two latter valves correspond to the inlet and exhaust valves of the Otto cycle engine.

Compressed air is used in starting the engine, which is admitted to the cylinder through an auxiliary valve which is operated by a starting cam on the cam shaft. By this mechanism, high pressure air is furnished to the cylinder during a portion of the working stroke, turning it over on the first few revolutions as a common air engine. As soon as the engine picks up speed, the starting valves are thrown out of operation, and the engine proceeds on its regular working cycle with the oil fuel.

When used for marine purposes in sizes over 100 horse-power, where it is not possible to use reverse gears, the Diesel engine whether of the two stroke cycle or four stroke cycle type must be made reversible. This may be accomplished by either of two methods, first, by changing the angular position of the cams in regard to the piston position, and second by using two sets of cams, one being for right hand rotation and the other for left hand. When a single cam is used, the relation of the cam shaft on which the oil pump cams and oil valve cams are located, is advanced or retarded in respect to the crank shaft by means of sliding the two spiral gears that drive the cam shaft, over one another, in a direction parallel to their axes. The spiral gears are moved back and forth by a hand controlled reverse lever. This type is used principally on the two stroke cycle type of engine as there are not so many factors to contend with as on the four stroke cycle.

With double cams, the system almost invariably used with the four stroke cycle engine, the cams may be mounted either on one shaft, or the ahead cams on one cam shaft and the reverse cams on another. When two shafts are used they are arranged so that either set of cams may be swung under the valve lifters by swinging the shafts in a radial direction by brackets. The single type of cam shaft is usually moved back and forth in a direction parallel to its axis, the ahead cams coming under the valve lifts at one position, and the reverse cams at the other. In the four stroke cycle Diesel it is evident that not only the relations of the oil pump and oil valves must be changed in respect to the piston position but the relations of the air inlet and exhaust valves must be changed as well. This necessitates double cams for the inlet and exhaust valves in order to reverse rotation.

Compressed air for starting and injection is generally supplied by a three stage air compressor or a compressor in which the pressure is built up in three different steps, the second cylinder taking the air from the discharge of the first, and the third cylinder taking the air from the second, and compressing it to about 250 pounds above the compression pressure of the engine. Perfect scavenging is possible with this engine because of the large excess of air supplied during the suction stroke and the period of injection. On the marine type the air pumps and water circulating pumps occupy about the same amount of space as the condenser and circulating pumps of a steam engine having the same outputs. In a recent test made with an Atlas-Diesel engine it was found that 11 per cent of the output was lost in driving the air pumps or more than 50 per cent of the total loss by friction and impact.

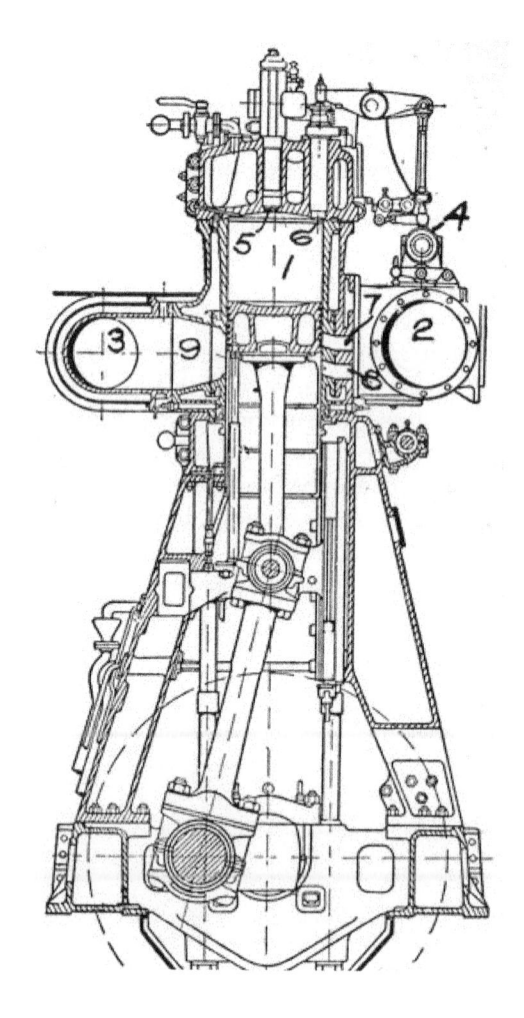

Fig. 67. Cross-Section Through the Working Cylinders of the M. S. Monte Penado Two Stroke Cycle Diesel Engine. From the *Motor Ship*, London.

Unlike the ordinary gasoline engine in which an increase of speed increases the output in an almost direct proportion, the output of the Diesel engine decreases when the speed rises beyond a certain limit due to imperfect combustion at speeds much over 350 revolutions per minute. Because of this fact it has been practically impossible to apply the type to automobile service which ordinarily requires a speed of from 400 to 800 revolutions per minute under ordinary conditions. In addition to the speed limitations, the Diesel engine weighs approximately 70 pounds per horse-power against an average weight of 17 pounds per horse-power with the ordinary type of gasoline automobile motor. Of course these objections may be overcome in time, as the engine is only in its infancy, and the two stroke cycle Diesel has not yet been fully developed, but at the present time it does not seem probable that this engine will ever be an active competitor of the gasoline automobile motor, at least from the standpoint of flexibility.

As the Diesel engine depends entirely upon compression for its operation, it is necessary that all of the parts such as the pistons, valves, etc., shall be perfectly fitted and air tight under extremely high pressures. The careful workmanship required for such fitting and the adjustments make the Diesel much more expensive to build than the ordinary type of gas engine, and for this reason the first cost and overhead charges cut into the fuel item to a considerable extent. A description of the Diesel engines will be found in the chapter devoted to oil engines.

(63) Diesel Engine (Marine Type).

As a practical example of a Diesel engine, which was described in Chapter III, we will give a brief description of the two 850 horse-power Diesel engines installed in the cargo vessel "M. S. Monte Penedo," which were built by Sulzer Brothers of Winterthur, Switzerland. We are indebted to the *Motor Ship*, London, for the details.

The engines are of the two stroke cycle, single acting type, with four working cylinders, a double acting scavenging pump cylinder, and a three stage ignition compressor cylinder. The bore of the working cylinders is 18.8 inches, and the stroke 27 inches. While the crank case is of the enclosed type, there are two sets of covers which can be easily removed for inspection while the engine is running, for as the scavenging pump performs the work of the crank case of the ordinary two stroke cycle engine there is no need of a tight case to retain the compression.

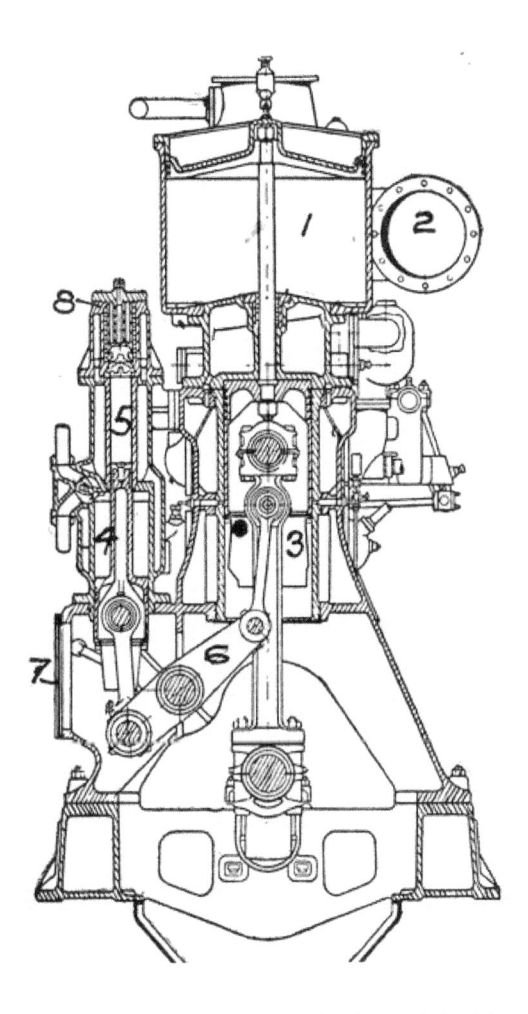

Fig. 68. Cross-Section Through the Air Cylinders of the Two Stroke Diesel Motors on the M. S. Monte Penado.

The scavenging pump is mounted on one end of the engine and is driven from the crank-shaft, the cross-head of the pump forming one piece with the piston of the low pressure cylinder of the injection air cylinder. All of the compressor stages are water cooled and fitted with automatic valves. The double acting scavenging pump has a piston valve driven by a link motion for reversing it when the engine is reversed. The air enters the pump through the top valve chamber from a pipe leading into the engine room. The air discharges a pressure of about 3 pounds per square inch in a header that passes in front of all four working cylinders. By means of a valve the air entering the low pressure stage of the compressor can be taken either from the atmosphere or from the discharge of the scavenging pump; taking the air from the latter allows of a greater weight of air taken by the compressor and consequently a higher compression for use in emergencies.

As in the ordinary type of two stroke cycle engine, two independent sets of exhaust ports are used, one set being for the scavenging air and the other for the exhaust gases, both sets being at the end of the stroke as usual. The air inlet ports are divided into two groups, however, one group being controlled by the piston of the working cylinder, and the other group by an independent piston valve driven from the cam-shaft. Both sets of ports connect with the main scavenging air header. By means of the valve controlled ports it is possible to admit scavenging air even after the other ports are closed by the piston, which greatly increases the scavenging effect. With the air at 3 pounds pressure the air from the valve controlled ports throw the scavenging air to the top of the cylinder even after the exhaust ports are closed. This valve is provided with a reverse mechanism. A single cam is used for operating the fuel inlet valve and the air starting valve, and the reversal of the engine is obtained by turning the cam shaft through a small angle relative to the crank-shaft, which of course also reverses the lead of the fuel valve. Starting is accomplished by compressed air, with the air valve lever on the cam, and the fuel valve lever off. After turning through a few revolutions, the air valve levers are raised, and the fuel levers dropped back on the cams which results in the engine taking up its regular cycle.

By moving the tappet rod of the fuel valve out of or into a vertical position, the time of the fuel valve opening is regulated and the amount of air is controlled. This movement is normally performed by a compressed air motor, but in an emergency hand wheels may be used.

One of these serves to rotate the camshaft through the required angle in order to set the cams in the positions for astern or ahead running and also reverses the link motion of the scavenging pump valve by the rotation of shaft, as mentioned above. The other auxiliary motor operates the fuel and starting air valves by moving the small spindle longitudinally to bring the tappet lever of the air valve about the required cam for ahead or reverse and also lifts this or the fuel valve tappet rod off its cam, according as it is desired to run on fuel or air.

The spindle on which the valve levers are pivoted is in two parts, divided at the center. This is to allow two of the cylinders to run on air whilst the other two are running on fuel, and, as can be seen from the dial where the pointer indicates the position, in starting up, whether astern or ahead, first two cylinders are put on air, then four on air, next two on air and two on fuel, and finally all four on fuel. This allows very rapid attainment of full speed.

The amount of fuel entering each cylinder can be regulated separately by small hand wheels.

Below the fuel pumps are arranged three auxiliary pumps, two of these being oil pumps for the oil circulation, whilst the other is of the piston cooling water. On the left of the engine and driven in a similar manner from the cross-head by links are three other pumps, one for the circulating water and the other for the general water supply of the ship.

Lubrication for the cylinders is furnished by 8 small pumps, just above the water pumps, two oil pumps being provided for each cylinder. As the supply pipe is divided into two parts, the oil reaches the cylinder at four points in its circumference. Four oil pumps are provided for the air compressor.

Four steel columns are provided for the support of each cylinder in addition to the cast iron frame of the base, and by this means the explosion stresses are transmitted directly to the bed plate. The cast iron columns provide guide surfaces for the cross-head shoes. The guides are all water cooled.

(64) The M.A.N. Diesel Engine.

The Maschinenfabrik Augsburg-Nürnberg, A. G., a German firm have built some remarkably large Diesel engines both of the vertical and horizontal types. The peculiar merits of the horizontal type of Diesel engine of which the M.A.N. company are pioneers are still open to discussion at present, but there is no doubt but what this type will be the ultimate form of very large engines when certain alterations are made in the design.

Fig. 69. Horizontal M. A. N. Diesel Engine at the Halle Municipal Plant.

Fig. 70. High Speed Mirlees-Diesel Engine.

In Fig. 69 is shown a 2,000 brake-horse-power horizontal M.A.N. Diesel engine of the four stroke cycle type which is installed at the Halle Municipal

Electricity Works, Halle, Germany. It is of the double acting type with twin-tandem cylinders giving four working impulses per revolution. This engine was installed in addition to the six producer gas engines already in place to take the peak load of the station at different times during the day, the gas engines meeting the normal steady demand.

This firm has built many thousands of the vertical type of Diesel engine of all sizes, and has recently installed 13 engines of 4,500 brake horse-power for operating the Kreff tramways. The company is now building cylinders giving outputs of from 1,200 to 1,500 brake horse-power per cylinder, giving outputs of from 5,000 to 6,000 horse-power in tandem twin type engines. As will be seen from the cut, the horizontal Diesel engine is remarkably free from complicated valve gear.

(65) Mirlees-Diesel Engines.

The Mirlees-Diesel engine is built by the English firm, Mirlees, Bickerton and Day both for stationary and marine service. A generating plant consisting of two, 200 horse-power Mirlees engines direct connected to Siemens generators has been installed in the municipal plant at Dundalk as shown by Fig. 71. On test these units consumed 0.647 pounds of oil per horse-power at full load and 0.704 pounds per horse-power at half load with a regulation of 3.24 per cent from full load to no load. All of the engines built by this firm are of the four stroke cycle type.

Fig. 71. Mirlees-Diesels at Dundalk.

(66) Willans-Diesel Engines.

The Willans-Diesel engines built by the Willans and Robinson Company of Rugby, England, are in sizes up to 400 brake horsepower, and run at speeds up to 250 revolutions per minute. They are all of the four stroke cycle type and are applied principally to the driving of electric generators. The cut shows one of the four, 280 horse-power units supplied to the Alranza Company and the Rosario Nitrate Works in South America.

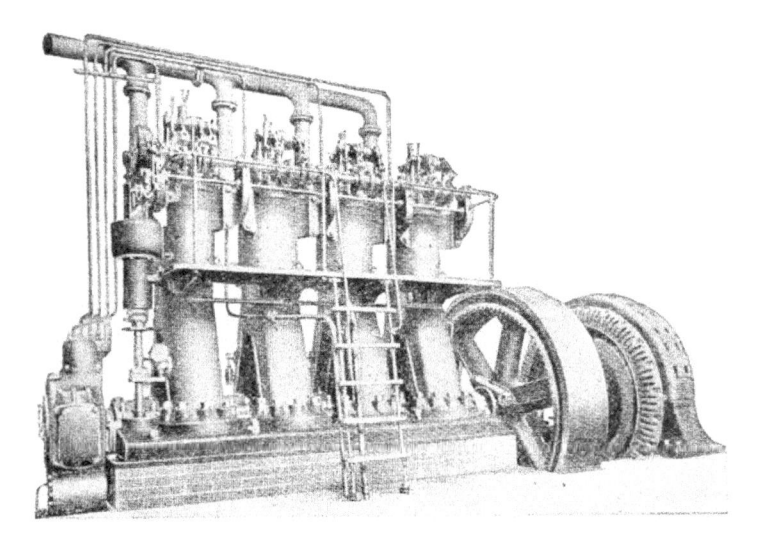

Fig. 72. Willans Vertical Diesel Engine.

Unlike the Mirlees engine, the Willans has an individual frame for each cylinder as in steam engine practice. Like the steam engine frame, the bottom is left open for the inspection of the connecting rod ends and the main bearings which is a most desirable feature. The air compressor and pumps are arranged in a most compact form at the left end of the crankshaft from which the pipes may be seen issuing to the four cylinders. The valves and over head gear are of the conventional type, which, with the exception of a few minor details are the same as those on the recently developed Sulzer-Diesel. The individual grouping of the cylinder units has many desirable features and should, we believe, be more extensively copied.

(67) Installation and Consumption of Diesel Plant.

An English gas-electric station was completed at Egham, England, that is a good example of the changes that have been made recently in the electricity supply abroad, with Diesel power.

The generating plant comprises two 94 K. W. Diesel engines built by Mirrlees, Bickerton and Day, direct connected to single phase alternators generating at 2,000 volts. The exciters are direct connected to the main shaft, and the plant is capable of generating an overload of 10 per cent for two hours. Space has been left for the installation of two more units of a larger size.

The following fuel consumption was guaranteed for a load of unity power factor, and the official tests show slightly better figures than the guarantee.

Full load	0.68 lb. oil per K. W. H.
Three-quarter load	0.72 lb. oil per K. W. H.
Half load	0.79 lb. oil per K. W. H.
Quarter load	1.15 lb. oil per K. W. H.

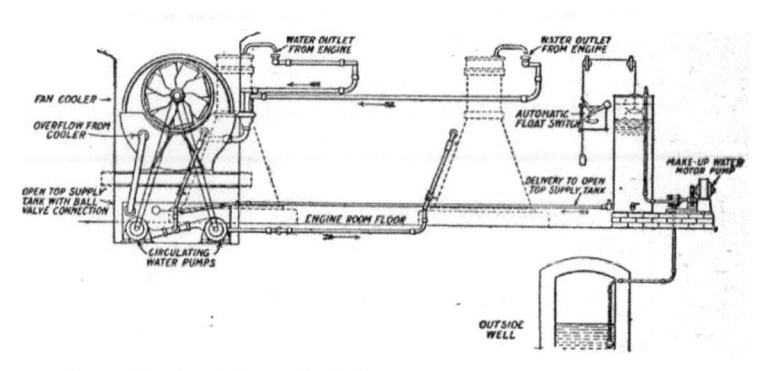

Cross-Section Through Egham, England Municipal Plant.

Particular attention has been given to the water supply for the jackets of the engines; the circulation being by two electrically driven, direct connected centrifugal pumps, one of which is a spare. A Little Company's cooler has been installed, which consists of a horizontal cylindrical chamber, the lower part of which contains water. In the tank are arranged a number of concentric metal cylinders spaced about ¼-inch apart, and in several sections, that are carried on a slowly revolving shaft, driven from the fan shaft. The cylinders are all of the same length, and are open at both ends.

The lower half of the cylinders dips into the water in the casing, and as they revolve, a thin film of water on each side of the plate is carried into the upper portion of the casing where it meets a blast of cold air from the fan. The fan is driven from the circulating pumps, and passes the air through the chamber in a direction opposite to that of the water, baffles being placed so that correct circulation is maintained.

The small loss is made up by connecting the ball cock in the tanks with another tank charged from the works well by means of a self-starting rotary pump, electrically driven. Very little power is required for the pumps and cooler. Fuel oil is stored in a tank outside the building, the oil being supplied to the tanks from an oil wagon by means of a small hand pump.

Oil is taken from the tanks and forced into the engine room by a rotary pump, from which it enters two graduated tanks located in the roof of the station. The graduations on the tanks allow the consumption of oil to be carefully recorded by alternately filling and emptying the two auxiliary fuel tanks.

The entire building is electrically heated, and the kitchen of the flat above the station is equipped with an electric cooking-stove for the use of one of the engineers who make it his residence.

DIESEL HORSE-POWER FORMULA

P. A. Holliday, in the *Engineer*, derives a new formula for computing the horse-power of the four stroke cycle, single-acting engine. For each horse-power developed by these engines about 21,000 cubic inches of displacement is necessary, per minute.

$$D = \text{Cylinder bore in inches.}$$
$$S = \text{Stroke in inches.}$$
$$M.P.S. = \text{Mean piston speed in feet per minute.}$$
$$R = \text{Ratio of stroke to bore.}$$
$$N = \text{Revolutions per minute, then}$$

$$D = \frac{\sqrt{B.H.P.} \times 2220}{M.P.S.}$$

Knowing the value of D, N = $\dfrac{6 \text{ M.P.S.}}{S}$

For high speed, low ratio (R), four stroke cycle engines, approximately 22,000 cubic inches displacement per minute is required.

D = $\dfrac{\sqrt{2{,}330 \text{ B.H.P.}}}{\text{M.P.S.}}$

In both formulae, the air compressor for fuel injection is included.

(32) Semi-Diesel Type Engine.

In the "Semi-Diesel" Type Engine the oil is injected into the cylinder at the point of greatest compression in the same manner as in the Diesel engine, and like the Diesel it compresses only pure air. In regard to the compression pressure, however, it stands midway between the pressure of the Diesel engine and that of the ordinary "aspirating" type oil engine, as the compression averages about 150 pounds per square inch. While this is a much higher pressure than that carried by the ordinary kerosene engine which compresses a mixture of kerosene vapor and air, it is not sufficiently high to ignite the oil spray by the increase in temperature due to the compression, but ignites the charge by means of a red hot bulb or plate placed in the combustion chamber.

This type of engine is built both in the two stroke and four stroke cycle types, the events occurring in the same order as in the two stroke and four stroke Diesel types, that is, pure air is drawn into the cylinder on the suction stroke (four stroke cycle) or is forced in at the beginning of the compression stroke (two stroke cycle), and is compressed in the combustion chamber. At the end of the compression stroke, the fuel is injected against the red hot bulb or plate by which the charge is ignited. Expansion follows on the working stroke after the fuel is cut off, and release occurs at the end of the stroke.

Fuel oil is supplied to the spray nozzles by a governor controlled pump having a variable stroke or by compressed air as in the Diesel engine, making the supply of fire proportional to the load. A separate pump is generally supplied for each cylinder, which is capable of developing a pressure of about 400 pounds per square inch. Several of the Semi-Diesel type engines have water sprayed into the cylinder for the purpose of cooling the cylinder and piston, and as an aid in the combustion. This water spray increases the output of a given size cylinder by the amount of the steam formed by the heat of the cylinder and piston walls, and by the increased rate of combustion. The amount of water supplied to the cylinder is equal, approximately to the amount of fuel oil. The water connection is made in the air intake pipe so that the water spray and the intake air are drawn into the cylinder at the same time.

There is very little difference in the efficiency of the Diesel and Semi-Diesel in favor of the true Diesel type for both have accomplished records of a brake horse-power hour on .45 pound of crude oil in units of the same capacity. Neglecting the question of efficiency the Semi-Diesel has many advantages which are due principally to the differences in compression pressures. Valve and piston perfection in regard to leakage is not as essential with the semi-type as with the Diesel, as the former is not dependent on compression for its ignition. This means that the Semi-Diesel has a lower first cost and a lower maintenance expense. Its low compression pressure makes starting possible without the use of compressed air with engines of a considerable horse-power. As the explosion pressure is much lower than with the Diesel type there is less strain on the working parts and lubrication is much more easily performed.

Compared with the ordinary type of kerosene engine the Semi-Diesel is much more positive in its action as the oil is sure to ignite when sprayed on the hot surface of the bulb or plate when under the comparatively high compression. In the engine where the air is mixed with the vaporized fuel before it is drawn into the cylinder, it is difficult to obtain perfect combustion because of the uncertain mixtures obtained on varying loads by the throttling method of governing. At light loads the only difficulty encountered with the Semi-Diesel type is that of keeping the igniting surface hot enough to fire all of the charges.

In the majority of cases the two stroke cycle type of Semi-Diesel engines compress the scavenging air in the crank chamber in the same way that a two stroke cycle gasoline motor performs the initial compression, although there are several makes that compress the air in an enlarged portion of the cylinder bore by what is known as a "trunk" piston. This initial compression determines the speed of the engine, the pressure limiting the time in which the air traverses the cylinder bore and sweeps out the burnt gases of the previous explosion.

(68) De La Vergne Oil Engines.

Two types of four stroke cycle oil engines are built by the De La Vergne Machine Company, which differ principally in the method and period of injecting the fuel into the cylinder. While both types compress only pure air in the working cylinder, the oil is injected in a heated vaporizer during the suction stroke in the smaller engine (type HA), and is injected directly into the combustion chamber of the larger engine (type FH) at the point of greatest compression. This fuel timing classifies the type FH as a semi-Diesel, while type HA comes under the head of that class of engines known as aspirators.

76-a. Elevation of De La Vergne Oil Engine, Semi-Diesel Type. Class F H.

Semi-Diesel (Type FH)

During the suction stroke, air is drawn into the cylinder through the inlet valve located on the top of the cylinder head, and on the return, or compression stroke, the air is compressed to about 300 pounds per square inch in the combustion chamber. The compression heats the air to a high temperature which is still further increased by contact with the hot walls of a cast iron vaporizer D, shown by Fig. 76-b. At the completion of the compression, the fuel is injected in a highly atomized state by compressed air through the spray nozzle F, the spray being thrown into the vaporizer.

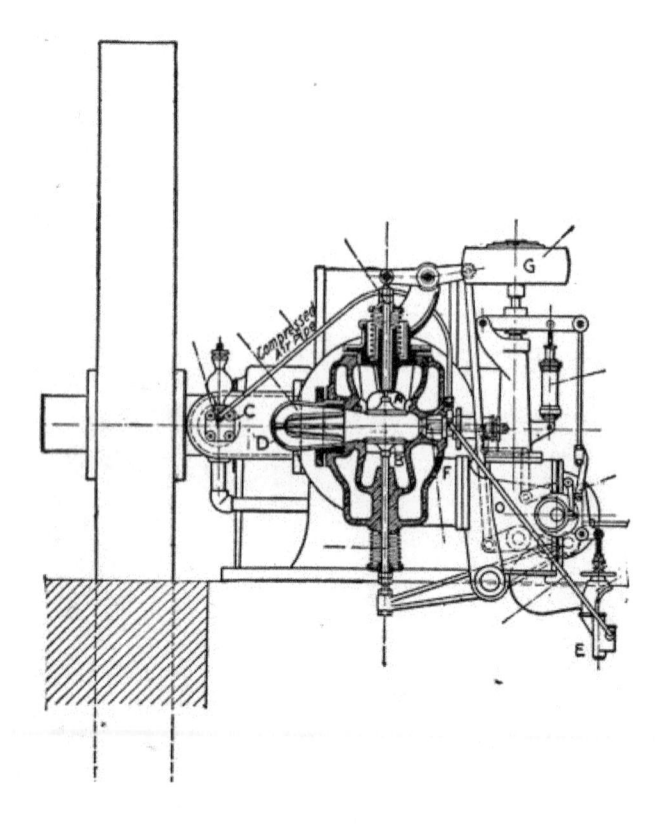

76-b. Cross-Section of Type F H, De La Vergne Oil Engine.

The vapor formed by the contact of the spray with the walls of the vaporizer mixes with the compressed air in the combustion chamber and is ignited at the instant of fuel admission by the combined temperatures of the vaporizer and compression pressure.

As the fuel is not injected until the proper instant for ignition, it is possible to obtain a relatively high compression without danger of the charge preigniting. The oil is supplied to the nozzle by a fuel pump under pressure. The atomizing air takes the oil at pump pressure and performs the actual injection. Details of the spray valve are shown by Fig. 76, in which the oil and air are entered at a pressure of about 600 pounds per square inch.

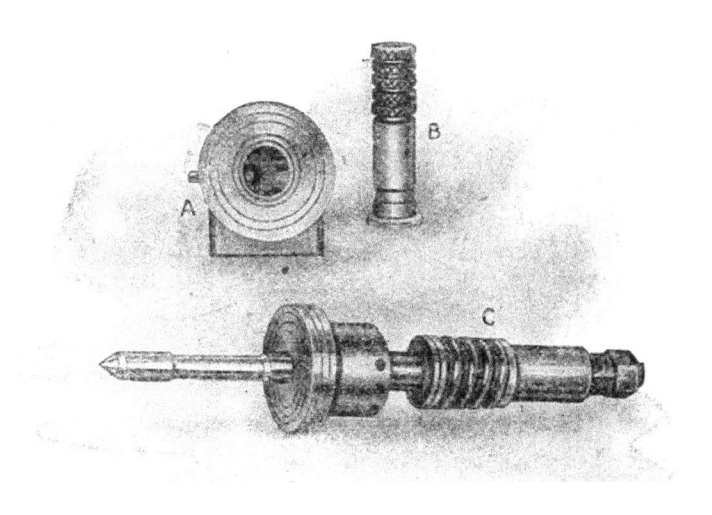

Fig. 76. De La Vergne Spray Nozzle.

The air and oil enter the nozzle at opposite sides of the cylinder B which fits snugly into the valve body A. As the air and oil proceed side by side along the outside of B, they are forced to pass through a series of chambers connected by a system of fine diagonal channels on the surface of B which results in a very fine subdivision and intimate mixture. The charge is admitted to the cylinder by a sort of needle valve about one-half inch in diameter which is provided with a spring that holds it closed on its seat as shown by C, in Fig. 76. The needle is so constructed that it may be readily removed at any time for inspection. The spray valve is located on the right hand side of the valve chamber directly opposite the vaporizer and is operated by an independent cam on the camshaft.

Fig. 76-c. De La Vergne Governor and Fuel Pump.

The vaporizer consists of an iron thimble having ribs on the inside to increase the radiating surface. In starting, the vaporizer is heated for a few moments until it reaches the temperature necessary for vaporizing the fuel, but after the engine is running, the blast lamp is removed and the temperature is maintained by the heat generated by the combustion of the successive charges. Since the fuel is ignited at the instant that it makes contact with the vaporizer, it is possible to accurately adjust the point of ignition by adjusting the position of the fuel cam on the camshaft.

Air for spraying the fuel is supplied by a two stage air compressor that is driven from the crankshaft by an eccentric. The air compressed by the first stage is stored in tanks at about 150 pounds pressure for starting the engine. The second stage compresses the air to about 600 pounds pressure, but is correspondingly small in volumetric capacity since it handles only enough air to spray the oil which amounts to about 2 per cent of the cylinder volume. A governor controlled butterfly valve in the air intake pipe regulates the amount of air taken in on the second stage to suit the varying charges of oil injected at each load.

In starting by compressed air, a quick opening lever operated valve on the cylinder head is used to admit air from the tanks to turn the engine over until the first explosion takes place. If the vaporizer is sufficiently heated by the torch, the explosion occurs during the first revolution of the crank shaft. At a point about 85 per cent of the expansion stroke, the exhaust valve is opened, and the products of combustion are expelled into the atmosphere. When starting, the compression may be relieved by shifting the starting lever from the exhaust cam to the auxiliary starting cam provided for that purpose.

Speed regulation is affected by a Hartung governor, driven from the camshaft, which actuates the oil supply pump through levers by shifting the point of contact between the pump levers and its actuating cam. This lengthens or shortens the stroke of the pump in accordance with the requirements of the load. The type FH engines are built in both single and twin cylinders ranging from 90 to 180 horse-power in the single cylinder type to 360 horse-power in the twin.

Since the fuel injection of the smaller engine type HA differs from that just described, it will be described separately in the following section.

The De La Vergne Oil Engine (Type HA)

In the small four stroke cycle De La Vergne Oil Engine, the fuel is injected into a heated vaporizer during the suction stroke in such a way that the vapor and intake air do not form a mixture in the cylinder proper. On the return stroke of the piston, the compression of the pure air takes place which forces the air into the vaporizer and into intimate contact with the oil vapor. This forms an explosive mixture which ignites and forces the piston outwardly on the working stroke. The release and scavenging are performed in a similar manner to that of a four stroke cycle gas engine. Both the inlet and exhaust valves are of the mechanically operated poppet type, and as both the inlet and exhaust gases pass through the same passage, the entering air is heated to a comparatively high temperature.

The injection pump receives the fuel from a constant level stand pipe or tank, located near the engine and injects the fuel into the vaporizer through a spray nozzle. The vaporizer is a bulb shaped vessel that is connected with the cylinder through a short post and really forms a part of the combustion chamber. Since no water jacket surrounds the vaporizer, it remains at a high temperature and vaporizes the oil at the instant of its injection. Because of the residual gases remaining in the chamber, ignition does not occur until air is forced through the passage by the compression. The air inlet valve and the fuel injection valve are opened at the same instant by a cam lever that also operates the pump.

On the compression stroke, the air which is at a pressure of approximately 75 pounds per square inch enters the vaporizer, and ignition occurs, partly because of the increased heat due to the compression and partly because of the supply of additional oxygen. Internal ribs provided in the vaporizer greatly increase the heat radiating surface and add to the thoroughness with which the atomized oil is vaporized. Since no mixture of air and fuel takes place in the cylinder proper, sudden changes in the load do not affect the ignition of the charge as the heated surfaces are surrounded with comparatively rich gas under all conditions.

Before the engine is started, the vaporizing chamber is heated to a dull red heat by means of a blast torch in order to vaporize the oil for the first stroke. As soon as the engine is running, the lamp is cut out and the temperature is maintained by the heat of the successive explosions. The combustion attained by this method is very complete even with the heaviest fuels, and whatever carbon deposit is formed occurs in the vaporizer from which it is easily removed. The contracted opening of the vaporizer passage effectually prevents the solid matter from working in the bore or valves.

A Porter-type fly ball governor maintains a constant speed at varying loads by regulating the quantity of fuel supply to the vaporizer, the air intake remaining constant. A by-pass valve, controlled by the governor divides the oil supplied by the pump, into two branches, one of which leads to the spray nozzle and the other to the supply tank. In the case where all of the oil is not supplied to the vaporizer because of a light load, the by-pass valve will return the surplus to the tank, thus maintaining a constant pressure at the spray nozzle.

When operating under ordinary loads, the governor opens only the small inside valve which regulates the amount of oil injected into the vaporizer. But should the engine speed up, due to a sudden change in the load, the governor will not only open the small valve but also the large concentric valve, in which case all of the oil will return to the tank. The makers guarantee the following speed variation limits under the different loads.

Ordinary Variation	$2\frac{1}{2}$	per cent.
Full load to one-quarter load	4	per cent.
Full load to no load	5	per cent.

(69) Operating Costs of the Semi-Diesel Type.

As the semi-Diesel type engine will operate successfully on the lowest grades of crude oils, with an efficiency that compares favorably with the true Diesel type, the operating expenses are very much lower than with the gas or gasoline engine. With the same fuels, the semi-Diesel will show greater net saving than the Diesel with a low load factor, as the fuel saving is not eaten up by the high first cost, and overhead charges of the true Diesel. Western crude oils with a specific gravity of .960 (16° Beaumé) are being used daily with this type of engine while nearly every builder of the semi-Diesel type will guarantee results with oils up to 18° Beaumé (.948 Specific Gravity). Fuel of this grade will cost anywhere from 1½ cents to 3½ cents per gallon in tank car lots, depending on the distance of the engine from the wells or refinery.

With fuel oil weighing 7½ pounds per gallon, an engine consuming .65 pounds per horse-power hour (a usual guarantee) at full load, the cost of a horse-power hour delivered at the shaft will be .26 cent with fuel at 3 cents per gallon. This the lowest fuel expense of any prime mover even with steam or gas units of great power. In a twenty-four hour test of a De La Vergne oil engine running on 19° Beaumé oil, the consumption was considerably below the figure assumed above, being .508 pounds per horse-power hour. Even the engine was exceeded in a test made on a 175 horse-power engine by Dr. Waldo, which gave a consumption of .347 pounds of oil per horse-power hour with oil of .86 Specific Gravity.

The following is a tabulation of reports received by the De La Vergne Machine Company from the Snead Iron Works, giving the cost of power at their plant under actual working conditions extending over a period of twenty-four months. The plant consisted of a 17 × 27½ inch De La Vergne semi-Diesel type engine of 180 horse-power rated capacity, the load factor being 54.2 per cent. The total power produced during the record was 552,217 horse-power hours, with a working period of 588 days. Fuel = 28.8° Beaumé = 7.35 pounds per gallon.

TABULATION

Items	Total Cost	Cost per K.W. Hour	Cost per H.P. Hour
Fuel Oil, 38,211 gallons	$859.75	$.00232	$.00155
Lubricating	228.72	.00061	.00041

Oil			
Miscellaneous Stores and Repairs	123.20	.00032	.00022
Labor and Attendance	1361.42	.00368	.00246
Total		$.00693	$.00464

Fuel oil used = .761 pounds per K. W. hour = .508 pounds per horse-power hour. Computing from the load factor of 54.2 per cent, the cost of power produced under the above conditions would be $9.30 per horse-power year, or $13.98 per kilowatt year. This result is obtained by assuming that the horse-power hours would be increased from 552,217 to 1,077,354, or in proportion to the actual load factor, the period, of course being the same in both cases.

(70) Elyria Semi-Diesel Type.

A type of semi-Diesel type oil engine has been recently developed by the Elyria Gas Power Co., Elyria, O., that presents many features of interest. It operates on the two stroke cycle principle, and with the exception of the spray nozzle has no valves in the working cylinder. The principle of the semi-Diesel type cycle as distinguished from the true Diesel engine, was described in Chapter III, as having the following characteristics. (1) Fuel injection. (2) Medium compression pressure. (3) Hot plate ignition. (4) An efficiency approximating that of the true Diesel type.

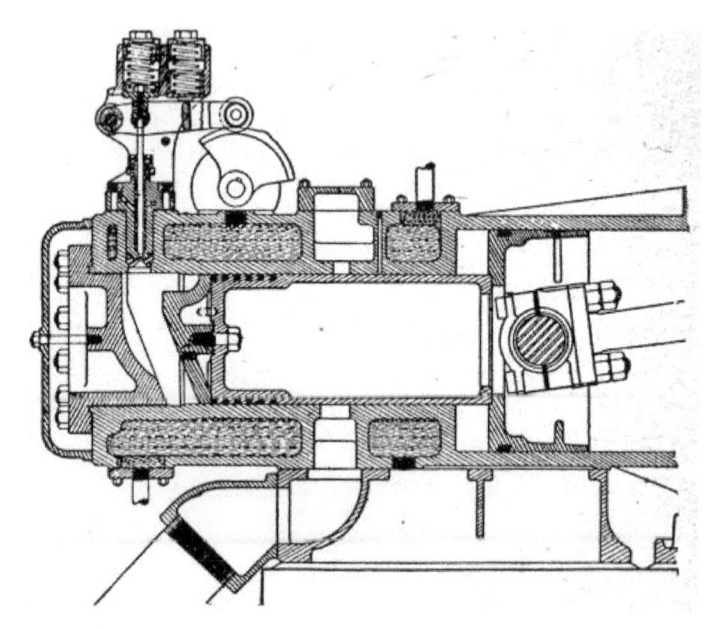

Fig. 77. Working Cylinder of Elyria Oil Engine.

It is claimed that the change from the ordinary four stroke cycle Diesel cycle has been accomplished with practically no loss of thermal efficiency, and that the elimination of the many moving parts of that type has done away with many of the operating difficulties. By the introduction of a false piston end and an unjacketed cylinder head, the loss of efficiency due to the lower compression is compensated by the reduction of heat loss to the jacket water. Because of the high temperature it is possible to burn the heaviest fuels with a maximum pressure not exceeding 400 pounds per square inch, and without trouble due to missed ignition at light loads. With a given cylinder capacity this heating effect has increased the output about 75 per cent. The loss due to the friction of the scavenging apparatus causes a fuel consumption of approximately 10 percent more than a standard four stroke Diesel.

Unlike the Diesel, this engine automatically controls the quantity of injection air admitted to the cylinder at different loads, the air corresponding with the amount of fuel injected. This is in marked contrast with the Diesel engine which admits a constant volume of air at all loads. In place of the usual crank-case compression of the scavenging air met with in the ordinary two stroke cycle engine, the initial compression in the Elyria engine is performed by a "differential piston" which acts in an enlarged portion of the cylinder bore. This construction increases the volumetric efficiency from 70 percent, in the case of the marine type, to well over 90 percent, and it also does away with the bad effect of the compression on the lubrication of the main crank shaft bearings.

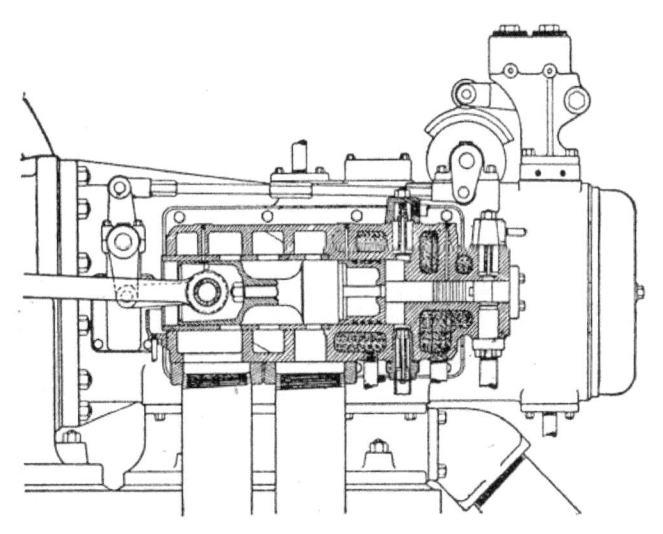

Fig. 78. Compressor Cylinder of Elyria Oil Engine.

The working piston and differential piston as shown by Fig. 77 is separate castings fastened together by four studs, and the piston pin is carried by the differential piston which acts as a cross-head, taking all of the side thrust from the main piston. The working piston is easily taken from the cylinder by removing the cylinder head and the four nuts that fasten it to the differential piston casting. The displacement of the differential piston is approximately 1.9 times the displacement of the working piston which is more than enough for thoroughly scavenging the cylinder and supplying air for combustion. The air suction is controlled by a piston valve which eliminates much of the loss encountered in the marine type of two stroke cycle.

In the figure may be seen the separate or auxiliary piston head which is bolted to the piston proper, a construction that greatly increases the working temperature, and allows a symmetrical form of piston. By removing the cap over the inlet port, it is possible to inspect the condition of the six piston rings with removing the piston from the cylinder. Because of the clean burning of the fuel lubrication is easily effected by the force pump which supplies oil at three points around the cylinder wall.

Three stages of compression are employed for providing the air for fuel injection, the first stage being accomplished by the differential piston, and the remaining two stages by a separate air pump driven by an eccentric from the crankshaft. This cylinder also supplies the air for starting the engine, the air being taken from the second stage and piped to the storage tank. The suction of the second stage pump which receives its air from the differential pump (first stage) is controlled automatically so that it is possible to keep the supply tank at any desired pressure regardless of the pressure or amount of air used for the fuel injection. Air from the tank (at approximately 200 pounds pressure) is piped to the suction side of the third stage air pump. In this suction line is a valve, controlled by the governor, which regulates the amount of air admitted to the injection nozzle, and also the amount. This pressure at the nozzle will vary from 500 pounds per square inch to 1000 pounds depending on the load and the nature of the fuel. The high pressure air travels directly from the pump to the fuel valve casing, and is equipped with a safety valve and pressure gauge.

The fuel pump is driven by a Rites Inertia Governor located in the flywheel which varies the stroke of the pump plunger and gives a correct proportion of fuel to the load. This type of governor has been extensively used on high speed engines and is exceeding accurate. The fuel pump may be disconnected from the governor drive, and operated by hand when it is necessary to provide fuel for starting. The spray or injection valve is operated by a cam, which lifts the valve at the proper moment in a very simple manner. The valve proper is made of a single piece of steel with openings of ample size, so that there is no danger of clogging with the heaviest fuels. As the valve only lifts 1 16 of an inch, the amount of work required to operate the valve is very small.

Starting is accomplished by spraying cold gasoline into the cylinder through the fuel valve in the same manner that the heavier oil is fed during operation, and the ignition is performed by a high tension coil and batteries. No spark time device is used, so that a continuous shower of sparks is thrown into the mixture during the starting period. Within a minute after the engine is started, the ignition switch may be opened, the gasoline cut off, and the heavy oil turned on for continuous running on full load. Starting by an electric spark avoids the inconvenience and danger of torch starting with a retort.

Cooling water is admitted around the compressor cylinder from which point it goes to the working cylinder, and is there discharged. Less water is required for this type of engine than for the ordinary gasoline engine, for with the water entering at 60°F, only 3 gallons per horse-power hour is used. With fuel oil weighing 7.33 pounds per gallon the makers claim a fuel consumption of .65 pounds per horse-power at the rated load. The amount of cylinder oil used does not exceed 1 pint per 100 horse-power hours, while the loss of the bearing oil is extremely small because of the return system.

(71) Remington Oil Engine.

The Remington Oil Engine is a vertical oil engine operating on the three port, two stroke cycle, and is an oil engine in the strict meaning of the word, the oil consumed being introduced into the combustion chamber as a liquid and gasified within this chamber.

The method of gasifying and igniting the charge of oil in the Remington Oil Engine is unique. Only clean air unmixed with any charge, is taken into the crankcase. This air is afterwards passed up into the cylinder and compressed until its temperature has raised to a point high enough to vaporize the oil which is injected into it. The charge of oil is then atomized into this hot compressed air and turns immediately into a vapor, which finds itself well mixed with the charge of air, comes in contact with a firing pin recessed in the head, ignite and burns. This method of having the oil well gasified and mixed with air before ignition begins, prevents the formation of carbon which is formed when oil not well gasified and mixed with air comes suddenly in contact with very hot surfaces.

This perfect system of gasifying the oil has the effect not only of preventing the formation of carbon in the cylinder, but also of increasing the mean effective pressure and therefore decreasing the amount of fuel necessary for doing a certain amount of work. The engine passes through its cycle of operations smoothly, and does not have to be constructed with excessive weight.

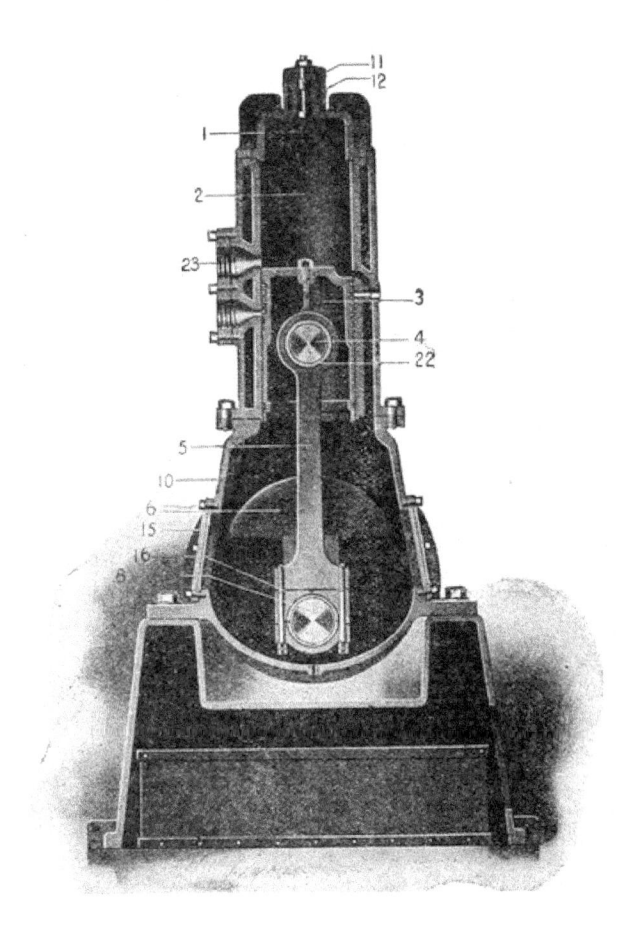

Fig. 79. Cross-Section of Remington Oil Engine.

The Remington Engine is of the valveless type, delivering a power impulse in each cylinder for each revolution of flywheel. The gases are moved in and out of the cylinder through ports uncovered by the movement of the piston, which itself performs also the function of a pump.

On the up stroke of the piston a partial vacuum is created in the enclosed crankcase, causing air to rush in when the bottom of the piston uncovers the inlet port seen directly under the exhaust port (23), Fig. 79. On the next down stroke this air is compressed in the crankcase to about four or five pounds pressure per square inch. Meanwhile the mixture of oil vapor and air already in the cylinder is burning and expanding. When the piston approaches the end of its down stroke, it uncovers the exhaust port (23), permitting the burnt charge to escape, until its pressure reaches that of the atmosphere. Directly afterward the transfer port on the opposite side of the cylinder is uncovered by the piston, thereby allowing a portion of the air compressed in the crankcase to rush into the cylinder, where it is deflected upwards by the shape of the top of the piston and caused to fill the cylinder, thereby expelling the remainder of the burnt charge. The piston now starts upward, compressing the fresh charge of air into the hot cylinder head. Near the end of the stroke, a small oil pump, mounted on the crankcase and controlled by the governor, injects the proper amount of oil through the nozzle (13), into the compressed and heated air.

Fig. 80. Remington Spray Nozzle.

This oil is atomized in a vertical direction through a hole near the end of the nozzle. It is therefore vaporized and gasified before there is a possibility of its reaching the cylinder walls.

The spray of oil is ignited by the nickel steel plug (12), which is kept red hot by the explosions because the iron walls surrounding it are protected from radiation by the hood (11). By the burning of the oil spray in the air the pressure is gradually increased and the piston forced downward, this being the power or impulse stroke. Near the end of the down stroke, the exhaust port is again uncovered and the burnt gases discharged.

Fig. 81. Fuel Pump and Mechanism of Remington Oil Engine.

The operations above described take place in the cylinder and crankcase with every revolution. Each upstroke of the piston draws fresh air into the crankcase and compresses the air transferred to the cylinder. Each down stroke is a power stroke, and at the same time compresses the air in the crankcase preparatory to transferring it to the cylinder by its own pressure at the end of the stroke.

The same volume of air enters the cylinder under all conditions, and the power is regulated by modifying the stroke of the oil pump, which may be done by hand or automatically by the governor in the flywheel. A separate fuel pump is provided for each cylinder when multiple cylinders are used, making it absolutely certain that each cylinder shall receive the same amount of fuel for a position of the control lever.

When starting the engine, the hollow cast iron prong rising from the cylinder head is heated by a kerosene torch, and when hot, a single charge of oil is admitted to the cylinder by working the hand pump. The flywheel is now turned backward, thereby compressing the charge which ignites the fuel before the piston reaches the highest position. After being started the engine, the torch may be extinguished.

Fig. 82. Two Cylinder Remington Oil Engine Direct Connected to Dynamo.

The governor is of the centrifugal type. It has an L-shaped weight, pivoted to the piece attached to the flywheel. As the engine speed increases, the weight tends to swing outward toward the flywheel rim, and thereby moves the arm attached to it so as to shift the cam along the crankshaft toward the left.

This cam turns with the shaft, and operates the kerosene oil pump. According to the position of the cam on the shaft, it will impart to the pump plunger a long or a short stroke, thereby injecting more or less oil into the cylinder. The lever pivoted on the bracket moves with the cam and is used for controlling the engine's speed by hand. To stop the engine the handle of the lever is pulled towards the flywheel, thereby interrupting the pump action altogether.

The handle of the control lever can be fitted with an adjustable speed regulator when required. This device is for use on marine engines to enable the operator to slow down the engine. The speed regulator does not interfere with the action of the governor but acts in conjunction with it. Whatever the speed of the engine may be, it is under the control of the governor. The engine can be controlled from the pilot house if such an arrangement is desirable.

The fuel pump is made of bronze. The valves are made of bronze and are designed with very large areas. The plunger is made of tool steel. A bronze cup strainer is attached to the lower end of the pump to prevent sediment or foreign matter from reaching the pump valves. As a result of the care used in its construction, the fuel pump is not only very sensitive in measuring the oil required by the governor, but is also very strong and durable.

The nozzle through which the fuel is atomized into the cylinder is thoroughly water jacketed to prevent the formation of carbon within the nozzle. It is so constructed that the water jacket spaces and fuel spaces can be opened for inspection.

Lubrication of all the important bearing joints is effected by a mechanical force feed oiler, pressure feed oiler or by gravity sight feed oilers, depending upon the service for which the engine is designed. Oil is fed in this manner to the piston, the main bearings and the crankpin bearings. The oil for the crankpin is dropped from a tube into an internally flanged ring attached to the crank by which it is carried by centrifugal force to a hole drilled diagonally through the crank and crankpin to the centre of the bearing. This insures that all the oil intended for the crankpin shall reach it. This feature, as well as the use of the sight feed oiler itself, is in line with the best modern high speed engine practice, and is an important factor in the reliability of the engine.

CHAPTER VIII
IGNITION SYSTEMS
(73) Principles of Ignition.

It is the purpose of the ignition system to raise a small portion of the mixture to the combustion temperature, or the temperature at which the air and fuel will start to enter into chemical combination. When combustion is once started in a compressed combustible gas it will spread throughout the mass no matter how small the original portion inflamed. The rate at which the flame spreads through the combustion chamber depends upon the compression pressure, the richness of the mixture, the nature of the fuel and upon the number of points at which it is ignited.

In practice perfect ignition is seldom realized. This is due not only to the ignition system itself but to poor mixture proportions, imperfect vaporizing of the fuel, and low compression; all of which tend to a slow burning mixture with the attendant losses.

The best ignition system will be that which will cause the ignition to occur invariably at the point of highest compression and which will supply ample heat to start the process of combustion with a cold cylinder, imperfect mixtures, and low compressions. An efficient and reliable ignition system is without a doubt the most important unit in the construction of a gas engine. As ignition systems have improved and become more reliable, so has the gas engine become more widely used and appreciated, and in almost a direct proportion to these improvements.

Many ingenious ignition systems have been proposed, but only two of these have met with any degree of success in practice; i. e., electrical ignition and ignition by means of the hot tube.

Sponge platinum has the peculiar property of igniting jets of hydrogen gas, or hydrocarbons, without the aid of heat; this is due to the condensing effect of the platinum on these gases.

It was proposed to ignite the gaseous charge of the gas engine by means of the platinum sponge (catalytic ignition) but the system proved a failure because of the clogging of the pores in the sponge by fine particles of soot.

Dr. Otto employed an open flame which was introduced into the mixture by means of a slide valve. This met with only a fair measure of success.

Cerium, Lanthum and several other rare metals cause a considerable spark when brought into contact with iron or steel. The objection to this method was the expense of the Cerium plugs which required frequent renewal.

The writer remembers a quaint attempt at firing the charge by means of a piece of flint and steel; the failure of this is obvious.

The Diesel Engine, a great success from a thermodynamic standpoint, is fired by means of the heat produced by the compression of air, the fuel being sprayed into air which is compressed to several hundred pounds pressure.

Mr. Victor Lougheed proposes ignition by means of a platinum wire rendered incandescent by a current of electricity. The plan sounds feasible, but we are still waiting to be shown.

Electric ignition is applicable to all classes of engines; in fact this system made the variable speed engine as used on automobiles, etc., a possibility, as accurate timing with the electric spark covers the range from the lowest possible speed to speeds of 4,500 revolutions per minute and over.

(74) Advance and Retard.

While the combustion of the mixture is extremely rapid under favorable conditions, there is, nevertheless, a perceptible lapse between the instant of ignition and the final pressure established by the heat of the combustion. For this reason it is necessary that ignition should be started a certain length of time before the pressure is required if we are to expect a maximum pressure at a definite point in the stroke of the piston. The amount by which the time of ignition precedes that of combustion is called the **ADVANCE**, and is usually given in terms of angular degrees made by the crank in traveling from the time of ignition to time of maximum pressure. Since the pressure is always required at the extreme end of the compression stroke, the degree of advance is given as the angle made by the center line of the cylinder with the center line of the crank at the instant of ignition. Should the advance be given as 10°, for example, it is meant that the crank is still 10° from completing the compression when ignition occurs.

Owing to variations in the richness of the mixture, and changes in the compression pressure, due to throttling the incoming charge, the rate of inflammation varies from time to time under varying loads. To keep the maximum pressure at a given point under these conditions it is necessary to vary the point of ignition to correspond with the increase or decrease of inflammation. This variation of advance to meet varying loads is approximated by the governor in some engines, and manually in others. The advance of an automobile is an example of manual ignition control. Should the point of ignition vary from the theoretical point it will result in a loss of fuel and power, and for this reason the ignition should be under at least an approximate control. A wide variation in engine speed has a very considerable effect on the ignition point as there is less time in which to burn the mixture at high piston speeds, and consequently the ignition must be further advanced to insure complete combustion at the end of the stroke. This fact is evident to those who have driven automobiles.

Should the ignition occur too early, so that combustion is complete before the piston reaches the end of the stroke, there will be a loss of power due to the tendency of the pressure to reverse the rotation of the engine. When starting an engine, over-advanced ignition will throw the crank over in the reverse direction from which it is intended to go, and will not only prevent the engine from coming up to speed but will prove dangerous to the operator.

Due to the effects of inertia and self induction in several types of ignition apparatus, a greater advance will be required than that demanded by the combustion rate of the mixture. This sluggishness of the apparatus in responding to the piston position is called ignition **LAG**. The total advance required to have the combustion complete at the end of the stroke is equal to the advance required by the burning speed plus the ignition lag. Since lag is principally due to inertia effects, it is much greater at high speeds than at low, and it therefore causes an additional advance at high speeds. Causing the ignition to occur before the crank reaches the upper dead center is called **ADVANCED IGNITION**, causing it to occur after the piston has reached the upper dead center, or when on the outward stroke, is called **RETARDED IGNITION**.

Ignition is retarded when starting an engine to prevent it from taking its initial turn in the wrong direction. As the combustion takes place after the compression, with the piston moving on the working stroke, in retard, it is impossible for the pressure to force the piston in any direction but the right one. Excessively retarded ignition will cause a power loss and will also cause overheating of the cylinder and valves as the combustion is slower.

(75) Preignition.

Preignition which is in effect the same as over-advanced ignition as due to causes within the cylinder such as incandescent carbon deposits or thin sharp edges in the cylinder that have become incandescent through the heat of the successive explosions. Preignition is very objectionable since it causes heavy strains on the engine parts and causes a loss of power in the same way as over-advanced ignition. Any condition that causes the preigniting of the charge should be removed immediately.

(76) Misfiring.

The failure of the ignition apparatus to ignite every charge is called **MISFIRING**. This missing not only causes a waste of fuel and a loss of power but it also causes an increased strain on the engine parts because of the violence of the explosion following the missed stroke. The heavy explosion is due to the fact that the stroke following the "miss" is more thoroughly scavenged by the two admissions of the mixture than the ordinary working stroke, and consequently contains a more active charge.

(77) Hot Tube Ignition.

A combustible gas may be ignited by bringing it into contact with surface heated to, or above the ignition temperature. It is upon this principle that hot tube ignition is based.

In practice this surface is provided by the bore of a tube which is in communication with the charge in the cylinder, the outer end of the tube being closed or stopped up. Around this tube is an asbestos-lined chimney which causes the flame from the Bunsen burner to come into contact with the tube and also prevents draughts of air from chilling it.

A Bunsen burner is located near the base of the tube and maintains it at bright red heat. The gas for the burner is supplied from a source external to the engine. When the fuel used is gasoline, a gasoline burner is used, which is fed from a small supply tank located five or six feet above the burner.

During the admission stroke, the hot tube is filled with the non-combustible gases remaining from the previous explosion, therefore, the fresh entering gases cannot come into contact with the hot walls of the tube and cause a premature explosion, before the charge is compressed.

As the compression of the new charge proceeds, the fresh gas is forced farther and farther into the tube and at the highest point of compression it has penetrated far enough to come into contact with the hot portion. At this point the explosion occurs.

The tube being of small bore, does not allow of the burnt gases mingling with the fresh within the tube; the waste gases in the tube acting as a regulating cushion. The distance of travel of the new mixture is proportional to the compression, hence the explosion does not occur until a certain degree of compression is attained.

The length of the tube required for a given engine is a matter of experiment, as is also the location of the heated portion. High compression naturally forces the mixture farther into the tube than low, therefore the flame should come into contact with the tube at a point nearer the outer end with high compression than with a low compression.

Shortening the tube causes advanced ignition, as the mixture reaches the heated portion sooner, or earlier in the stroke, because of the decreased cushioning effect of the residue gases in the tube.

The length of tube and location of maximum heat zone should be so proportioned that combustion will take place at the highest compression. Moving flame to outer end of the tube retards ignition. Moving the flame toward the cylinder advances it.

While the hot tube is the acme of simplicity in construction, it is not the easiest thing to properly adjust, as the adjustment depends on compression, temperature of the tube, and the quality of the mixture. Any of these variables may cause improper firing.

The hot tube is rather an expensive type of ignition with high priced fuel, as the burner consumes a considerable amount of gas, and is burning continuously during the idle strokes as well as during the time of firing.

It is practically impossible to obtain satisfactory results from a hot tube on an engine that regulates its speed by varying the mixture or compression, as engines running on a light load will not have sufficient compression to cause the mixture to come into contact with the hot surface, the engine misfiring on light loads.

The tubes are made of porcelain, nickel steel alloy, or common gas pipe, and are of various diameters and lengths.

All of these materials have their faults. Porcelain being very brittle, is liable to breakage. Gas pipe burns out and corrodes rapidly. Nickel alloy is not liable to breakage, is not so susceptible to corrosion as iron, but is far from being a permanent fixture.

Timing valves are a feature of some systems of hot tube ignition, which correct to a certain extent the irregularity of firing of the plain type of tube.

The timing valve is introduced in the passage connecting the cylinder and tube, and prevents the gas in the cylinder from coming into contact with the heated surface until ignition is desired.

The valve is operated by means of mechanism connecting it with the crank shaft. It is evident that with sufficient compression in the cylinder, the time of ignition can be obtained with certainty.

This mechanism is rather complicated, and subject to wear, and the advantage gained by the fixed point of ignition is offset by mechanical complication and consequent trouble.

The action of hot tube igniters is erratic and their use is not advisable unless under unusual conditions. The open flame used in heating the tube is a constant menace, as it is surrounded by inflammable vapors. This feature alone condemns it in the eyes of the insurance underwriters; in many places the use of the hot tube is prohibited both by the underwriters and city ordinances.

The above inherent defects of hot tubes are supplemented by breakage, "blowing," and clogging of the tube or passage with soot and products of corrosion, each factor of which will cause misfiring.

In case of misfiring, after determining that the tube is not broken or clogged with soot or dirt, see that the engine is being supplied with the proper mixture; that you are obtaining the proper compression; and that the Bunsen burner is delivering a bright blue flame on the tube at the proper point. Never allow the burner to develop a yellow sooty flame. A yellow flame indicates that insufficient air is being admitted to the burner. Remember that an overheated tube is quickly destroyed, and will cause misfiring as surely as an underheated tube. Regulate the gas supply to the burner.

A small leak near the outer end of the tube will destroy the cushioning effect of the burnt gas, and hence will cause premature firing of the charge. Procure a new tube.

Many engines are provided with a sliding burner and chimney which allows of some adjustment of the flame on the tube. In cases of persistent misfiring, move the chimney one way or the other. It may improve the ignition.

(78) Electrical Ignition.

Ignition by means of an electric spark is by far the most satisfactory method as it makes accurate timing and prompt starting possible. It is the most reliable of all systems and is easily inspected and adjusted by anyone having even a rudimentary idea of electricity or the gas engine. For this reason electric ignition is used on practically all modern engines (with the exception of the Diesel types). The spark is caused by the current jumping an opening or gap in the conducting path of the current, and the ignition of the charge is obtained by placing this cap in the midst of the combustible mixture to which the spark communicates its heat.

The method of producing the spark gap, and the method by which the current is forced to jump the gap, divides the electrical ignition system into two principal classes:

- (1) The **MAKE AND BREAK**, or **LOW TENSION** system.
- (2) The **JUMP SPARK** or **HIGH TENSION** system.

In either system the spark is produced by the electrical friction of the current passing through the high resistance of the gas in the spark gap. The incandescent vapor in the gap formed by this increase of temperatures causes the flash that is known as the spark. The temperature of the gap depends principally upon the current flowing through it, the amount of heat developed being proportional to the square of the current.

There is of course a practical limit to the amount of current used in the ignition apparatus to produce spark heat. The limit is generally set by considerations of the life of the battery furnishing the current, expense of generating the current, and the life of the contact points between which the spark occurs.

The heat developed by an electric current is proportional to the amount of resistance offered to its flow and the strength of the current employed. The greater the resistance, the more heat developed.

The resistance of copper wire (the usual conducting path), being very low causes little rise in temperature, but the air in the opening or break has a resistance of many thousands of times the resistance of the copper; hence the current passing across the opening spark or gap raises the air to an exceedingly high temperature.

With a comparatively heavy current flowing across the break, the temperature developed is high enough to boil or vaporize any metal in contact with the spark or flame, rendering the metallic vapors incandescent. With sufficient current, the ends of the wires which constitute the break may be melted away.

For the successful and continuous operation of the engine it is imperative that ends of the conducting path or terminals be made of a metal of a high fusing point in order to withstand the heat of the spark and also that the current be kept to as low a value as possible.

In actual construction the spark gap terminals are generally made of platinum or platino-iridium, or an alloy of high fusing point. Iron is sometimes used, but deteriorates rapidly. Nickel steel lasts longer than common iron or steel but is not as durable as platinum or its alloys.

As the temperature of the electric spark or arc is approximately 7,500° F., and the ignition temperature of an ordinary rich gas at 70 lbs. compression is 1,100° F., it is evident that the quantity of current for ignition may be kept to an exceedingly low value. High compression increases the resistance of the spark gap, and requires higher electrical pressure to force a given current across a gap of given length.

(79) Sources of Current.

The electric current that causes the ignition spark is usually generated or supplied by one of the three following methods:—

1. By the primary battery which converts the chemical energy of metal, and some corroding fluid, into electrical energy, by chemical means.

2. By the magneto or dynamo that converts mechanical work or energy into electrical energy through the method of magnetic induction.

3. By the storage or secondary battery which acts as a reservoir or storage tank for current that has been generated by either of the two above methods. A storage battery simply returns electrical energy that has been expended on it by an external generator. A storage battery does not really generate electricity but as it is often used as a source of current for an ignition system, we will consider it as a generator.

Current producers that convert chemical or mechanical energy into electrical energy are called primary generators, and are represented by the primary battery and dynamo. The above methods are used for generating current for either the high or low tension systems.

Electricity may also be produced by friction, but as such current is without heat value it is not used for ignition purposes. Electricity produced by friction is called static electricity.

Primary and storage batteries always deliver a direct or continuous current of electricity, that is a current which flows continually in one direction. Dynamos are usually made to furnish a direct current, but can be built to deliver either direct or alternating.

Alternating current, unlike the continuous current, changes the direction of its flow periodically; flowing first in one direction and then in the other, the flow alternating in equal periods of time.

Magnetos being a special form of dynamo can furnish either class of current, but with few exceptions are built for generating alternating current.

Either current may be used for ignition purposes for either high or low tension systems.

Alternating current has several advantages not possessed by the continuous current, when used for ignition purposes. The principal advantages are:

1. Alternating current does not transfer the electrode metal of contact points, and consequently causes less trouble with vibrators and "make" and "break" ignitors.

2. Magnetos generating alternating current are less complicated, have fewer parts to get out of order, and are cheaper to keep in repair.

3. Alternating current is not liable to burn out spark coils or overheat with an excessive voltage.

4. Alternating current generators can be used at any speed without the use of governors.

43-a. The Esselbé Rotary Aero Motor. Four Pistons are Contained in the Ring Shaped Cylinder at the Left Which are so Connected with Cranks and Gears in the Gear Box that the Pistons and the Cylinder Rotate in Opposite Directions. As the Pistons Rotate they also Oscillate Back and Forth in Regard to One Another, so that the Working and Compression Strokes are Performed. From Aero London.

When installing an ignition system give due consideration to the reliability of the source of current. The gas engine is no more reliable than its source of current. Failure of the current means the failure of the engine.

(80) Primary Batteries.

Current is produced in a primary battery by the chemical action of a fluid known as an **ELECTROLYTE** upon two dissimilar metals or solids known as the electrodes. One of the electrodes, the negative, is usually made of zinc which is more readily attacked by the electrolyte than the positive electrode. As the metal of the negative electrode is dissolved and passes into the solution during the process of current generation, the electrolyte is also exhausted. The production of current is accompanied by the liberation of hydrogen gas from the electrolyte from which it is displaced by the zinc taken into solution.

When the electrodes are immersed in the electrolyte, and the outer ends of the electrodes are connected with a wire, a current will flow from the positive electrode to the negative through the wire, and from the negative to the positive electrode through the fluid. It will be seen that to complete the circuit between the electrodes it is necessary that the current flows through the electrolyte.

Electrical energy is actually generated in the primary battery by the chemical combustion of the negative electrode in the same way that heat energy is developed by the burning of a fuel.

By connecting the binding posts of the electrodes to the two wires of the external circuit, a current will flow through the circuit as long as the electrodes remain undissolved, or until the positive electrode is covered with hydrogen gas bubbles.

The bubbles of gas tend to insulate the positive electrode from the electrolyte or fluid, thus breaking the circuit through the fluid, and stopping the flow of current. This action is known as polarization.

When a battery is polarized, the only remedy is to disconnect it from the circuit and allow it to rest or recuperate. The greater the current drawn from a battery, the more rapid the polarization, and it is evident that if the battery is to be used for long periods, polarization must be eliminated, or the current must be considerably reduced in volume. A battery that delivers a small current has a much greater capacity in ampere hours than a battery that has a higher rate of discharge. The greater the discharge rate the longer must be the rest periods.

A battery that is designed for continuous service, or for delivering heavy currents of long duration, is called a closed-circuit battery. Polarization is eliminated in closed circuit batteries by various methods, the usual methods being to place some substance in the electrolyte that will destroy the hydrogen film; or by packing some solid oxidizing material around the positive electrode that will absorb the hydrogen; or by making the positive electrode of some material that will destroy the hydrogen as soon as it is developed.

Batteries that are capable of being operated only for short periods, on account of polarization, are called open circuit batteries. Open circuit batteries are cheaper and more simple than closed circuit batteries. For ignition purposes, a battery is made that is a compromise between the closed and open circuit cells, this being a battery in which the polarization is only partially suppressed. As the demand for current on an ignition battery is small with comparatively long rests between contacts, the compromise battery answers the purpose and is fairly cheap.

All primary batteries are in reality wet batteries, for the reason that it would be impossible to cause a chemical reaction and a current with a dry electrolyte. The action of dry and wet batteries is identical.

There are many types of wet battery in use for various purposes, but few of them are adapted for gas engine ignition because of a tendency to polarize or because of the cost of maintenance.

All wet batteries are not suitable for portable or automobile engines because of the slopping of the liquid electrolyte and the danger of breaking the containing jars. Their weight and bulk is also a drawback.

If the electrolyte or the electrodes be made of impure material local currents will be generated. These currents decrease the life of the cell without producing any useful current in the ignition circuit. Due to the deteriorating effects of the local currents, batteries standing idle for several months will often be found to be completely discharged and worthless without having done any useful work. In the better grade of cells this loss is reduced to a minimum.

A type of wet battery using a solution of caustic soda for an electrolyte, and having zinc and copper oxide electrodes, is extensively used for stationary ignition purposes, and is the most satisfactory type of wet cell for continuous work with this class of engine. The caustic soda battery is of the **CLOSED** circuit type, and is capable of furnishing a strong uniform current without danger of polarization.

The hydrogen bubbles which cause polarization are oxidized or eliminated by the copper oxide electrode as soon as they are formed. The hydrogen combines with the oxygen of the copper oxide forming water.

The copper oxide is gradually reduced to metallic copper by the reaction with the hydrogen, and in the course of time requires renewal. The copper oxide element is rather expensive and cannot be obtained as readily as the electrodes used in other cells.

It will be noted that both electrodes are consumed in the caustic battery, the consumption of the zinc furnishing the current, and the reducing of the oxide furnishing the chemical energy for depolarizing the cell.

(81) Dry Batteries.

Dry batteries are by far, the most convenient and economical form of primary battery to use, for there is no fluid to slop and leak, the first cost is low, the output is large for the weight, and last but not least, the cell can be thrown away when exhausted without great monetary loss. This does away with the expense and annoyance of changing wet cells, a factor that will be appreciated by those that are far from a source of chemical supplies. Since the advent of the automobile the use of dry cells has extended so that they may be obtained in almost any country town or village.

While the cell is not dry, strictly speaking, the solution is held in such a way that it cannot slop around in the cell nor leak out of the seal. The only fault of a dry cell is its tendency to deteriorate with age because of the constant contact of the electrolyte with the electrodes.

The negative electrode of the dry cell (zinc) is in the form of a cup which serves as a containing vessel for the electrolyte and the depolarizer.

The electrolyte is usually composed of a solution of ammonium chloride, with a small percentage of zinc sulphate, this fluid being held by some absorbent material such as blotting paper, or paper pulp.

The electrolyte is applied to the electrodes by means of the saturated blotting paper, which is also used to line the zinc container, thus providing insulation between the electrodes.

A rod of solid carbon which forms the positive electrode is placed in the center of the container, and the space between the rod and the zinc is packed solidly with granulated carbon, the blotting paper lining preventing contact of the zinc with the carbon.

Pulverized manganese dioxide is mixed with the granulated carbon for a depolarizer.

Brookes Four Cylinder Gasoline Engine Direct Connected to Dynamo.

After the zinc container is filled with the electrolyte and pulverized carbon, the top of the container is closed hermetically by means of sealing wax. Granulated carbon is used for it presents a large surface to the electrolyte, reduces the internal resistance of the cell, and therefore increases the current output of the battery.

As soon as the battery starts generating current, polarization begins, with the liberation of hydrogen gas. If the cell is discharged at a high rate, the manganese dioxide will be unable to absorb all of the gas, and consequently pressure will be erected within the cell. The greater the rate of discharge, the greater will be the amount of hydrogen set free, and the higher the pressure.

If a short circuit exists for any length of time, the pressure of the excess hydrogen will speedily ruin it, as the cell will puff up, or even burst under the pressure. If the rate of discharge be kept so low that all of the gas will be absorbed by the manganese, as soon as generated, the cell will furnish a steady current until the elements of the cell or the electrolyte are exhausted.

The steady current limit, or non-polarizing limit is about one-half ampere and if long life of the cell is expected, the current drain should be less than this amount. A good spark coil will develop a satisfactory spark on a quarter to one-half ampere, so that the demand of a good coil is well within the safe limits of battery capacity. The voltage of the average dry cell when in good condition is 1.5 volts on open circuit. When the cell is old or exhausted, the voltage falls rapidly when any demand for current is made on the cell, and the voltage also varies with the rate of current flow, the voltage decreasing with an increase of current.

As there is not much difference in voltage between a new and old cell when on open circuit, it will be seen that the ammeter giving the current output will give a more accurate determination of the condition of the battery. The voltage is independent of the size of cell.

The battery showing the greatest amperage is not necessarily the best for general use, as cells having an unusually high current capacity are generally short lived. The strong electrolyte used in high ampere batteries causes them to burn out or deteriorate rapidly when not in use.

Under ordinary conditions, a correctly proportioned No. 6 ignition cell should show a current of from fifteen to twenty amperes on short circuit when the cell is new, although higher results may be obtained safely with some makes of cells.

While the voltage is the same for all sizes of batteries, and depends on the material used in the construction, the amperes increase with the size of the cell, and the area of the electrodes.

If a cell does not show more than ten amperes on short circuit, it should be thrown out and another substituted for it, as the cell is liable to go out of commission at any minute when reaching this point of exhaustion.

A small battery testing voltmeter or ammeter should be in the kit of every gas engine operator using a battery for ignition, as the exact condition of a vital part of the power plant can be determined quickly and with accuracy. For dry batteries an ammeter is preferable; for storage batteries a voltmeter must be used.

When buying dry batteries insist on having new, fresh cells, as any battery depreciates in value with age. Never take a cell without testing it, as it is the practice of dealers to work off their old stock on unsuspecting customers. Examine the battery closely for the makers' dates, and if the battery is several months old, it is probable that the electrolyte is dried up or that the electrodes are wasted through long continued local action. As heat stimulates chemical action in the cell, they should be stored in a cool place to retard the wasting action as much as possible. Under all conditions, the cell should be kept dry, since the moisture that is deposited on the cell forms a closed circuit for the current which soon exhausts the battery. Cold retards chemical action in the cell and consequently reduces the output in zero weather to such an extent that starting is frequently impossible.

Multiple cylinder engines exhaust a battery quicker than those with a single cylinder, as there are more current impulses in a given time and consequently more current is used. A battery may be compared with a bottle that holds a certain quantity of fluid. If the water is allowed to drip out slowly it will last for a long time, but if allowed to flow in a continuous stream will soon be exhausted.

With badly designed or poorly adjusted spark coil, the demand on the batteries is greater than with one that is in proper condition. An engine that runs continuously exhausts a battery faster than one that is run at long intervals. Always open the battery switch when the engine is to be idle for any length of time, as the engine may have stopped with the igniter in contact, allowing the battery to expend its energy uselessly.

Test batteries immediately after a run, as the batteries will recover after standing a while, and will show a fictitious value.

A weak, partially exhausted battery will cause a poor spark that will result in misfiring or a loss of power. It is poor economy to attempt running an engine on a weak battery. An engine may run on a weak battery for a short time, and then gradually decrease in speed until it comes to a full stop. Misfiring is generally in evidence as the engine dies down. In case of an emergency, weak batteries may be made to run an engine of an automobile or boat to its destination, by stopping the engine frequently and allowing the batteries to recuperate during the idle periods. A battery that is temporarily weakened by hard service or by a temporary short circuit will usually revive or partially recover its strength if allowed to "rest" for a short time until the hydrogen is absorbed by the depolarizing material. The life of a dry cell can be extended for a few hours by punching a hole in the sealing wax on the top of the battery, and pouring water, or a solution of water and sal-ammoniac into the cell. This will reduce the internal resistance and increase the current. The batteries will run under these conditions for a short time only, and new cells should be procured at the earliest possible moment. No old cell can be made as good as new by any method. Never drop the cells on the floor nor subject them to hard usage mechanically, for if the active material is loosened, the current output will be reduced. A short circuit through a closed switch with the engine stopped or a loose dangling wire will put the cells beyond repair.

If the binding screw on the carbon electrode does not make good contact with the carbon, tighten it to decrease the resistance. Fasten the connecting wires firmly under the binding screws and keep the connections clean.

In the absence of an ammeter, a rough estimate of the condition of the cell may be made by fastening a short wire tightly in the zinc binding post, and touching the carbon surface lightly and intermittently with the free end of the wire. When contact is made with the free end of the wire, a small puff of smoke will arise and a red spark will be seen if the cell is in good condition.

Sometimes the contact made on the carbon will produce only a small black ring on the surface of the electrode. This indicates a battery that is nearly exhausted, and one which is good for only a few more hours of service.

When a number of cells are connected together in such a way that they collectively form a single source of current, the group is called a battery, and the resulting voltage and amperes of the group depends on the way in which the cells are interconnected.

It is possible to connect the cells of a battery in such a way that total voltage of the group or battery is equal to the sum of the voltages of the individual cells. A battery connected in this manner is said to be connected in series. While the voltage of a battery is increased, by series connection, the number of amperes is the same as that given by a single cell, the same current flowing through the set.

(82) Series and Multiple Connections.

Fig. 86 shows the cells connected in series, the carbon terminal of one cell being connected to the zinc terminal of the second. The carbon of the second cell is connected to the zinc of the third, and so on throughout the series, the two remaining terminals of the battery being connected with the ignition circuit. The number of watts or power developed by the group is equal to the sum of the outputs of the separate cells. If the voltage of each cell shown in diagram is 1.5 volts, the total voltage of the group of five cells will be $1.5 \times 5 = 7.5$ volts, and if the current of a single cell is 15 amperes, the current output of the group will be 15 amperes, or the same as that of a single cell. Almost all ignition apparatus now on the market requires six volts for its operation, so with cells having a voltage of 1.5 volts such apparatus would call for four cells in series, as $6 \div 1.5 = 4$.

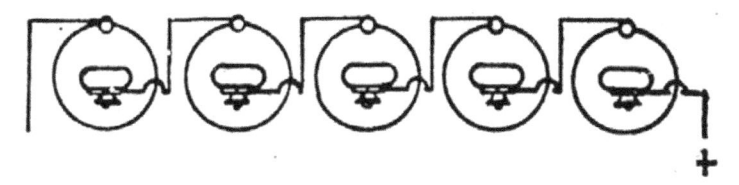

Fig. 86. Five Cells in Series.

Owing to the increase of internal resistance caused by series connections it is usual to add one more cell than is theoretically required, making a group of five cells to supply the six volts required. A large number of cells will give a hotter spark than a smaller, but the excessive current causes the contact points of the igniter or vibrator to burn off rapidly and also hastens the destruction of the cells themselves.

Batteries connected in such a way that the total amperes of the group is increased without increased voltage are said to be connected in multiple or parallel. When batteries are connected in multiple, the total current in amperes is equal to the sum of the amperes delivered by the separate cells; and, while the current in amperes is increased by multiple connection, the voltage of the group remains equal to that of a single cell.

If each cell connected in multiple has an electromotive force of 1.5 volts, and can deliver 15 amperes, the total current delivered by this system of connection will be $15 \times 5 = 75$ amperes with five cells, and the electromotive force will be 1.5 volts as in the case of the single cell. By connecting batteries in multiple, the resistance is reduced, allowing a maximum flow of current. The demand on the individual cells is reduced by multiple connection, as each cell only furnishes a small part of the total current. The greater the number of cells, the less will be the current required per cell, with a given total current. As the life of a battery depends entirely upon the rate at which it is discharged, it is necessary, for economical reasons, to keep the current per cell as small as possible, therefore the multiple system would prove of value as it reduces the load to the smallest possible limit. Enough cells should be placed in multiple to reduce the current to less than a quarter of an ampere per cell. The cells shown will not have sufficient voltage to operate ordinary ignition apparatus requiring a potential of six volts, hence the multiple system must be modified in order to have an increased voltage, and at the same time secure the advantages of multiple connections.

(83) Multiple-Series Connections.

A compromise is affected by the multiple series system of connections in which are combined the advantages of both the series and multiple systems of connection.

This arrangement allows sufficient voltage to operate 6 volt apparatus and at the same time reduces the rate of discharge on the individual cells. The series-multiple battery shown in the diagram 88 consists of four groups of batteries connected in multiple, each group of which consists of five cells that are connected in series. The current and voltage in the various branches is shown in the diagram. The series-multiple system is adapted for use with multiple cylinder engines, as engines with more than one cylinder cause a severe drain on the ignition system. Arranging the series groups in parallel increases the life and efficiency of the cells. If an efficient coil is used, the drain of a single cylinder is not too great to be met with a single set of series cells. If possible the set should be provided with a duplicate, so that the load could be transferred from one set to the other at proper intervals by means of a double throw switch.

With a single set of batteries in series the working life of the cells will be approximately twenty hours under ordinary conditions. With four groups of four cells in series, the life of the cell will be approximately 160 hours, or eight times the life of the single set under similar conditions.

While the cost of the cells will be only four times that of the single set, it will be seen that the cost of battery upkeep is halved by reducing the demand on the cells.

Sometimes duplicate sets of series multiple connected batteries are used for heavy duty engines, the engine running on one set for a while and then on the other, allowing the first set to thoroughly recuperate before it is again thrown in service, by means of the double throw switch.

When batteries are multiple or series-multiple connected they should be of the same size and make and of the same voltage. If the cells are of different voltages useless local currents will circulate among the cross-connections, shortening the life of the battery and reducing the output.

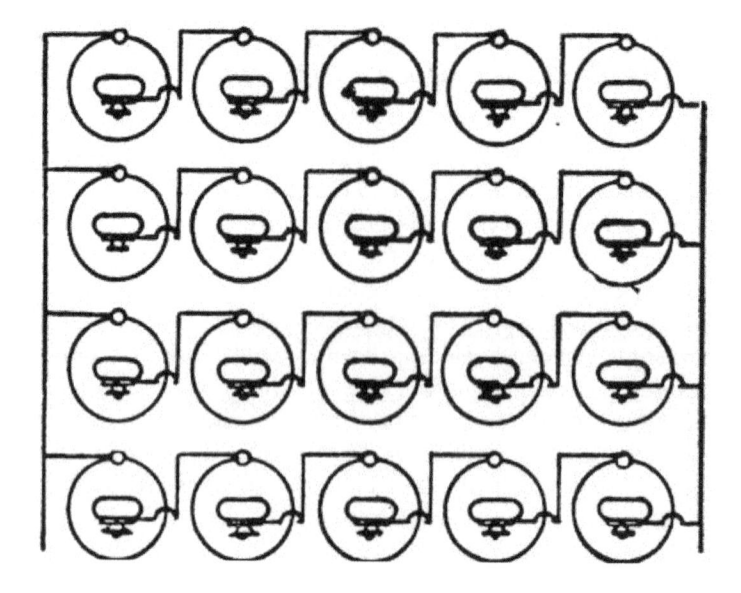

Fig. 88. Cells in Multiple Series.

In connecting a dry cell use a good grade of rubber insulated wire, preferably wire with a stranded conductor, as it is less liable to break or loosen at the binding screw of the battery. Carefully remove the insulation from the end of the wire that is to be fastened under the binding screw of the battery. Scrape it until it is bright and perfectly free from dirt before fastening it in the battery terminal. Never allow a dirty or corroded connection or a loose wire to exist. An open battery circuit or loose connection will stop engine suddenly, or will prevent starting.

The battery connections should be screwed down tight with the pliers, care being taken that the screws are not broken by the tightening process. See that frayed ends of the wire do not project beyond the binding screw to which they are connected and make contact with other cells or metal objects. Be sure that no insulation gets between the contact braces of the binding screw.

(84) Operation of Dry Cells.

The following hints should be observed to obtain the best results with dry cells.

(1) Never remove the paper jackets from the cells.

(2) Never lay tools or other metallic objects on top of the cells for this will cause a "short" that will quickly exhaust them.

(3) Do not connect old and new cells together, especially with the multiple-series system of connections, for the old cells will limit the output of the new, or else will cause cross-currents that will exhaust all of them.

(4) When trouble develops in the battery, test each cell separately and remove the faulty cells. Do not reject all of the battery because of one or two dead cells.

(5) Place the cells in a wooden box that will protect them from dirt or moisture, and if possible divide the box off into pigeon holes with a cell in each hole. For the best protection against moisture, the box should be boiled in paraffine.

(6) Provide a battery switch on the box that will cut both leads from the cells completely out of circuit when the engine is stopped.

(7) Never place a dry cell in a box that has contained storage cells unless the box has been thoroughly washed out, for the residual acid of the battery will destroy the zinc elements.

(8) Make all connections firmly with well insulated wire and take care that the wire does not make contact with any part of the battery except that to which it is connected.

(9) Keep the battery dry.

(85) Storage Batteries.

The purpose of the storage battery is to store or accumulate the current generated by a dynamo until so that the current will be available when the dynamo is not running. A storage cell does not "store" current in the same way that water is held in a tank, but returns the energy expended on it through the chemical changes caused in the cell by the current.

When the charging current passes through the storage battery chemical changes are produced in the electrodes and electrolyte, and the energy expended on the cell is in the form of latent chemical energy, in which state it remains until the electrodes are connected with one another by a wire or some other conducting medium. When the electrodes are connected through an external circuit, the electrolyte acts on the electrodes causing them to assume their original composition. As they pass into their previous chemical condition the latent chemical energy is converted into electrical energy. The current thus produced may be used in the same way as in a primary cell.

When discharging, the action of a storage battery is similar to that of a primary battery, the current being produced by the action of a fluid on two dissimilar electrodes. Instead of supplying new elements when the battery is discharged, as in the case of the primary cell, the elements are brought back to their original state by passing a current through the cell in the opposite direction to that of the discharge.

There are several combinations of materials which may be used in the making of storage battery electrodes and electrolytes, but with the exception of the lead sulphuric battery and the new Edison battery none have proven a commercial success.

The most common type of storage or secondary cell is the lead-sulphuric type in which the electrolyte is dilute sulphuric acid and the electrodes are lead plates, covered with a chemical composition known as the active material. These plates usually consist of a lead grid, or lattice frame in the pockets of which is pasted the active material. The pockets or lattice bars of the plates are for the purpose of supporting the active material which is of a weak and spongy nature. The active material on the positive plate is usually litharge, while that on the negative plate is red lead.

After charging, the active material on the positive plate is changed to lead peroxide by the action of the current, and the active material on the negative plate is changed into spongy metallic peroxide. The composition of the active material on the plates determines the direction of flow of the discharge, or secondary current. The current flows from the positive plate to the negative through the external circuit.

When fully charged, and in good condition, the positive and negative plates may be readily distinguished by their colors, the positive plate being a dark brown or chocolate color, and the negative a slate or grey color.

The positive active material is hard, while the negative may be easily cut into by the finger nail. The density of the material changes slightly with the charge, as the material expands during the discharge.

The problem of holding the active material securely to the plates during expansion and contraction has been a hard one to solve, each manufacturer having some favorite form of grid or material plug to which he pins his faith. While great improvements have been made in this direction, it is certain that we have not yet reached perfection. Loose active material will cause short circuits and will reduce the output of the cell; loose active material frequently ruins a cell.

The current capacity of a storage battery depends on the area of the plates or electrodes, and in order to increase the capacity of a battery, and consequently the area, it is usual to use a number of plates connected in parallel. A number of small plates of a given area are to be preferred to two large plates of the same area, as the battery will be of a more convenient size.

Customarily there is one more negative plate than positive, so that the extreme end plates in a cell are negative, as the positive and negative plates alternate with each other when assembled.

An ignition battery usually consists of two negative plates and one positive. Cells used for power purposes have as high as sixty plates.

A single cell of storage battery should show about two volts when fairly well charged. If more than two volts are desired more cells should be connected in series. The total voltage will be equal to the number of cells, in series, multiplied by the voltage per cell. The voltage per cell should never be allowed to drop below 1.7 volts, as the cell is likely to be destroyed when operated with a low voltage. Recharge as soon as the voltage drops to 1.8 volts.

The ordinary six volt ignition battery consists of three separate cells connected in series, which are encased in one protecting box.

The plates are prevented from touching each other within the cell by means of a perforated sheet of hard rubber that is inserted in the space between the plates. The perforations allow the liquid to circulate between the plates.

The storage battery is furnished as standard equipment with several well known gas engine builders, and its use is advocated by nearly all. When used in connection with a low tension direct current magneto two independent sources of current are at hand, either of which will ignite the engine in an emergency.

With the magneto-storage battery combination, it is possible to obtain a few small lights at any time, whether the engine is running or not, and the engine is always ready to start on the first "over" with the storage battery and a good mixture.

If a magneto is not used, difficulty is sometimes experienced in obtaining a suitable source of charging current, as many localities do not possess direct current plants. Batteries may be charged from the direct current exciter in an alternating current station, or may be charged by an alternating current rectifier such as is used by automobile garages.

The principal objections to the storage cell are: inconvenience of charging; sulphating of cell when standing without a charge; ease with which the cell is ruined by short circuits; the damage caused by the spilling of the electrolyte; and the fact that the cell gives no warning of failing or discharged condition.

Since the composition of the plates depends on the direction in which the current flows through the cell, it is obvious, that an alternating current which periodically changes its direction of flow will first charge the plates and then discharge them alternately. The result of an attempt at charging with alternating current would be that the plates would be in the same or a worse condition in a short space of time than they were at the beginning. In charging a storage cell care should be taken to determine the character of the current, especially when the cell is to be charged from a magneto. When under charge, the cell is connected to the charging circuit in such a way that the current flows backwards through the cell or in a direction opposite to that when the cell is discharging.

(86) Care of the Storage Cell.

The storage battery should never be left in an uncharged condition with the acid electrolyte in the cell, for the solution will quickly attack the uncharged plates and combine with them to form lead sulphate. As lead sulphate has a high electrical resistance and is insoluble in the electrolyte the sulphate coating will reduce the output or if present in excess, ruin the cell. The sulphate appears as a white coating on the surface of the plates. The only remedy for this condition at the hands of the average engine operator is a prolonged charge, or over charge, at a slow rate. There are several chemical processes but they are too complicated for the average man.

As sediment collects on the bottom of the battery jars, and is liable to cause a short circuit, the plates should be held about half an inch from the bottom of the jar. Care should be taken that the cells of the stationary type of battery are kept dry and clean. Do not allow dirt to drop into the solution as it is liable to destroy the cell.

A volt meter should be used to determine the condition of the battery, and should be used frequently. An ammeter should never be used on a storage battery, as it is of very low resistance, and would probably cause a rush of current that would destroy both the battery and the instrument.

Never short circuit a storage battery, even for an instant, as excessive current will cause the plates to buckle, or will loosen the active material on the plates.

The plates are immersed in the electrolyte, which should cover the entire plate or active surface. If the solution does not cover the plate, the capacity of the cell will be reduced. Plates that are partially covered with solution deteriorate rapidly from "sulphating." This is caused by the air and acid acting on the damp inactive portion of the plate.

Usually the electrolyte consists of a dilute solution of sulphuric acid and water, but in some ignition cells the solution is "solidified" by some substance to about the consistency of table jelly. The object of this thickened solution is to prevent the solution from slopping and leaking when the battery is being transported.

The solution used in a storage battery is exceedingly corrosive in its action, and if spilled on metal or wood will destroy it immediately. Care should be taken in handling the electrolyte.

A cell should never be discharged below 1.7 volts for below this point, the plates are likely sulphate. When the solution is replaced by fresh, or water is added for the purpose of restoring the electrolyte to its original level, use only distilled water, free from metallic salts and suspended matter.

Many people "test" their cells by snapping a wire across the terminals to "see if there is a good spark." Nothing could be more injurious to the battery, and as this test indicates nothing, the practice should be discontinued. Make all your tests either with a hydrometer or a voltmeter, the latter is preferable in the average case.

The electrolyte is a solution containing approximately 10% of chemically pure sulphuric acid and 90% of distilled water. The specific gravity of the fluid should be from 1,210 to 1,212 in all cases. A standard battery hydrometer should be used by all storage battery users to ascertain the exact density of the solution as the specific gravity is a direct index to the condition of the cell. A gasoline hydrometer is useless for a storage battery.

When mixing the electrolyte it should be placed in a glass or porcelain jar, and the process should never be performed in the battery jar in the presence of the plates. The solution is very active chemically and should not be brought into contact with metallic or organic substances because of the danger of contaminating the fluid. The acid should always be poured into the water in a thin stream while the mixture is being stirred with a glass or porcelain rod. Pouring the water into the acid is likely to produce an explosion and should therefore be carefully avoided.

As the acid heats the water during the mixing the hydrometer reading should not be taken until the heat caused by the first addition of acid has been reduced to that of the room. Taking a reading with a hot solution will give inaccurate results, unless, of course, the reading is reduced to normal by the method described in a previous chapter. When the reading has been taken and found to be correct and the solution has been reduced to the temperature of the room, the electrolyte may be poured into the cell through the filler openings in the top of the cell. Pour into each cell sufficient fluid to cover the plates but avoid filling the cell to the top, or flooding it.

At the end of the charging time given by the maker, withdraw a sample of the electrolyte by means of a syringe and test the specific gravity. This should not be over 1,290 for a fully charged cell, and if the solution exceeds this amount, pure water should be added until the proper point is reached. Always correct the specific gravity in this way every time the battery is charged as evaporation and internal chemical changes cause the density to change from time to time. The voltage of a good storage battery will be about 2.1 volts when fully charged. Overcharging is wasteful and finally destroys the cell, the effects being similar to those caused by excessive discharges, that is, buckled plates and loosened active material. Overcharging a sulphated battery may cure the trouble, a little overcharging at intervals being better than a long continued overcharge.

An increase in the specific gravity of the electrolyte of from 30 to 50 degrees, with a corresponding rise of voltage, shows that the cell is fully charged.

After the charging is completed remove all of the solution spilled on the battery, preferably by washing, and wipe bone dry. If the solution is higher in the air, remove the excess with the syringe.

(87) Make and Break System (Low Tension).

When a circuit carrying a current is opened or broken at any place in its length, an electric spark will occur at the point at which the wires or contacts are separated. This is due to what might be termed the "momentum" of the current which causes it to persist in its course even to the extent of jumping over a short distance of the highly resistant air in the gap. The size and heat of the spark may be increased by placing a coil of copper wire in series with the circuit that has an iron core in the center of the turns. This coil increases the tendency of the current to jump the gap, or in other words increases the momentum of the circuit.

Each separation of the terminals of the circuit causes but a single spark, so that in order to obtain another the terminals must be again brought into contact and the current reestablished in the circuit before the circuit is again opened. Thus the function of the make and break igniter is to alternately make and break the circuit in the presence of the combustible mixture. To obtain the greatest spark and most certain ignition, the contact points should be opened with the greatest possible speed, an action that is accomplished in the actual engine by springs and triggers.

A typical cylindrical make and break coil consisting of an iron wire core surrounded by a coarse copper wire core is shown by Fig. 91. At one end of the coil will be seen the two terminal screws by which it is connected with the circuit. Another make and break coil is shown by Fig. 92, which has the same type of winding, but differs in having the core wire coil extended beyond the winding and heads. By closely examining the cut, the iron wires will be seen in the projecting core tube at the left end of the coil. A flat base is also provided for fastening it to a stationary foundation.

A typical make and break igniter is shown by Fig. 93, together with the usual circuit consisting of a primary coil and battery. In this figure, A and C are the two electrodes provided with platinum contact points N and O respectively. The electrode A is stationary and is insulated from the iron casing K by the insulating washer H, and the insulating bushing or tube I. The electrode C is oscillated intermittently by the engine through its shaft E, and the trigger G, the springs S serving to snap the platinum contact O away from N at the proper moment. This electrode (C) is in electrical connection with the shell K, and the engine frame at all times, and is provided with a brass bushing F for a bearing surface. The outer containing casing K is bolted to the combustion chamber of the engine by the bolts LL, so that the electrodes A and C project into the combustion chamber.

Fig. 91. Kingston Cylindrical Make and Break Coil.

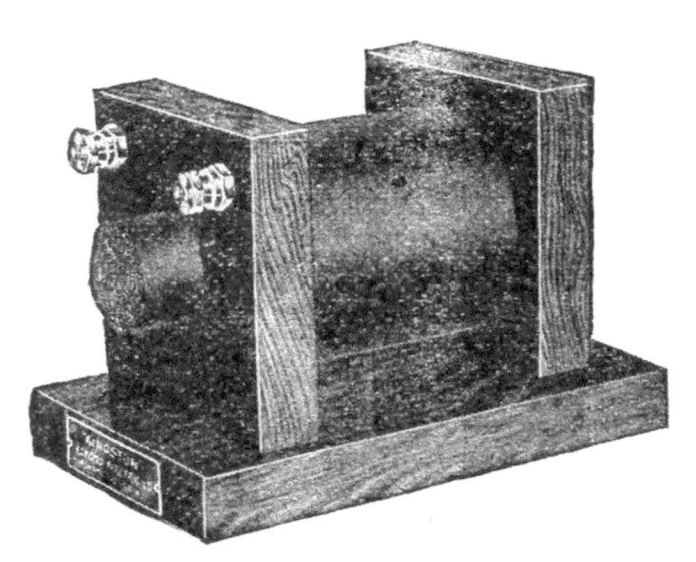

Fig. 92. Kingston Make and Break Coil. Short Type.

Current from the battery R passes through the coil winding P to the coil terminal U from which it passes from V to the igniter binding post J. From J it flows along the rod D to the stationary electrode A. Since the rod D is surrounded by the insulating washers and tube H, T and I, the current cannot escape directly to the casing K. With the two platinum points N and O in contact, the current flows through C to the shell K from which point it flows back to the battery R through the conducting path V, completing the circuit. The greater portion of the path V consists of the engine frame. When the electrode is moved in the direction of arrow B, the current is opened and a spark occurs at the point of separation M, in contact with the gas in the combustion chamber. The electrode C being connected with the engine frame is said to be "grounded." If the stationary electrode A were not insulated from the casting K, the current would pass directly from the terminal J back to the battery R without passing through the contact points at all, and consequently no spark would be produced on the separation of the points.

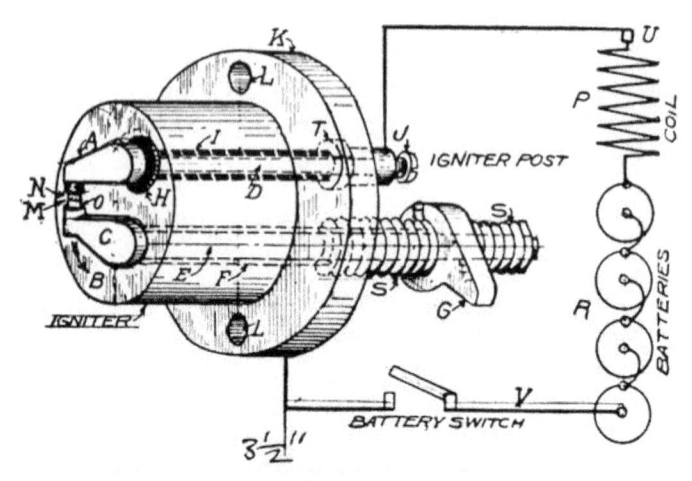

Fig. 93. Diagram of Igniter and Connections.

A push rod which is actuated by a cam on the engine, engages with the trigger G, and causes the spark to occur when the piston is on the end of the compression stroke. In nearly all engines, the relation between the time of the spark and the piston position can be regulated to suit the requirements for advance and retard. This adjustment is necessary in order that the spark may be varied to meet the difference between the starting and running requirements.

While the ignition should be considerably advanced while running, it is necessary to retard it when starting, as the engine is liable to "kick back" with an advanced spark.

This advance and retard device should be accessible while the engine is running, and the operator should be able to control the point of ignition at all times. Many men have been seriously injured by the lack of this device or by neglecting to use it.

The contact points make contact only for a short time before the spark is required in order to reduce the amount of current to the minimum, and therefore increase the life of the batteries.

The duration of the "make" or contact should be as short as possible. Prolonged contact weakens the batteries and causes them to run down rapidly. For the same reason the electrodes should remain separated until the make is actually required.

A certain period of contact is necessary, however, to allow the spark coil to "build up," but with a properly designed coil the time required is very short.

Some engines provide a device that cuts out the ignition current altogether during the idle strokes. This adds materially to the life of the batteries.

The igniter should be located near the inlet valve, as the cold incoming gases tend to keep it cool and clean, besides insuring the presence of combustible gas around the igniter electrodes. Improper placing of the igniter will greatly reduce the efficiency of the engine. Avoid placing the igniter in a pocket, or in the path of the exhaust gases.

The make and break ignition system has many good features, but cannot successfully be applied to engines running over 500 revolutions per minute, nor can it be applied to engines of less than 3 H. P. as the parts would be too small and delicate to be durable.

The make and break igniter produces the largest and "hottest" spark of any type of ignition, and is especially desirable for large or slow running engines. Being operated at a low voltage, it is not as easily affected by moisture, poor insulation, or dirt as the high tension or jump spark system, nor is it liable to give the operator such a violent "shock."

Engines governing by the "hit and miss" system have a device that cuts out the current during the "missed" power strokes. This effects a considerable saving in battery current, especially on light loads when the engine misses a great number of strokes.

While possessing many points of merit, the make and break system is open to several serious objections:

1. Due to the high combustion temperature there is excessive wear of the working parts in the cylinder, this wear causes a change in the ignition timing.

2. The low voltage used in the make and break system calls for perfect contact of the electrodes in the cylinder. This contact is often interfered with or entirely prevented by the accumulation of carbonized oil and soot deposited on the surfaces.

3. The wear of the operating spindle or shaft, which passes through the cylinder wall causes leakage, which in turn causes a loss of compression in the cylinder.

4. The wear of the external operating mechanism produces a change in the timing. The edge of the fingers, wiper blades, etc., tend to cause an advance in the ignition as a general rule, with the attendant danger of broken crank shafts.

5. The system is mechanically complicated, correct operation calling for constant care as to adjustment.

All ignition apparatus wears in the course of time and changes the timing of the engine. The electrodes and push-rods wear and require readjustment. Generally the tendency of worn parts is to advance the ignition. This change in timing occurs so gradually that the operator does not notice it until the engine begins to pound, or until the efficiency has been considerably reduced.

When the engine is new it is well to mark the ignition mechanism in such a way that the relative positions of the crank and igniter will be shown at the time when the igniter trips. It will then be possible for the operator to refer to the marks at any time to tell whether his ignition is occurring at the proper time. Always mark the half-time gears when taking the engine apart for the difference of one tooth when reassembling will be sufficient to throw the engine out of time.

The usual method of marking the gears, is to center punch, or scratch one tooth on the small gear, and then mark the two teeth of the large gear that lie on either side of it. With these marks it is possible to replace the gears in their original and proper positions.

The igniter should trip, causing the electrodes to separate just before the end of the compression stroke is reached, or just before the crank reaches the inner dead center. The distance lacking the exact dead center represents the instant of time between the time of ignition and the actual pressure established by the combustion.

As most engines have the ignition considerably retarded when starting, the igniter will trip later with the lever in the "start" position than when in the "running" position. Never fail to retard spark when starting nor forget to advance it when engine is up to speed.

The actual advance given to an engine depends on the character of the fuel and on the speed.

An engine is said to have an advance of $10°$, if the crank lacks $10°$ of having made the inner dead center at the time of ignition.

The most economical point of ignition is easily determined when the engine is running on a steady load, by varying the point of ignition and noting the position assumed by the governor.

(88) Operation of the Make and Break Igniter.

To keep the igniter in order, and to obtain the best results with the least trouble, the following hints should be observed:

(1) Clean the igniter frequently, and remove all deposits of oil and carbon. For cleaning, the igniter must be removed from the cylinder, care being taken to avoid injury to the packing or gasket. Graphite dusted on the gasket will prevent it from sticking to either the igniter or cylinder.

(2) If the contact points are rough, pitted, or covered with a carbon deposit, the scale should be removed, and the points smoothed down with a fine file, taking care that the two faces are filed parallel with one another.

(3) Insulating washers and tubes should be removed and washed in gasoline. The hole through which the igniter rod passes should be scraped free from any deposit for much trouble can be caused by a tight working shaft.

(4) Examine the hole or bushing through which operating spindle passes, for wear. A worn spindle or bushing may cause a serious loss of compression; replace worn bushing at once.

See that the insulation of the stationary electrode is not broken. If it is injured in the slightest degree, replace it with new.

(5) Often the sparking points may be cleaned temporarily without removing the igniter from the cylinder by pulling upon the outside finger or trigger until the points come together, and then pushing in towards the cylinder several times on the movable electrode, which slides them one on the other, scraping off the deposit. This method is only a make shift.

(6) After removing igniter, replace all wires, screwing them firmly into place. The ends of wires and connecting screws should be perfectly clean when the connection is made; to insure perfect contact, the surfaces should be scraped or sand-papered until bright and shining. See that no foreign matter of any kind gets between the wires and the metal of the binding screws. Wherever possible connections should be soldered.

(7) A small coil of the wire should be made at the point of connection; i. e., the wire should be a trifle longer than necessary to reach the binding screw, the excess wire being coiled up on a pencil. This coil allows of removing igniter, allows for broken wire ends and reduces the tendency to loosen the connection.

(8) Ground wires, or wires connected with the frame of the engine should receive careful attention. They are generally fastened under some screw or bolt on the engine which may become loose or fail to make contact, thus

opening the entire circuit and causing the engine to stop. The ground wires are generally connected in inaccessible places, and require all the more attention for this reason.

(9) For the primary of low tension wiring, use only the best grade of stranded rubber covered wire. A special wire for ignition purposes is on the market. It is rather expensive but is just the thing for the service.

Never use cotton covered or waxed wire. This covering affords absolutely no protection against moisture or abrasion.

(10) As the voltage of a primary circuit, or circuit for make and break is very low, and the current comparatively high, it is well to have the copper as large as possible. It should never be less than number 14 gauge. Don't use solid wire if you can obtain stranded conductor. (Stranded wire is made up of a number of fine wires which are twisted into a cable or rope of the desired size.)

(11) Oil destroys rubber insulation and should be kept off the wiring. Try to locate the conductors so that they will be out of range of oil thrown by the moving parts.

(89) Jump Spark System (High Tension System).

Due to its simplicity and the light weight of its moving parts, the high tension ignition system is applied to practically all small, high speed engines running 500 R.P.M. or over. The high tension system is also desirable from the fact that it has no moving parts in the cylinder of the engine.

The principal objection to the high tension system is the ease with which the high voltage current leaks or short circuits, moisture being fatal to the operation of a jump spark engine.

Instead of producing the spark by breaking the circuit of a low tension current, the spark is produced by increasing the voltage to such a point that the current will jump directly across a fixed gap. To cause the current to jump through the air requires an extremely high voltage, and as the battery current is very low it is necessary to introduce a device known as a "transformer" to stop the current up to the required tension. In addition to the voltage required at atmospheric pressure (about 50,000 volt per inch of spark) we must also furnish sufficient pressure to overcome the increased resistance due to the compression in the cylinder.

Unlike the spark coil used on the low tension make and break system, the induction coil or transformer coil has two separate and distinct coils, that are thoroughly insulated from each other. One coil has a few turns of heavy copper wire which is called the primary. The other consists of many thousands of turns of very fine copper wire, and is called the secondary. Both coils are wound around a bundle of soft iron wire called the core, from which they are carefully insulated. When a battery or magneto current flows through the primary coil, the core is magnetized, and throws its magnetic influence through the turns of the secondary coil.

In Fig. 94 the primary coil and the low tension battery and magneto circuit are represented by heavy lines. The secondary coil, and high tension circuit are represented by light lines.

In order to obtain a continuous discharge of sparks it is necessary to make and break the current in the primary coil very rapidly. This is done by means of the interrupter or vibrator, which is indicated in the diagram by V. The interrupter consists ordinarily of a spring A on which is fastened a soft iron disc D and a platinum contact point B. When the core is magnetized it attracts the iron disc D which is pulled toward the core, bending the spring A and breaking the contact between the platinum point B and C. When the contact points are separated, and the current broken, the core loses its magnetism, and the spring assumes its normal position, which brings the platinum points B and C into contact once more, and reestablishes the current through the primary. The core is again magnetized and the primary current is again broken, and so on. This make and break of the current is thus accomplished automatically, the current being broken many thousands of times per minute, the vibrator moving so fast as to cause a continuous hum.

As soon as the current starts flowing, the magnetic force spreads out through the secondary coil and threads through the turns of which it is composed. The instant that the current ceases, the magnetic force decreases and the turns are again threaded by the magnetic field on its return to the core.

Thus two magnetic waves are sent through the secondary coil, one when the circuit is "made," and one when the circuit is "broken."

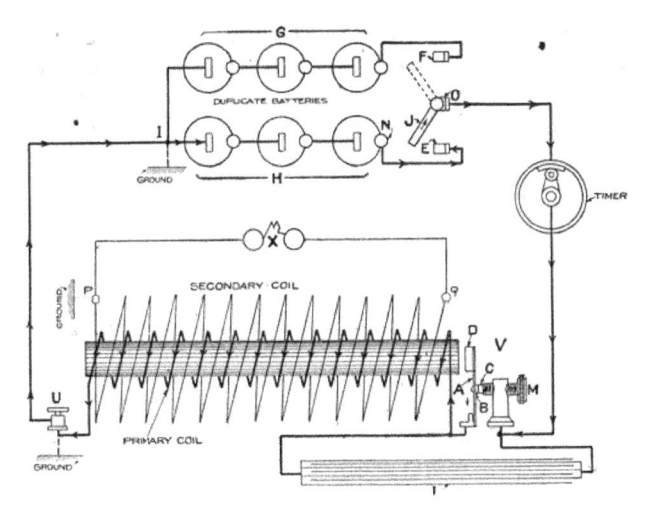

Fig. 94. Diagram of High Tension Coil.

When a magnetic wave threads or spreads through the turns of a coil of wire, a current of electricity is generated in the coil, the quantity and pressure or voltage of which is proportional to the intensity of the magnetism, and to the number of turns of wire in the secondary coil.

Thus it will be seen that at every make and break of the low tension current in the primary coil, a current is generated in the secondary. As the voltage generated in the secondary is roughly proportional to the number of turns in the secondary, and as there are many thousands of turns, it is evident that the voltage in the secondary will be very high. Thus by the use of the induction coil, the low tension battery current is transformed into a high tension current of sufficient voltage to break down the high resistance of the spark gap.

The condenser is shown at L which has one wire leading to the vibrator spring A, and one wire to the contact screw M. The function of the condenser is to absorb the spark produced at the vibrator points so that the break is made quickly, producing a maximum spark. The intensity of the spark depends upon the quickness with which the primary current is broken, and if it were not for the condenser the length and intensity of the spark would be greatly reduced. This device consists of alternate layers of paper and tin foil, every other leaf of foil being alternately connected to the vibrator spring and to the contact screw.

A method of using two independent sets of battery is shown in the diagram, so that either set may be thrown into circuit by means of the double throw switch O. When handle J is in contact with E, the current of battery set H flows through the coil as shown by the arrows. When J is in contact with F, the battery C is thrown into circuit. The spark gap is shown by X, which represents the spark plug in the cylinder.

In practice, the portion of the circuit shown by I-U is generally formed by the frame of the engine, or is grounded. The terminal P of the high tension circuit is always grounded through the threaded shell of the spark plug, the grounded circuit being shown by the dotted lines. Grounding saves wire and many connections, for with P and U connected to ground it follows that one binding post will serve the place of one high tension and one primary post, making three coil connections instead of four.

In order that the spark will occur in the cylinder of the engine at the proper time, a switch must be placed in the primary circuit of the coil, that will open and close the circuit at proper intervals. Such a switch is called a timer, and is always driven by the engine. The timer is connected to the engine shaft in such a way that contact is made at, or slightly before, the time at which the explosion is required, and as soon as possible after spark occurs the current is cut off.

For multiple cylinder engines it is usual to provide one coil for each cylinder, the primaries of which are controlled by a single timer and battery. A high tension wire from each coil runs to the corresponding cylinder. Instead of having a number of coils with a battery system, there are two or three makes that operate with one coil in combination with a special device known as a distributor which controls the high tension current. The high tension distributor directs the current to the proper cylinder that is in the order of firing, the timing being performed by a timer similar to that used with multiple coils except that a single contact sequent is supplied.

(90) Vibrator Construction.

Since the efficiency of the high tension coil depends largely on the construction and efficiency of the vibrator, the different coil makers have developed various types of vibrators that differ greatly from the simple device shown in the coil diagram in details.

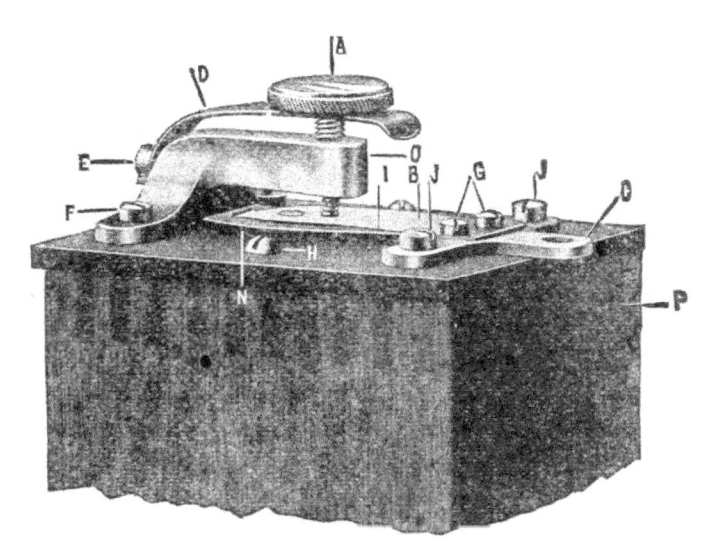

Fig. 95. Kingston Vibrator.

The main objects in view in the construction of a successful vibrator are:

1. To reduce the weight of the moving part as much as possible in order to increase the speed of vibration, and to make the trembler instantly responsive to the timer.

2. To cause the contact points to separate as rapidly as possible in order to cause the maximum spark.

3. To have the contacts as hard and infusible as possible to resist wear and the action of the spark between the contacts.

4. To make any adjustments that may be required, due to wear, as simple and accessible as possible.

The types of vibrators are legion, and we have not the space to go into the details of all the prominent makes, but will illustrate and describe two well known types.

The Kingston vibrator made by the Kokomo Electric Company, is a good example of a modern vibrator and is shown in detail by Fig. 95. All adjustments between the contact points are made by means of the contact screw A which carries a platinum point at its inner end. The retaining spring D keeps the contact screw from being jarred out of place by the engine vibration, without the use of lock nuts. Turning A against the vibrator, the tension of the spring B is increased, raising the screw decreases the tension. Increasing the tension screw increases the length and heat of the spark, and also increases the current consumption. At N is a separate thin iron plate which is acted on by the magnetized core, a rivet fastening the plate to the main vibrator spring is shown at the end of the spring. The current enters through the lug C, and from this point the circuit is the same as shown in the coil diagram.

(91) Operation of the Jump Spark Coil.

The spark produced by a coil in good condition should be blue-white with a small pinkish flame surrounding it, when the gap is ¼ of an inch or less. The sparks should pass in a continuous stream with this length of gap without irregular stopping and starting of the vibrator. Coils giving a sputtering, weak discharge that causes sparks to fly in all directions are broken down and should be remedied.

The secondary windings of coils are often punctured or broken down by operating the coil with the high tension circuit open, or by trying to cause long sparks by increasing the spark gap over ⅜ of an inch in the open air. Coils are also broken down by allowing excessive currents to flow in the primary coil. Never cause a spark to jump over ⅜ of an inch.

High compression in the cylinder shortens the jumping distance of a high tension spark. Coils that will cause a stream of sparks to flow across a gap of ½ an inch in the open air are often unable to cause a single spark to jump a gap of 132 of an inch under a compression of 80 pounds per square inch in the cylinder.

Remember that a hot spark causes rapid combustion, and will fire a greater range of mixtures and "leaner" charges, than a straggling, thin, weak spark. Spark coils that give poor results with a long spark gap under high compression are often benefited by the shortening of the spark gap. Shortening the gap will increase the heat of the spark, and will insure the passing of a spark each time that the timer makes contact. A good coil should have no difficulty in igniting a piece of paper inserted between the wires forming the spark gap in the open air.

Fig. 96. Kingston Dash Coil.

The adjusting screw affords a means of increasing or decreasing the tension of the vibrator spring, and the amount of battery or magneto current flowing through the primary coil. Increasing the tension of the spring requires stronger magnetization of the core to break the circuit of the contact points. This in turn calls for more current from the battery; hence in order to lessen the demand for current on the battery, the tension should be as little as possible to obtain the necessary spark. An increased tension produces more spark as the magnetization of the core is increased, but for the sake of your batteries decrease the tension as much as possible with a satisfactory spark.

Almost all operators have a tendency to run with too stiff a vibrator, and hence use too much current. An efficient coil should develop a satisfactory spark with ¼ to ½ of an ampere of current in the primary coil. I have often found coils that would work well with ½ ampere, that were screwed up so tight that the coils were consuming 4 to 5 amperes or 8 to 10 times as much as they should.

A battery ammeter used for testing the current consumed by coil will save its cost many times over in batteries and burnt points if used at frequent intervals in the primary circuit.

An automobile or marine engine should be tested for vibrator adjustment in the following way:

Adjust vibrator so that spring is rather stiff. Start engine and get it thoroughly warmed up and running at full speed, then slowly and gradually decrease the tension of the spring until misfiring starts in; then slowly increase tension until misfiring stops. Increase the tension no farther; this is the correct adjustment.

Poor vibrator adjustment is the cause of much trouble and expense as it uses up the batteries and wastes fuel. The principles of correct adjustment are simple, the adjustment easily made, and there is no possible excuse for the high current consumption and rapid battery deterioration met in every day practice. The usual practice of the average operator is to tighten the vibrator until the spark (observed in the open air) is at its maximum. This is commonly known as "adjusting the coil;" shortly after you hear of him throwing out his batteries as no good. After once getting the vibrator in proper trim the ear will give much information as to the adjustment.

A vibrator adjusted too lightly will cause "skipping" or misfiring with the consequent loss of power.

Never attempt to operate a coil that is damp; the coil will be ruined beyond repair. Above all, do not place the coil in a hot oven to dry, as the box is filled with wax, and if this is melted it will run out and reduce the insulation of the coil. Dry coil gradually.

If the batteries are new or too strong the vibrator may be held against the core of the coil so that the vibrator will not buzz. If this is the case loosen the screw until it works at the proper speed. If the batteries are weak, the coil may not be magnetized sufficiently to draw the vibrator and break the circuit. If this is the case tighten the screw. If the vibrator refuses to work with the battery and wiring in good condition, and if you are sure that the current reaches the coil, look for dirty or pitted contacts on the vibrator.

Should the contact points be dirty, clean them thoroughly by scraping with a knife or sandpaper. Water on the points will stop the vibrator, as will oil or grease.

If contact points are of a uniform gray color on their contact surfaces, and are smooth and flat without holes, pits or raised points, they are in good condition. If pits, discolorations or projections are noted, the contact surfaces should be brought to a square, even bearing by means of a small, fine file. The point should not come into contact on an edge, but should bear on each other over their entire surface. Do not use sand paper to remove pitting, as it is almost impossible to secure an even, flat surface by this means.

It is best to remove the contact screw and vibrator blade for examination and cleaning, as it is much easier to file the points square and straight when removed from the coil.

Be careful not to bend the vibrator spring when cleaning, as the adjustment will be impaired. When replacing contact screw and vibrator blade in coil, be careful that they are in exactly the same relative position as they were before removing. Also be sure that the contacts meet and bear uniformly on their surfaces.

(92) Primary Timer.

The duty of the primary timer is to close the primary circuit of the spark coil at, or a little before the time at which the explosive of the charge is required. The exact time at which the timer closes the circuit depends on the load, the speed, and the nature of the fuel. The lapse of time between the instant that the timer closes the circuit and the instant at which the piston reaches the end of the compression stroke is called the "advance" of the timer. When the timer closes the circuit after the piston reaches the end of the stroke, the timer is said to be "retarded." The timer is constructed so that the time of ignition or the advance and retard can be varied between wide limits. Advancing the spark too far will cause hammering and power loss as the piston will work against the pressure of the explosion.

Retarding the spark will cause a loss of power, as the compression will be less when the piston starts on the outward stroke; and also for the reason that more of the heat will be given up to the cylinder walls as the combustion will be slower. The pressure in the cylinder is less with retarded ignition. Greatly retarded ignition often causes overheating of the cylinder walls, especially with air cooled engines, and also overheats and destroys the seat and valve stem of the exhaust valve. Do not expect the engine to develop its rated horse-power or run efficiently with a late, or retarded spark.

When the engine is installed, and before the timer wears or has a chance to get out of adjustment, look it over carefully and see whether the maker has left any marks relating to the timing of the spark. If there are no marks, it is well to determine the relation between the position of the piston and the timer, as the efficiency of the engine depends to a great degree upon the firing point.

Timers are advanced and retarded by partially rotating the housing either in one direction or the other. When the timer is mounted directly on the cam shaft with the cam shaft traveling in a direction opposite to that of the crank shaft, the timer will be retarded by moving it in the same direction as the cam shaft travels, moving it against cam shaft rotation advances the spark.

Timers for two stroke cycle engines rotate at crank shaft speed, and the direction of advance and retard varies with the methods adopted for driving the timer.

(93) Timer Construction.

Fig. 97 shows a typical timer and circuit arranged for a four cylinder engine. The device can be arranged for any number of cylinders, however, by changing the number of sectors, the sectors being equal to the number of cylinders. There are timers on the market that differ from the one shown in the diagram but the principle of operation is the same with all. The shaft E is usually connected to the cam shaft and is electrically grounded to the engine frame at L by means of the bearing in which the shaft rotates.

The lever F mounted on the shaft E carries the pivoted arm H which is free to move on the pivot to a limited extent to allow for wear on the walls W-W-W-W. At one extremity of H is the roller I which rotates on the pin J, as the roller runs around W-W-W-W. At the other extremity of H is fastened the spring S, which forces I into contact with the walls. A-B-C-D are metallic contact sectors whose connections lead to the four spark coils.

When the metal roller I comes into contact with one of the sectors as at B, the sector is grounded to the engine frame by the roller, the current traveling through the roller and its pin, through lever H and its pin, through the lever F and shaft E to ground at L, the course of the current being indicated by the arrows.

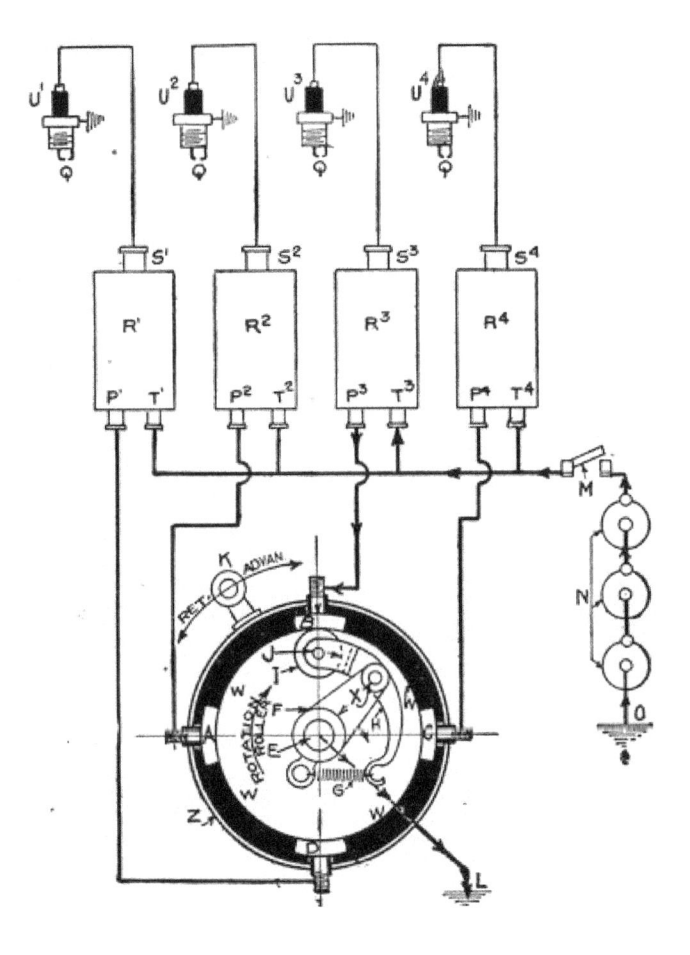

Fig. 97. Timer Diagram.

As the shaft E rotates and carries with it roller I, the roller makes contact with the sectors in order B-C-D-A, if rotated in the direction shown by arrow, which rotation grounds the primary coils of the spark coils R^3-R^4-R^1-R^2 in succession; the connection from the timer to the primary being to the primary binding posts P^3-P^4-P^1-P^2. A high tension spark occurs at each contact of the roller with the sectors, as the contact allows current to flow through the primary of the coils. The high tension binding posts S^1-S^2-S^3-S^4 are connected with the spark plugs or spark gaps U^1-U^2-U^3-U^4 by means of high tension cables. As soon as the timer grounds a coil, the coil produces a high tension spark in its corresponding spark plug.

It is evident from the foregoing that the timer not only determines the time at which a spark will take place, but it also determines the cylinder in which the spark will be produced, providing of course that a spark coil is provided for each cylinder.

The contact sectors A-B-C-D are insulated from each other by the insulating walls W-W-W-W, the inner surface of which provides a path on which the contact roller I revolves.

The contact sectors and insulating walls are encased by the protective housing Z, to which they are rigidly fastened.

The housing Z can be moved back and forth on the shaft E for advance and retard, by means of the lever K.

The current flows from the battery terminal V (with the roller in the position shown) through the switch M, through coil R^3, post P^3 to sector B, from which it passes through the roller I, levers H and F to ground. From the ground on the engine frame the current flows back to its source, the battery O, thus completing the circuit. When the roller makes contact with sector C, the coil R^4 is energized, contact with D energizes R^1, and so on. No two coils can be thrown on simultaneously as only one coil is grounded at a time. The high tension current flows from each coil to its plug as soon as the current passes through the primary of that coil.

In some timers, the current is taken from the revolving arm through a separate connection to ground instead of grounding the shaft through the bearings. With these timers, the connection is not affected by worn bearings or an oil film that tends to insulate the shaft from the bearings.

(94) Operation of Timers.

Timers frequently cause misfiring which is generally due to dirt or oil getting between the contacts, or to the wear of the insulating walls W-W-W-W, or to the wear of the moving parts.

Dirt or gummy oil will prevent the contact coming together and completing the circuit, or will clog up the rollers or levers so that they cannot perform their functions properly. This will of course interfere with production of the spark.

The contacts and moving parts of the timer should be kept as clean as possible, all dirt and heavy oil being removed by means of gasoline at regular periods. Make a practice of cleaning out the timer at intervals not greater than one month; oftener if possible.

Parts subject to wear, such as the roller pin J and the bearings should be well lubricated, none but the lightest oil being employed for this purpose. Heavy grease will gum the contacts and cause trouble. There should be no rough places or shoulders on the contact sectors or on the walls W-W-W-W as roughness will cause the roller to jump over the high places which in turn result in misfiring. The remedy is to machine the surfaces of the sectors and walls by grinding or turning in the lathe. Care should be taken in this operation to have the interior perfectly smooth and the sectors perfectly flush with the walls. Repair black or burnt sectors immediately by grinding or sand paper.

Burnt spots or blackened surface on the contact sectors prevent good contact between roller and sector, sectors should show a bright, shining metallic surface.

Sometimes the insulation warps or swells above the contacts so that the roller jumps over the contacts without touching them, or if for any reason that contact is made under these conditions, it is of a short period and results in a poor spark.

Timers often make good contact when starting, or at low speed, and misfire badly at high speed. This will be caused generally by the contact sectors or insulation projecting beyond one another, the roller has time to make good contact at low speed but jumps over the sector at high.

The roller I may become rough or develop a flat stop which will cause it to jump over the contact occasionally, or it may become loose on its bearing pin J, causing intermittent misfiring.

The wearing or loosening of pins J and X result in poor contact. Should pin J fall out of the lever H, the roller would drop out of the fork and cause serious damage. This has happened in two cases to my knowledge.

Should the spring G weaken or break, contact will be made intermittently at high speed, and no contact at low. In this case it would probably be impossible to start the engine. In case the spring breaks, a rubber band may be used temporarily. Wire connections to the timer should be examined frequently as the continual back and forth movement tends to twist and loosen the wire. Use stranded or flexible wire for these connections, if possible.

Before removing the timer mark the hub and the shaft so that the hub can be properly replaced. If this is not done the engine will be out of time with the usual results of hammering or power loss.

Should the gears which drive timer shaft be removed, be sure and mark the teeth of both gears in such a manner that there will be no mistake possible in reassembling them. Mark a tooth on the small gear by scratching or with a center punch (the tooth selected should be in mesh with the large gear). Then mark the two teeth of the large gear that lay on either side of the marked tooth of the small gear. Thus it will be easy to locate the proper relative position of the two gears at any time.

(95) High Tension Spark Plug.

The high tension spark plug is a device that introduces the spark gap and spark into the combustion chamber, and at the same time insulates the current carrying conductor from the cylinder walls. Since the voltage of the jump spark current is very high it is evident that the insulation of the plugs must be of a very high order and that this insulation must be capable of withstanding the high temperature of the combustion chamber. A cross-section of a typical plug is shown by Fig. 98, together with its connections and the course of the current, the latter being shown by the arrow heads.

The electrode B through which the current enters the cylinder is thoroughly insulated from the walls by the porcelain rod C.

The porcelain forms a gas tight joint with the threaded metal bushing F at the point P, the tension caused by the electrode B and the nut I holds the porcelain firmly on its seat at P.

The nut is supported by the porcelain shell H which rests in the top of the metal bushing F. A washer L is inserted between H and F to insure against the leakage of gas from the plug should a leak develop at P. L being a soft washer (usually asbestos) allows the porcelains C and H to expand and contract without breaking. A packing washer or gasket is also placed at the point where the electrode B passes through the porcelain H. This is the washer Q, held in position by the nut I. This washer is elastic and reduces strain on porcelain caused by the expansion.

The cylinder wall G has a threaded opening R into which the plug is screwed, the threads of the opening corresponding with the threads on the metal sleeve E. The plug may be removed from the cylinder for examination without disturbing the adjustment of the electrode and porcelains by unscrewing it at R.

Allowing the current to jump from the electrode to the cylinder wall via the metal sleeve saves one wire and connection, the cylinder and the frame of the engine serving as a return path for the current. This simplifies the wiring and minimizes the danger of high tension short circuits.

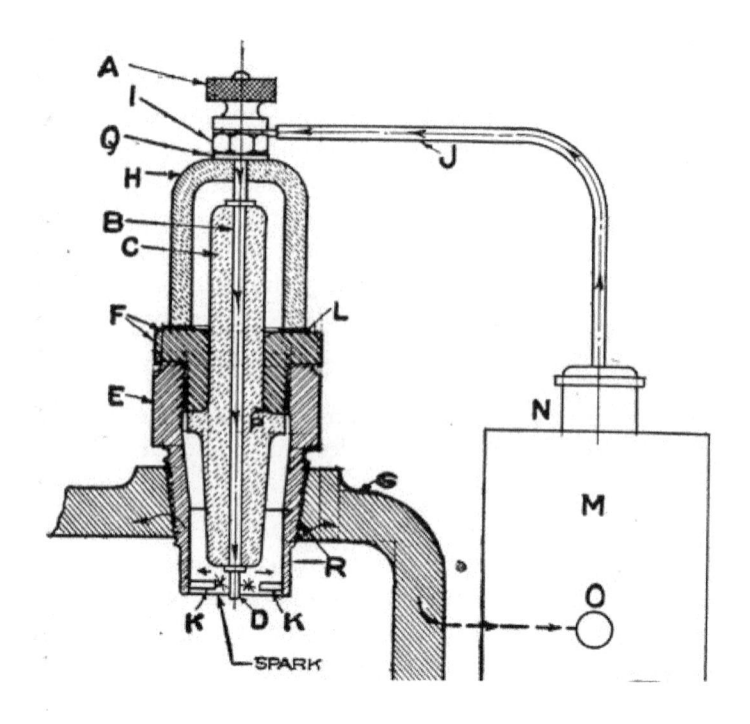

Fig. 98. Cross-Section of Typical Spark Plug.

By unscrewing the threaded metal bushing F it is possible to examine the condition of the porcelain rod C at the point where it is exposed to the heat of the cylinder. This inspection can be made without disturbing the packed joints at L or Q.

In the high tension, or jump spark system, the spark gap D-K is of fixed length, hence there are no moving parts or contacts within the cylinder to wear, to cause leakage of gas, or to cause a change in the timing. This advantage is offset to some degree by the difficulty experienced in maintaining the insulation of the high tension current.

The high tension current leaves the spark coil M at the binding screw N, flows along the wire J, and enters the spark plug at the binding screw A. From the binding post the current follows the central electrode B to its terminal at D. At D a break in the circuit occurs which is called the spark gap. It is at this point that the spark occurs, the current jumping from D to point K through the air. Point K is fastened in the threaded metal sleeve E which is in turn screwed into the cylinder wall G or ground. From the ground the current returns to its source through binding post O to the coil.

The spark therefore occurs inside of the cylinder wall and in contact with the combustible charge, at the point marked "spark" in the cut.

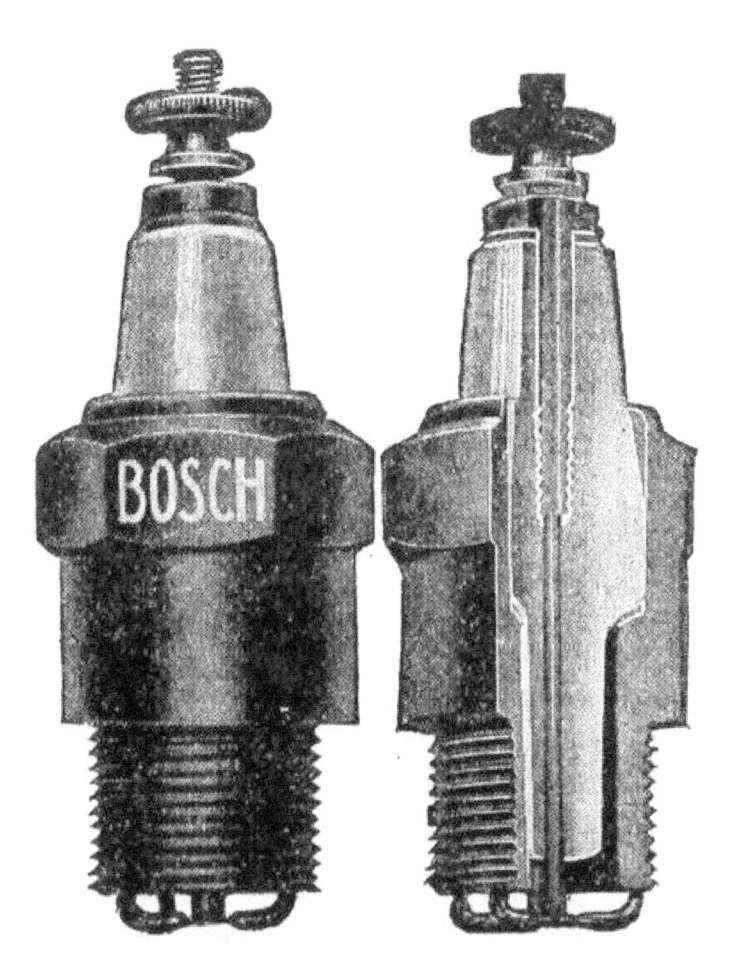

Fig. 99. Bosch Spark Plugs.

If the fuel, lubricating oil, and air are not supplied in proper proportions, soot will be deposited on the lower surface of the porcelain, and as soot is an excellent conductor of high tension current, the current will follow the soot rather than the high resistance of the spark gap, a condition that will result in misfiring or a complete stoppage of the motor. Carbonized lubricating oil or moisture have the same effect.

Preventing the deposits of soot, moisture and carbonized oil is the chief object of plug manufacturers, many of whom have brought out designs of merit. In fact the problem of elimination of soot is the principal cause of the many types of plugs now on the market.

While many plugs differ in minor refinement of detail from the typical plug shown, the connections and general construction are the same in all types, the spark being produced in a gap of fixed length which is insulated from the cylinder.

A well known form of plug, the Bosch, is shown by Fig. 99 a-b. In this plug a special material known as Steatite is used instead of the usual porcelain. The three external electrodes surrounding the center electrode is a particularly efficient arrangement, especially for magnetos. A peculiar form of pocket minimizes the soot problem.

As porcelain is brittle and is easily broken by the effects of heat or blows, mica insulation is often used in place of the porcelain. The central core of a mica plug is formed by a stack of mica washers, which are held in place by the central electrode and the upper lock nuts.

A poorly constructed mica plug is easily destroyed by a weak, stretching, electrode, or by an overheated cylinder. The latter causing the washers to shrink and admit oil between the layers of mica washers causes a short circuit. As soon as the mica washers loosen and separate, they should be forced together by means of the mica lock nuts on the top of the plug.

If by any reason the mica core becomes saturated with oil, it is best to obtain a new one, as it is almost impossible to remove the oil by simple means open to the average operator.

The chief value of a mica plug lies in its toughness and mechanical strength, a good mica plug being practically indestructible.

When heated, porcelain does not expand at the same rate as the metal sleeves, hence in poorly designed or imperfect plugs, heavy strains are thrown on the delicate porcelains which causes them to crack. When a crack develops it provides a lodging place for soot and carbon which of course causes a short circuit. Should a compression leak occur through faulty packing between the porcelain and sleeve, it should be immediately tightened up for eventually it will leak enough to destroy the plug or reduce the output of the engine.

When ordering a plug be sure that you know the size and type required by your engine. Some engines require a longer plug to reach the combustion chamber than others. Never install a shorter plug than that originally furnished with the engine. Be sure that the plug is not too long as it may interfere with the action of the valves or may be damaged by them. Plugs are furnished with several threads and taps, i. e.:

- ½ inch pipe thread (Generally used on stationary engines).
- Metric Thread (Generally used on imported autos).
- ⅞ inch A. L. A. M. Standard (Used on Domestic automobiles).

Using a plug in a hole tapped with the wrong thread will destroy the thread in the cylinder casting and cause compression leaks.

(96) Care of Spark Plug.

Porcelains are often broken by screwing the plug too tightly in a cold cylinder, as the cylinder expands when heated and crushes the frail plug. A plug installed in this manner is difficult to remove as the expanded walls grip the thread. The plug should be screwed in just enough to prevent the leakage of gas. A short thin wrench should be used in screwing the plug home such as a bicycle wrench. A wrench of this type is so short that it will be almost impossible to exert too much force, and will be thin enough to avoid any possible injury to the packing nut. Bad leaks may be detected by a hissing sound that is in step with the speed of the engine, small leaks may be detected by pouring a few drops of water around the joint. If a leak exists bubbles will pass up through the water and show its location.

Plugs are more easily removed from a cold cylinder than a hot. If the plug sticks when the engine is cold and is impossible to remove with a moderate pressure on the wrench squirt a few drops of kerosene around the threads. Never exert any force on the porcelain or insulation. The high tension cables should be connected to the plugs by means of some type of "Snap Terminal," such terminals may be had from automobile dealers.

These terminals make a firm contact with the plug and do not jar loose from the plug by the vibration of the engine. They are easily disconnected when the inspection of the plug becomes necessary, and are generally a most desirable attachment.

The high tension cable should be firmly connected to the plug terminal under all circumstances. A loose connection will cause misfiring or will bring the engine to an abrupt halt. If snap terminals are not used the plug binding screw should be screwed down tightly on the wire. When making connections see that the wire is bright and clean, and that frayed ends of the wire do not project beyond the plug and make contact with other parts of the engine.

A large percentage of high tension ignition troubles are due to short circuits in the spark plug which are generally caused by deposits on the surface of the plug insulation. Soot or oil may be removed from the plug by scrubbing the porcelain and the interior of the chamber with gasoline applied by a tooth brush. Examine the plug for cracks, and if any are found, replace the porcelain or throw the plug away. A cracked porcelain is always a cause of trouble.

To test a plug for short circuits, remove it from the cylinder, reconnect the wire, and lay the sleeve of the plug on some bright metal part of the engine in such a way that only the threaded portion is in contact with the metal of the engine. Close the switch and see if sparks pass through the gap. If no sparks appear, and if the coil is operating properly, clean the plug. As an additional test for the condition of the coil, hold the end of the high tension cable about ¼ inch from the metal of the engine while the coil is operating. If a heavy discharge of sparks takes place between the end of the cable and the metal of the engine, the coil is in good condition.

If a partial short circuit exists, the spark at the gap will be weak and without heat; the result will be intermittent, or misfiring with a loss of power. Moisture in the cylinder is a common cause of plug short circuits, the moisture coming from leaks in the water jacket or from the condensation of gases in a cold cylinder. A drop of water may bridge the spark gap, allowing the current to flow from one electrode to the other without causing a spark.

If a cloud of bluish white smoke has been issuing from the exhaust pipe before the misfiring started, you will probably find that the trouble is due to sooted or short circuited plug.

The remedy is to decrease the amount of lubricating oil fed to the cylinder.

When a magneto is used the intense heat of the spark causes minute particles of metal to be torn from the electrodes and deposited on the insulation as a fine metallic dust. This will of course cause a short circuit and must be removed. Short circuits are sometimes caused by the magneto current melting the electrodes and dropping small beads of the metal between the conductors. All metallic particles should be removed from the plug.

While a spark plug may show a fair spark in the open air test, it will not always produce a satisfactory spark in the cylinder on account of the increased resistance of the spark gap due to compression.

Compression increases the resistance of the spark gap enormously and thin, highly resisting carbon films that would cause very little leakage in the open air will entirely short circuit the gap under high pressure, the current taking the easiest path which in the latter case is the carbon deposit.

In order to produce conditions in the open air test similar to those in the cylinder we must devise some method of increasing the resistance of the spark gap in the open air above any possible resistance that could be offered by the carbon film.

Placing a sheet of mica or hard rubber between the electrodes, or in the spark gap, will increase the resistance to the required degree. If the spark plug is in good condition the spark will jump from the insulated terminal to the shell when the mica is in the spark gap, but if a short circuit exists the current will go through it without causing a spark. It is assumed that the battery and coil are in good condition when making the above test.

If the electrodes or spark points are dirty they should be cleaned with fine sand paper, special attention being paid to the surfaces from which the spark issues. When reassembling the plug, see that all of the washers and gaskets are replaced and that the length of the spark gap is unchanged. A little change in the spark gap may make a great change in the spark. A good spark is blue white with a faint reddish flame surrounding it. When the discharge is intermittent or sputters in all directions, either the coil or the plug are partially short circuited. Always have a spare plug on hand.

Ordinarily the length of the gap or the distance between the electrodes should be about 132 inch for batteries, and a trifle less for magnetos. A silver dime is a good gauge for the gap. If the engine misfires with the coil and batteries in good condition, try the effects of shortening the gap a trifle, usually this will remedy the difficulty. Exhausted batteries may be made operative temporarily by closing up the plug gap to 164 inch or even less. Shortening the gap increases the heat of the spark and nothing is gained by having it over 132 inch.

Almost all high tension magnetos have visible safety spark gaps that show instantly the presence of an open circuit in the secondary or high tension circuit. If an open circuit exists, a stream of sparks will flow across the safety spark gap at low speed.

To determine the cylinder that is misfiring in a four cylinder engine proceed as follows:

Remove cover on spark coil, and hold down one vibrator spring firmly against the core while the engine is running.

If the engine speed is not decreased by cutting this coil out of action, it is probable that this is the coil connected to the misfiring cylinder. Now release this vibrator and proceed to the next coil, and hold its vibrator down. If this decreases the speed of the engine you may be sure that the first coil is in the defective circuit. If the vibrator buzzes on the coil under inspection the trouble will be found in the plug.

Cutting out a coil connected to an active cylinder decreases the speed of the engine. Cutting out the coil connected with a dead cylinder makes no difference.

(97) Magnetos.

A magneto is a device that converts the mechanical energy received from the engine into electrical energy, the electricity thus produced being used to ignite the charge in the engine. This appliance does away with all of the troubles incident to a rapidly deteriorating chemical battery and produces a much hotter and uniform spark. A magneto is especially desirable with multiple cylinder engines where the demand for current is almost continuous, as the amount of current delivered by the magneto has no effect on its life or upon the quality of the spark.

The principal parts of the generating system of the magneto are the magnets, the armature, the armature winding, and the current collecting device, of which the armature and its windings are the rotating parts. The production of current in the magneto is the result of moving or rotating the armature coil in the magnetic field of force of the magnets. When any conductor is moved in a space that is under the influence of a magnet a current is generated in the conductor which flows in a direction perpendicular to the direction of motion. The value of the current thus generated depends on the strength of the magnetic field, the speed with which it is cut, and the number of conductors cutting it that are connected in series. Roughly, the voltage is doubled, with an increase of twice the former speed, and with all other things equal, the voltage is doubled by doubling the number of conductors connected in series.

By employing powerful magnets, and a large number of conductors (turns of wire) on the armature it is possible to obtain sufficient voltage for the ignition system at a comparatively low speed. The number of amperes delivered depends principally upon the internal resistance of the armature and the external circuit, and not on the number of conductors, nor directly upon the strength of the field. For this reason, low voltage machines that are intended to deliver a great amperage have only a few conductors of large cross section, while high tension machines have a great number of conductors of small size. In all cases the magneto, or ignition dynamo must be considered simply as a generator of current in the same way that a battery is a source of current since the current generated by them is utilized in precisely the same way.

The class of ignition system on which the magneto is used determines the class of the magneto. The low tension magneto is used principally for the make and break system, although it is sometimes used in connection with a high tension spark coil or transformed in the same way that a battery is used with a vibrator coil. The high tension magneto is used exclusively with the jump spark system and high tension spark plug.

These classes are again subdivided into the direct and alternating current divisions, depending on the character of the current furnished by the magneto. Briefly a continuous current is one that flows continually in one direction while an alternating current periodically reverses its direction of flow. As the alternating current magneto is the most commonly used type, we will confine our description to this class of magneto. The alternating current magneto is much the simplest form of machine as it has no commutator, complicated armature winding, nor field magnet coils, and in some types the brushes and revolving wire are eliminated.

As the magnetic flux of an alternating magneto is changed in value, that is increased and decreased, twice per revolution, it follows that the current changes its direction twice for every revolution of the armature. Each change in the direction of current flow is called an alternation.

The voltage developed in each alternation or period of flow is not uniform, the voltage being low at the start of the alternation, rapidly increasing in voltage until it is a maximum at the middle, and then rapidly decreasing to zero, from which point the current reverses in direction. As we have two such alternations, in a shuttle type magneto, per revolution we have two points at which the maximum voltage occurs; that is in the center of each alternation. These high voltage points are called the peak of the wave and consequently the sparking devices should operate at the peak of the wave or at the point of highest voltage. The spark therefore should occur when the shuttle or inductors are at a certain fixed point in the revolution at which point the peak of the wave occurs. The peak of the wave occurs when the shuttle is being pulled or turned away from the magnets.

In what is known as the "shuttle type" alternating current magneto, the generating coil is wound in the opening of an "H" type armature. This iron armature core is fastened rigidly to the driving shaft and revolves with it. As the armature revolves, it is necessary to collect the current that is generated by means of a brush that slides on a contact button B, the button being connected to one end of the winding.

(98) Low Tension Magneto.

The winding of the low tension magneto consists of a few turns of very heavy wire or copper strip, one end of which is grounded to the armature shaft and the other passing through the hollow shaft from which it is insulated. The end of the insulated wire is connected to the contact button (B) on which the current collecting brush presses. As one end of the winding is grounded, one brush, and one connecting wire is saved as the current returns to the magneto through the frame of the magneto. As the shuttle revolves between the magnet poles the magnetism is caused to alternate through the iron of the armature, thus causing the current to alternate in direction and fluctuate in value.

Since there are only two points at which the maximum current can be collected during a revolution with the alternating current magneto, it is necessary to drive it positively through gears, or a direct connection to the shaft so that this maximum point of voltage will always occur at the same point in regard to the piston position. If it is driven by belt without regard to the position of the piston, it is likely that there will be many times that the voltage is zero or too low in value when the spark is required in the cylinder. Alternating current magnetos must be positively driven, and the armature must be connected to the engine so that the peak of the wave occurs at, or a little before the end of the compression stroke.

With this type of magneto the only point that is likely to give trouble is the point at which the brush makes contact with the contact button. If the brush should stick or not make contact, or if the button is dirty or rusty, the current will not flow; this point should always be given attention. Outside of this the only attention necessary is to keep the bearings oiled.

Fig. 101. Sumter Magneto Advanced.

Fig. 102. Sumter Magneto Retarded.

Fig. 103. Sumter Magneto on Horizontal Engine.

Fig. 101 and Fig. 102 show the Sumter low tension magneto as arranged for make and break ignition. The armature and its connections are of exactly the same type as that shown in the previous diagram. The magnets and frame are arranged to tilt back and forth so that the peak of the wave will occur at the advanced and retarded positions of the igniter. This arrangement allows the full voltage of the magneto to be obtained at any point within the range of the igniter, an important item when starting the engine or running at low speed. When mounted on the engine, as shown by Fig. 103, the magnets are provided with an operating rod that is marked "start" and "run." When the pin on the engine bed is engaged under "start," the magneto is retarded, when the pin is under "run" it is advanced. A number of intermediate points are provided at which the operating arm is held fast by tooth engagements as shown in the slotted handle. As shown in the illustration the magneto is fully advanced. The gears by which the magneto is driven are clearly shown in the cut, the ratio between the gear on the crank shaft and that on the magneto shaft being exactly 2 to 1. One lead is carried to the make and break igniter in the cylinder head, the current being returned through the bed of the engine. The same make of magneto is shown mounted on a vertical engine in Fig. 104. In this case the magneto is positively driven from the crank shaft of the engine by a chain. The single conductor running from the magneto to the cylinder heads is clearly shown. To start the engine, the igniter is set in the usual manner and the magneto tilted to starting position, as shown in the illustration. The engine is then started in the usual manner and, when running, the igniter is changed to running position, and the magneto is tilted outwardly. It is not important which is changed first, the magneto or the igniter. It is easy to remember the "starting" and running "position" of the magneto, the running position always being that in which the magnetos are tilted in the direction opposite to that in which the engine runs.

(99) Care of Low Tension Magnetos.

(1) Avoid setting a magneto on an iron or steel plate, unless stated otherwise in the manufacturer's directions, as in some makes the magnetism will be short circuit by iron or steel and will reduce the output.

(2) Do not jar magnets or magneto unnecessarily, for this tends to weaken the magnets.

(3) Never remove the magnets if it can possibly be avoided. If this must be done, mark the magnets and gears so that they may be replaced in exactly the same position. If your magneto refuses to generate after reassembling it is probable that they are reversed in position or that the magnetism has been knocked out of them while off of the magneto.

(4) As soon as the magnets are removed, or better before, place a plate of iron or steel across both ends of the magnet. Don't leave the magnets without this keeper for any length of time or they will lose their magnetism. The best plan is to leave the magnets alone.

(5) Remember that the running clearance between the magnets and armature is very small, only a few thousandths of an inch, and that any error in replacing the bearings in their proper position will cause the armature to bind in the tunnel. Handle armature carefully and do not lay it in a dirty place as a bent shaft or grit in the armature tunnel will fix it permanently.

(6) Most all magnetos are practically water proof, but don't experiment with the hose.

(7) Make all connections firmly and have the wire clean under the binding posts.

(8) Only a few drops of oil are needed at long intervals, don't neglect to oil them, but above all do not drown them with oil.

(9) Examine the brush occasionally and clean off all oil and dirt.

(10) When replacing the magneto on the engine after its removal see that the gears are meshed in the former position. Best to mark the teeth before removal.

(100) High Tension Magnetos.

The "true" high tension type magneto is complete in itself, requiring no jump spark coil nor timer, the high tension current being generated directly in the coils carried by the armature. This arrangement reduces the wiring problem to a minimum, as the only wires required are those leading directly to the spark plugs, and one low tension wire connecting the cutout switch used for stopping the engine.

Fig. 105. Single Cylinder High Tension Bosch Magneto.

The armature of this type of magneto carries two independent windings, one of a few turns of coarse wire called the primary coil, and the other consisting of thousands of turns of extremely fine wire called the secondary coil. It is in the latter coil that the high tension current is generated. The timer is connected directly to the armature shaft, and is an integral part of the magneto. All primary connections are therefore made within the magneto.

Belts or friction drives cannot be used with this type of magneto.

As there are no vibrators or independent coils used, the spark occurs exactly at the instant that the timer operates or breaks the primary circuit. It will be noted that the spark is produced with this magneto when the primary circuit is broken by the timer, instead of made as is the case with battery coils, or coils used with low tension magnetos. There is no lag and consequently the time of ignition is not affected by variations in the engine speed, which requires an advance and retard of the spark with batteries and vibrator coils.

When used with multiple cylinder engines the high tension magneto is provided with a distributor, which connects the high tension current with the different cylinders in their proper firing order. The timer determines the time at which the spark is to occur and the distributor determines the cylinder in which the spark is to take place.

Fig. 106. Connecticut High Tension Magneto.

The sparks delivered by the high tension magneto are true flames or arcs of intense heat, and exist in the spark gap for an appreciable length of time. It is evident that such flames possess a much greater igniting value than instantaneous static spark delivered by the high tension spark coil used with the battery or operated by the low tension magneto, and are capable of firing much weaker mixtures.

Like low tension magnetos, the true high tension type may be of either the inductor or shuttle wound class. All high tension magnetos are positively connected or geared to the engine in such a manner that there is a fixed relation between time of the current impulse produced by the magneto and the firing position of the engine piston.

The current is generated on the same principle as in the low tension shuttle type; that is, by a coil of wire revolving in the magnetic field established by permanent magnets.

During each revolution of the armature, two sparks are produced at an angle of 180° from each other.

The advance and retard of the spark is obtained by means of the timing lever which shifts the timer housing back and forth which results in the primary current being interrupted earlier or later in the revolution of the armature.

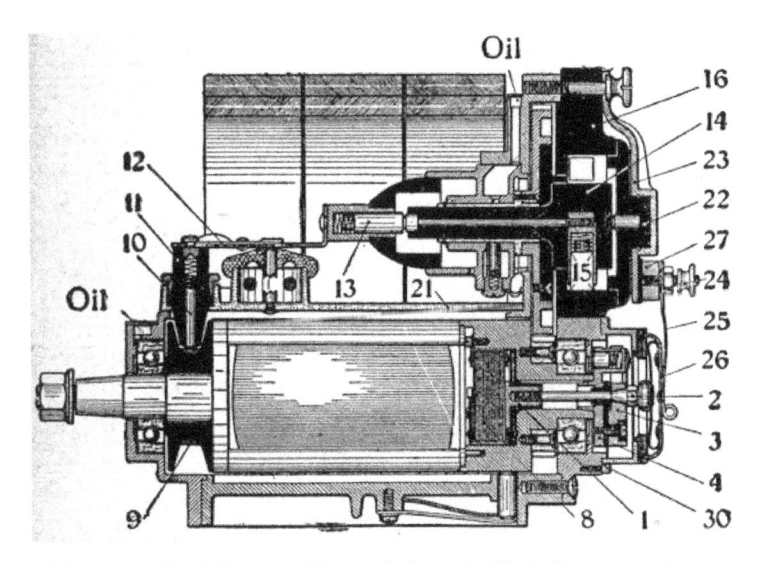

Fig. 107. Longitudinal Section Through Bosch High Tension Magneto.

The timing lever can turn through an angle of 40° measured on the armature spindle, and the angle of advance for multiple engines is as follows:

Advance for 1 cylinder   40°

Advance for 2 cylinders   40°

Advance for 3 cylinders   50°

Advance for 4 cylinders   40°

Advance for 6 cylinders   27°

A timer is used with the magneto on a "jump spark" system in the same way as with a battery, providing a vibrating coil is used.

In one type of magneto the Connecticut, the coil is part of the magneto, and is fastened to the magneto frame. This type of magneto uses a non-vibrating coil, and produces but a single spark each time the primary circuit is broken by the magneto timer. As the timer on this type is driven by the magneto shaft, it is evident that the magneto must be "timed" with the engine, or must have its armature shaft connected to the shaft of the engine in such a manner that the timer contact is broken, and the single spark produced at the instant that ignition is required in the cylinder.

Unlike the dynamo, the alternating current magneto cannot be used with a storage battery, the alternating current producing no chemical change in the electrodes of the battery.

Four Cylinder "D4" High Tension Bosch Magneto Showing Distributor.

The Bosch high tension magneto is a typical high tension magneto having the primary and secondary windings wound directly on the armature shaft, there being no external secondary coil. The end of the primary winding is connected to the plate (1) Fig. 107, which conducts the current to the platinum screw of the circuit breaker (3). Parts (2) and (3) are insulated from the breaker disc (4), which is in electrical contact with the armature core and frame. When the circuit breaker contacts are together the primary winding is short circuited, and when they are separated the current is broken and the spark occurs. The breaker contacts are simply two platinum pointed levers that are separated and brought together by the action of a cam as they revolve. A condenser (8) is provided for the circuit breaker to suppress the spark and to increase the rapidity of the "break."

The secondary winding of fine wire is a continuation of the primary winding, and the secondary is wound directly over the primary. The outer end of the secondary connects with the slip ring (9) on which slides the carbon brush (10), which conducts the high tension current from the armature. This brush is insulated from the frame by the insulation (11). From (10) the current is led through the bridge (12) through the carbon brush (13) to the distributor brush (15). Metal segments are imbedded in the distributor (16), the number of which corresponds to the number of cylinders. As the brush rotates, it makes consecutive contact with each of the segments in turn and therefore leads the current to the cylinders in their firing order. Wires from the cylinders are connected to sockets that in turn connect with the segments. The disc driving the distributor brush (15) is geared from the armature shaft in such a way that the armature turns twice for every revolution of the distributor, when four cylinders are fired, and three times for the distributors once when six cylinders are fired.

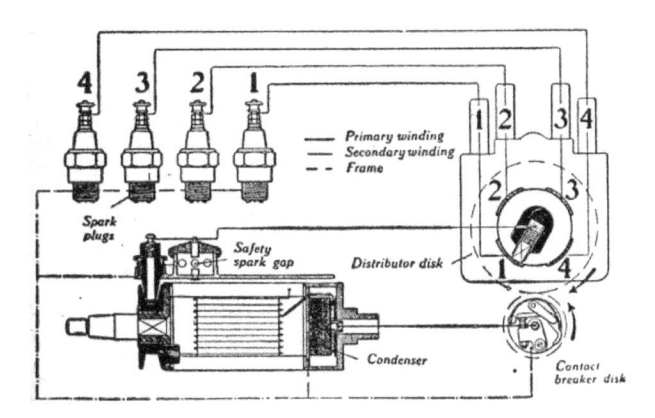

Fig. 108. Bosch High Tension Circuit.

The voltage of the current generated in the secondary coil by the rotation of the armature is increased by the interruption of the primary circuit caused by the opening of the contact breaker.

At the instant of interruption of the primary circuit the high tension spark is produced at the spark plug.

As the spark must occur in the cylinder of the engine at a certain position of the piston, it is necessary that the interrupter act at a point corresponding to a definite position of the piston, consequently this type of magneto must be driven positively from the motor by means of gears, or directly from the shaft.

These magnetos run in only one direction. This running direction should be given when magneto is ordered, as being "clockwise" or "counter-clockwise" when looking at the driving end of the magneto.

The magneto for the single and double cylinder engines has no distributor, the high tension current being led directly from the armature.

The circuit diagram of the Bosch four cylinder magneto is shown by Fig. 108, the winding and plug connections being clearly shown. When connecting the magneto care should be taken to have the distributor and plug connections arranged so that the cylinders will fire in the proper order.

(101) Bosch Oscillating High Tension Magneto.

The oscillating type of magneto is used on slow speed heavy duty engines that move too slowly for the ordinary type of magneto. In the oscillating type the armature is given a short angular swing by the action of a tripping device operated by the engine which results in an intense spark at the lowest speeds.

Magneto type "29" is constructed with two powerful steel magnets, while magneto type "30" is provided with three; an armature of the shuttle type is arranged to oscillate between their pole-shoes.

The magneto is actuated by a rotating cam or other suitable device, which moves the armature 30° from its normal position whenever ignition is required. To permit this movement, a trip lever is mounted upon the tapered end of the armature shaft, this trip lever being held in a definite position by the tension of the spring or springs 1. The trip lever is only supplied when specially ordered, but each magneto is provided with the necessary springs and spring bolts.

When the trip lever is moved from its normal position by the operating mechanism, the springs are extended, and when the operating mechanism releases the trip lever, the later returns the trip lever and armature to their normal position, this movement resulting in the production of a sparking current in the armature winding.

The winding of the armature is composed of two parts, one being the primary winding, which consists of a few turns of heavy wire, and the other the secondary winding, which consists of many turns of fine wire.

The tension of the current produced by the oscillation of the armature is increased by closing the primary circuit at a certain position in its movement, and then interrupting it by means of the breaker. At the moment of the interruption, an arc-like spark is formed at the spark plug and ignition occurs.

Fig. 109. Elevation of Bosch Oscillating Magneto for Slow Speed Engines. High Tension Type.

On cam shaft (c) two cams are mounted side by side. One of these cams (a) is to be used for starting the motor, or for the retarded spark position, while the second (b) is to be used for operation, or for the full advance position. These cams are mounted on a sleeve, which may be moved longitudinally on the shaft, so that the trip lever may be operated by cam (a) or cam (b) as desired. The sleeve is caused to rotate with the shaft by a key. Between the cam (b) and a fixed collar (f) a spiral spring is arranged, which tends to maintain the sleeves in the position when the cam (b) is in operation. A stop collar is also provided to limit the movement of the sleeve beyond this full advance position. Over this collar is fitted a hand wheel, which, in the position illustrated in the diagram, acts together with the collar as a stop. Around the collar is a circular key-way, and a brass bolt is located in the hand wheel to lock into this key-way when the hand wheel is pushed into the position indicated by the dotted lines. This movement of the wheel forces the cam sleeve forward, and brings the retarded cam (a) into the operating position to permit the engine to be started.

(102) The Mea High Tension Magneto.

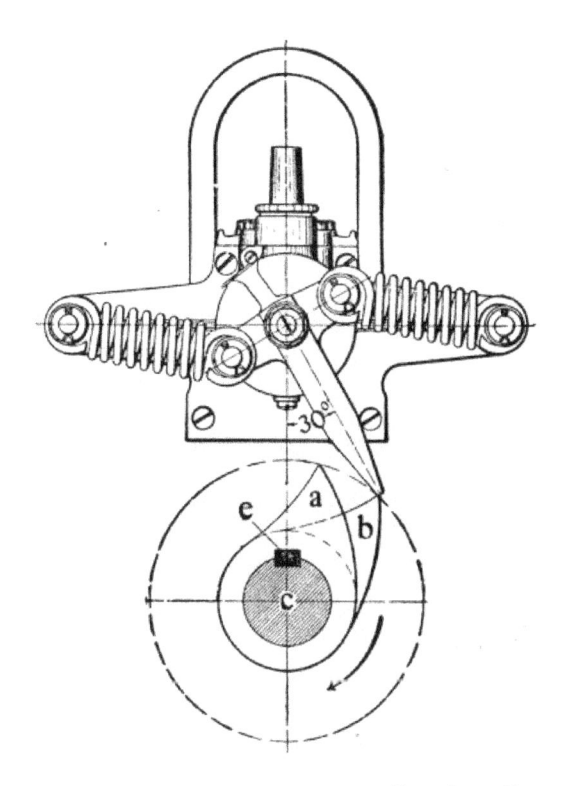

Fig. 110. Diagram of Oscillating Magneto, Showing Cam and Trigger Arrangement.

The low tension winding of the ordinary type of magneto is short-circuited by a breaker which opens at certain points of each revolution with the result that a high voltage is generated across the high tension winding at the moment of the break, and a spark produced across the spark gap in the cylinder to which it is connected. The quality of this spark, or in other words the heat value, depends among other factors upon the particular position of the armature in relation to the magnetic field at the moment the spark is produced. As the armature in this type of magneto is in a favorable position for obtaining a spark twice every revolution, two sparks can be obtained per revolution. The timing of the spark is accomplished by opening the breaker earlier or later, by shifting the breaker housing naturally with the unavoidable result that if the position of the magnetic field remains stationary, the relative position between armature and field at the moment of the break must vary. Since, however, as explained above, the quality of the spark depends upon this relative position, it is apparent that a good spark, can, with a stationary magnetic field, be produced only at one particular timing.

Fig. 111. Side Elevation of "Mea" Magneto, Showing the Magnets, and Cradle in Which the Magneto Swings When Advanced and Retarded.

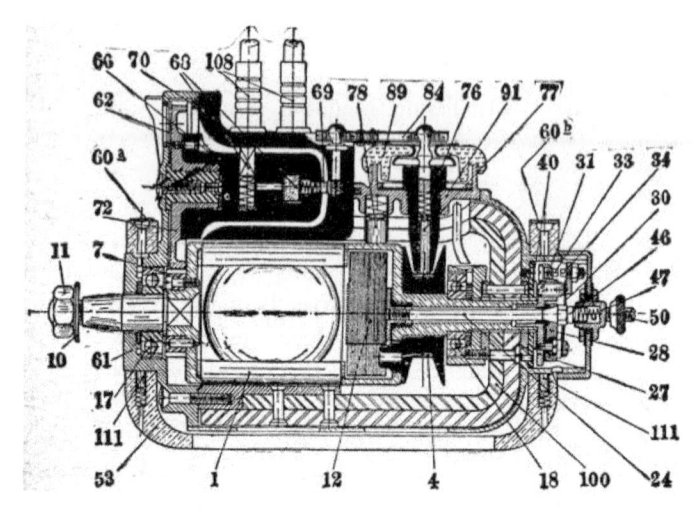

Fig. 112. Longitudinal Section of "Mea" High Tension Magneto.

The result of these conditions are known to everybody familiar with automobiles. They are the difficulty of cranking a motor on one of the average high tension magnetos, if the spark is fully retarded, and of operating the motor on the magneto at very low speed, particularly when it is overloaded, as for example, in hill climbing. Attempts have therefore been made to obtain the spark, independent of the timing, always at the same favorable position of the armature.

The distinct innovation and improvement incorporated in the Mea magneto consists in bell shaped magnets (Fig. 111) placed horizontally and in the same axis with the armature, instead of the customary horse-shoe magnets placed at right angle to the armature.

This at once makes possible and practicable the simultaneous advance and retard of magnets and breaker instead of the advance and retard of the breaker alone as the magnets may be moved to and fro with the breaker housing. It will be seen that as a result of this new departure the relative position of armature and field at the moment of sparking is absolutely maintained, and the same quality of spark is therefore produced, no matter what the timing may be. Furthermore, the range of timing, which with the horse-shoe type of magneto is limited to 10° or 15° at low speeds (i. e. at speeds at which a retarded spark is of value) becomes limited only by the necessity of supplying a suitable support for the magnets. With the standard types of Mea magnetos described in the following, this range varies from about 45° to 70°, but if necessary this range can be increased to any amount desired.

The bell-shaped magnets are fixed to the casing which is mounted on a base supplied with the magneto. The timing is altered by turning the casing and magnets together on the base.

Fig. 112 shows a longitudinal section of a four cylinder Mea magneto. The armature F with the ball bearings 17–18 rotates in the bell-shaped magnets 100, the poles of the magnets being on a horizontal line opposite the armature 1. The armature is of the ordinary H type iron core wound with a double winding of heavy primary and fine secondary wire. On the armature are mounted the condenser 12, the high tension collector ring 4, and the low tension circuit breaker 26–39.

The circuit breaker consists of a disc 27 on which are mounted the short platinum 33, the other contact point 34 is movable and is supported by a spring 30 which is fastened to the insulated plate 28 mounted on disc 27. Fiber roller 31 in connection with cam disc 40 which is provided with two cams is located inside the breaker. Revolving with the armature the roller presses against the spring supported part of the breaker whenever it rolls over the two cams which of course is twice per revolution.

Magneto of Roberts Motor in Advanced Position.

112-a. Advance and Retard Mechanism Used in the Roberts Motors. The Magneto is Driven by a Helical Gear from the Small Pinion. By Shifting the Gear Back and Forth on the Pinion, the Armature of the Magneto is Advanced or Retarded in Regard to the Piston Position. The Reason for this Change Will be Seen from the Cuts by Noting the Position of the Lower Helix.

Inspection of the breaker points is made easy by an opening in the side of the breaker box. The box is closed by a cover 74 supporting at its centre the carbon holder 47 by means of which the carbon 46 is pressed against screw 24. This latter screw connects with one end of the low tension winding while the other end is connected to the core of the armature. It will, therefore, be seen that the breaker ordinarily short-circuits the low tension winding and that this short-circuit is broken only when the breaker opens; it will also be apparent that when the screw 24 is grounded through terminal 50 and the low-tension switch to which it is connected, the low-tension winding remains permanently short-circuited, so that the magneto will not spark. The entire breaker can be removed by loosening screw 24.

The high tension current is collected from collector ring 4 by means of brush 77 and brush holder 76, which are supported by a removable cover 91 which also supports the low tension grounding brush 78 provided to relieve the ball bearings of all current which might be injurious. Cover 91 also carries the safety gap 89 which protects the armature from excessive voltages in case the magneto becomes disconnected from the spark plugs.

The distributor consists of the stationary part 70 and the rotating part 60 which is driven from the armature shaft through steel and bronze gears 7 and 72. The current reaches this distributor from carbon 77 through bridge 84 and carbon 69. It is conducted to brushes 68 placed at right angles to each other and making contact alternately with four contact plates embedded in part 70. These plates are connected to contact holes in the top of the distributor, into which the terminals of cables leading to the different cylinders are placed.

In the front plate of the magneto is provided a small window, behind which appear numbers engraved on the distributor gear which correspond to the number of the cylinder the magneto is firing. This indicator is of great value as it allows a setting or resetting after taking out, without the necessity of opening up the magneto to find out where the distributor makes contact.

The magneto proper is mounted in the base 53 which is bolted to the motor frame and the arrangement is such that the magneto can be removed from its base by removing the top parts 60a and 60b of the two bearings. The variation in timing is affected by turning the magneto proper in the stationary base which is accomplished by the spark lever connections attached to one of the side lugs 88. The spark is advanced by turning the magneto opposite to rotation and is retarded by turning it with rotation. One cylinder magnetos are similar to the four cylinder except that the distributor and gears are omitted.

(103) The Wico High Tension Igniter.

The Wico igniter produces a spark similar to that of the conventional high tension magneto except that the heat of the spark is independent of the engine speed. In other respects it is very different from the types described in the preceding pages for its motion is reciprocating instead of being rotary, and because all of the wire is stationary, the only movement being that of the iron core that passes through the center of the fields. The fact that the spark is of the same intensity at all speeds makes this device particularly desirable in starting the engine at which time the mixture is always of the poorest quality.

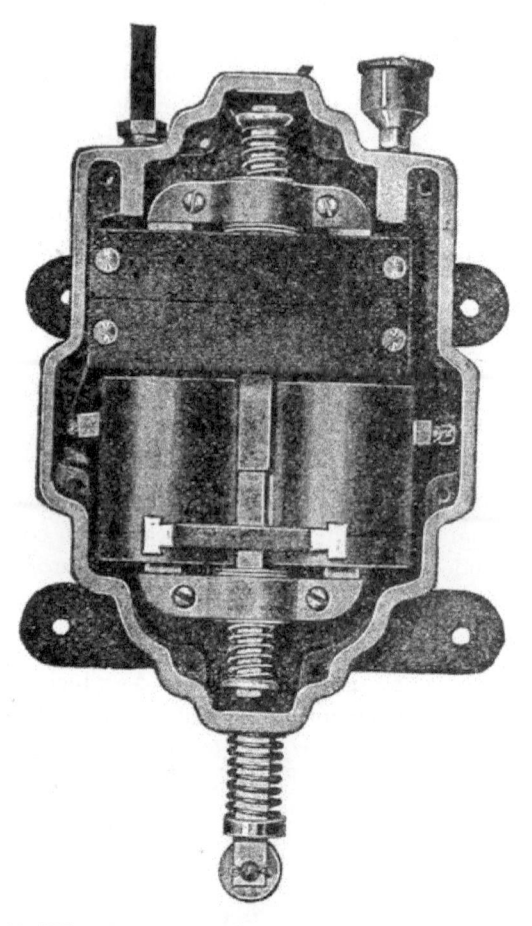

Fig. 113. Wico Igniter. High Tension Reciprocating Type.

It is very simple, and is without condensers, contact points or primary windings, and has no parts that require adjustment.

The current is generated by the reciprocating movement of two soft iron armatures shown as a bar across the bottom of the two coils, which move alternately into and out of contact with the ends of the soft iron cores. The movement of these armatures in the upward direction is produced by the motion of the engine and the speed of this movement is, of course, proportional to the speed of the engine. The downward movement, which produces the spark, is caused by the action of a spring, is much more rapid than the upward movement and entirely independent of the speed of the engine.

The magnets are made of tungsten steel, shown as two bars across the top of the coils, hardened and magnetized and are fastened by machine screws to the cast iron pole pieces, which serve to carry the magnetic lines of force from the poles of the magnets to the soft iron cores. The cores, which fit into slots milled in the pole pieces, are laminated or built up of thin sheets of soft iron, each sheet being a continuous piece, the full length of the core. Each core, extends from just below the top armature, down through the pole piece, and coil to just above the bottom armature.

Each armature consists of a number of laminations or sheets of soft iron mounted on a spool shaped bushing, which, in turn, is loosely fitted onto the squared end of the armature bar. The armature bar is supported with a sliding fit in a box shaped guide which is fastened in the case.

On the outer ends of the armature bar are spiral springs held in place by cup shaped washers and retaining pins, the combination making a self-locking fastening similar to the familiar valve spring fastening used almost universally on gas engines. These springs bear against the armatures and tend to force them against the shoulders of the armature bar.

The coils each have a simple high tension winding of many turns, thoroughly insulated and protected against mechanical injury. They are connected together in series by means of a metal strip, thus making one continuous winding. In the single cylinder igniter, one end of the winding is grounded to the case of the igniter, while the other end is connected to the heavily insulated lead wire. This lead wire passes out through a stuffing box, packed with wicking and thoroughly water tight, direct to the spark plug in the cylinder.

In the two cylinder machine no ground connection is used, but both ends of the winding are connected to lead wires passing out of the case to the spark plugs.

The action of the igniter is as follows:—As the driving bar, through its connection with the engine, is moved downward to its limit of travel, carrying the latch with it, the shoulder on the side of the latch snaps under the head of the latch block. As the motion reverses the latch carries the latch block and armature bar upward. The lower armature, being in contact with the stationary cores, cannot rise with the armature bar, but the lower armature spring is compressed between its retaining washer and the armature, while the bar rises and carries with it the upper armature, which bears against the upper shoulders on the bar.

As the driving bar continues its upward motion the upper end of the latch meets the lower end of the timing wedge and, as the wedge is held stationary by the timing quadrant, a further movement of the latch causes it to be pushed aside until the shoulder on the latch clears the latch block and releases it.

As the lower armature spring is at this time exerting a pressure between the armature bar and cores through the medium of the lower armature, the instant the latch block is released, the armature bar is quickly pulled downward, carrying the upper armature with it. Just before the motion of the upper armature is stopped by its coming in contact with the cores, the lower shoulders on the armature bar come in contact with the lower armature, and, as the bar has acquired considerable velocity, its momentum carries the lower armature away from the cores against the pressure of the upper armature spring, which thus acts as a buffer to gradually stop the movement of the armature bar. The armature bar finally settles in a central position.

The timing of the spark is accomplished by releasing the armature bar earlier or later in the stroke. This is done by shifting the position of the eccentric timing quadrant, which in turn varies the position of the wedge so that the latch strikes it earlier or later in the stroke. The timing quadrant is furnished with several notches into one of which the top of the wedge rests, thus holding the quadrant in the desired position.

The motion should preferably be taken from an eccentric on the cam shaft of a single cylinder four cycle engine, or the crank shaft of a single cylinder two cycle or a two cylinder four cycle engine. On a two cylinder four cycle engine, it is sometimes more convenient to drive the igniter from the cam shaft, using a two throw cam to produce the required number of sparks. In this case the shape of the cam should be such as to duplicate the motion of the eccentric. That is, it should start the driving bar slowly from its lower position, move it most rapidly at mid stroke and bring it to rest gradually at the upper end of the stroke, exactly as is done by the eccentric motion.

When an eccentric is already on the engine the motion may be taken from it to an igniter with a driving bar through a properly proportioned lever that will give the required length of stroke. Where a plunger pump is used on the engine the motion can usually be taken from the pump rod. Where an eccentric has to be provided especially for the igniter, the driving bar is generally used with its roller running on the eccentric.

(104) Starting On Magneto Spark.

A four-cylinder engine in good condition will come to rest with the pistons approximately midway on the stroke and balanced between the compression of the compressing cylinder and of the power cylinder. When the cylinders of such an engine are charged with a proper mixture, the engine will start by the ignition of the mixture contained in the compressing cylinder, for the pressure produced by the ignited gas will be sufficient to rotate the crankshaft.

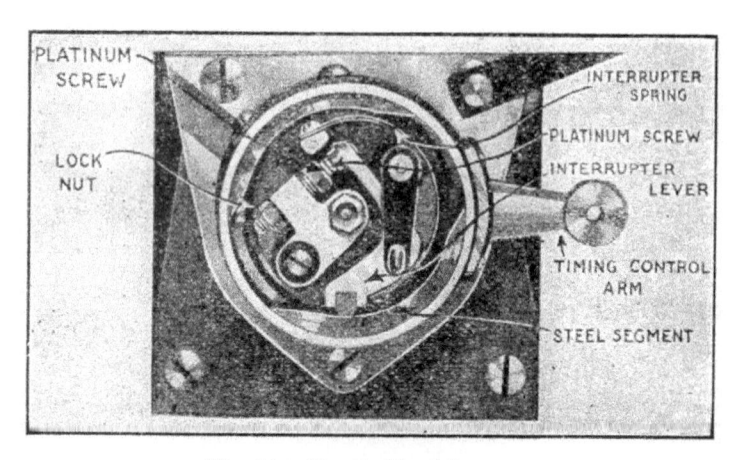

Fig. 114. Bosch Dual System.

It is essential, for the ignition system to be so arranged that a spark can be produced at any point in the piston travel, and in this the Bosch dual, duplex and two independent systems are successful.

The Bosch dual system, Fig. 114, is part of the equipment of many of the cars and engines marketed, and is composed of two separate and distinct ignition systems, one supplying ignition by direct high-tension magneto, and the other by a battery and high-tension coil. These two systems consist in reality of but two main parts; the dual magneto, incorporating a separate battery timer, and the single unit dual coil with its battery. The sparking current from either battery or magneto is brought to the magneto distributor, so that the only parts used in common are the distributor and the spark plugs; the common use of the latter for both magneto and battery systems is cause for the popularity of the dual system for motors having provision for only one set of plugs.

In both the magneto and the battery sides the spark is produced on the breaking of the circuit, and the coil is so arranged that by pressing a button when the switch is in the battery position, an intense vibrator spark is produced in the cylinder during that period when the circuit breaker is open, which will be the case during the first three-fourths of the power stroke. The current is transmitted to the distributor and passes through the spark plug of the cylinder that is on the power stroke.

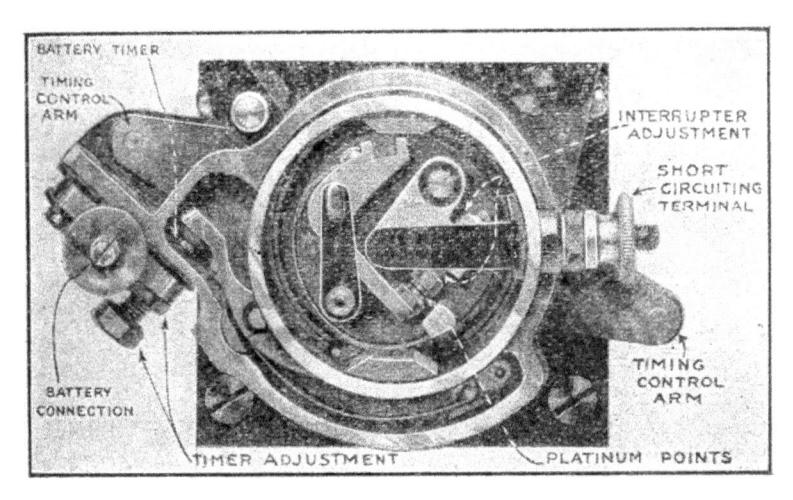

Fig. 115. Bosch Duplex Breaker.

Should the engine come to a stop in such a position that the battery timer is closed, it will not be possible to produce a vibrator spark by the pressing of the button, but the releasing of the button will produce a single contact spark that will ignite the mixture and thus start the engine.

Thus if the engine should stop in some odd position, and the spark is produced when the piston is slightly before top center, for instance, there will be a slight reverse impulse which will bring another cylinder on the power stroke and into the ignition circuit. The engine will thereupon take up its cycle in the proper direction.

In the Bosch duplex system the coil is in series with the magneto armature, but the spark is produced under the same condition, that is, on the breaking of the circuit. In consequence the Bosch duplex system will permit the production of a spark during the first three-quarters of the power stroke by the pressing of the push button set on the switch plate.

The Herz High Tension Magneto in Which the Magnets are Built up of Thin Steel Plates Without Pole Pieces (Four Cylinder Type).

The Bosch two independent system is composed of a separate Bosch battery system and a separate Bosch magneto. Although the operation of the coil is somewhat similar to that of the dual system, the nature of the battery system is such as to require arrangements for two separate sets of spark plugs. The coil is not unlike that supplied with the dual system in that by pressing a button located on the switch plate a series of intense sparks may be produced in the cylinder at all advantageous points of the power stroke.

CHAPTER IX
CARBURETORS
(105) Principles of Carburetion.

The carburetor is a device for converting volatile liquid fuels, such as gasoline, alcohol, kerosene, etc., into an explosive vapor. Besides vaporizing the liquid, the carburetor also controls the proportion of the fuel to the air required for its combustion. The mixture produced by the carburetor must be a uniform gas and not a simple spray to accomplish the best results for complete and instantaneous combustion. Proper combustion cannot be attained with any of the fuel in a liquid state as all of the fuel contained in a liquid particle cannot come into contact with the consuming air. It is of the utmost importance to have the air and fuel in correct proportions so that the fuel may be completely consumed without danger of interfering with the ignition by an excess of air.

With few exceptions modern gasoline carburetors are of the nozzle type in which the liquid is broken up into an extremely fine subdivided state by the suction of the engine piston. This fine spray is then fully vaporized or gasified by the heat drawn from the surrounding intake air that is drawn through the carburetor and into the cylinder on the suction stroke. Owing to the low grade fuels now on the market and to the constantly varying atmospheric conditions it is seldom possible to obtain a perfect vapor in the correct proportions, and for this reason much heat is lost that would be available were the mixture perfect.

Carburetors for automobiles and boats vary in detail from those used on stationary engines due principally to the difference in matters of speed. A stationary engine runs at a constant speed which makes adjustment comparatively easy, while automobile engines have a wide range of speeds and loads making it very difficult to maintain the correct mixture at all points in the range. The difference in the fuel and air adjustments for varying of speeds marks the principal difference between stationary and automobile carburetors. There are many types of successful carburetors on the market, so many in fact that we have room for the description of only three or four of the most prominent, but we will say that the well known carburetors are based on the same principles and differ only in matters of detail.

A cross-sectional view of the well known Schebler Type D carburetor is shown by Fig. 116, and is of the type commonly used on automobile motors and boats.

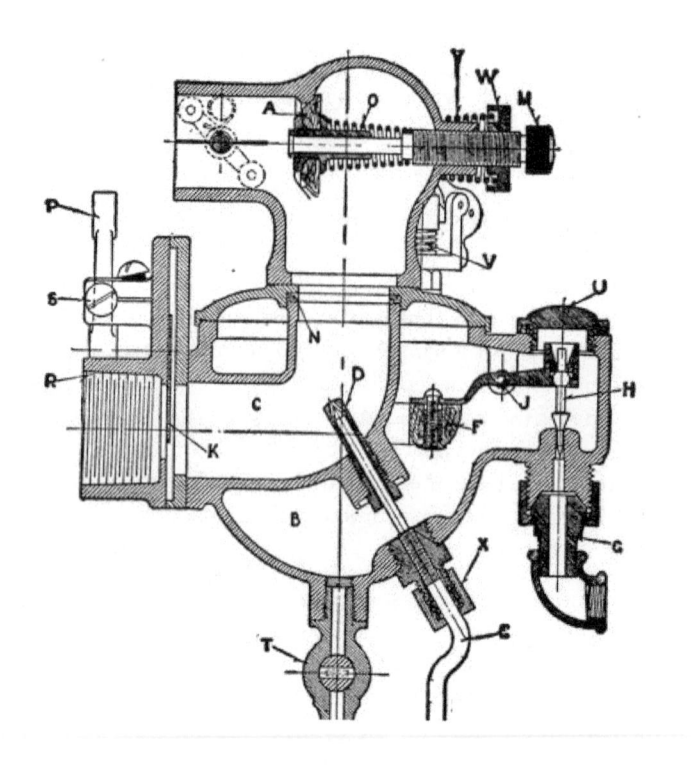

MODEL "D"

Fig. 116. Cross-Section Through Type "D" Schebler Carburetor.

(106) Schebler Carburetor.

The carburetor is connected to the intake of the engine by pipe screwed into the opening **R**, the gas passing from the carburetor to the engine through this opening.

D is the spray nozzle which opens into the float chamber **B**, the opening of the nozzle being regulated by needle valve **E** which controls the quantity of gasoline flowing into the mixing chamber **C**.

On the suction stroke of the engine, air is drawn through the upper left hand opening, past the partially open auxiliary air valve **A**, past the needle valve **D**, through the mixing chamber **C**, and into the engine through **R**.

The suction of the engine produces a partial vacuum in the mixing chamber **C** which causes the gasoline to issue from the nozzle **D**, in the form of a fine spray which is taken up by the air passing through the passage **H**, and is taken into the engine through **R**, thoroughly mixed. The amount of mixture entering the engine, and consequently the engine speed is regulated by the throttle valve **K**, operated by the lever **P**.

In order that the amount of spray given by the nozzle **D** be constant it is necessary that the level, or height of the gasoline in the nozzle be constant. The level is maintained by means of the float **F**, which opens, or closes the gasoline supply valve **H**, opening it and allowing gasoline to enter when the level is low, and closing the valve when the level is high.

The carburetor is connected to the gasoline supply tank, by pipe connected to the inlet **G**, through which the gasoline flows into the float chamber **B**. The float chamber carries a small amount of gasoline on which the float **F** rests. The richness of the mixture is controlled by opening or closing the nozzle needle valve **E**, which passes through the center of the nozzle **D**.

The float **F** surrounds the nozzle in order to keep the level of the liquid constant when the carburetor is tilted out of the horizontal by climbing hills, or by the rocking of the boat when used on a marine engine.

A drain cock **T** is placed at the bottom of the float chamber for the purpose of removing any water, or sediment that may collect in the bottom of the float chamber.

At low speeds, the auxiliary air valve **A** lies tight on its seat, allowing a constant opening for the incoming air through the space shown at the bottom of the valve.

When the speed of the engine is much increased, the vacuum is increased in the mixing chamber **C**, which overcomes the tension of the air valve spring **O** and allows the valve to open and admit more air to the mixing chamber.

The action of the auxiliary air valve keeps the mixture uniform at different engine speeds, as it tends to keep the vacuum constant in the mixing chamber.

When the engine speed increases, the flow of gasoline is greater, and consequently more air will be required to burn it; this additional air is furnished by the automatic action of the valve, and when once adjusted, compensates accurately for the different engine speeds.

The gasoline is generally supplied by a tank elevated at least six inches above the level of the fluid in the float chamber; although in some cases the gasoline is supplied by air pressure on a tank situated below the level of the carburetor.

In some types of Schebler carburetors, the float chamber **B** is surrounded by a water jacket that is supplied with hot water from the cylinder jackets of the engine. This keeps the gasoline warm so that it evaporates readily under any atmospheric conditions.

The quantity of air admitted to the carburetor is controlled by an air valve shown in the air intake by the dotted lines. This is adjusted by hand for a particular engine and is seldom touched afterward.

When starting the engine it is necessary to have a very rich mixture for the first few revolutions, this mixture being obtained by "flooding" the carburetor.

On the Schebler carburetor the mixing chamber is flooded by depressing the "tickler" or flushing pin **V**.

(107) Two Cycle Carburetors.

Nearly any type of carburetor can be used on a two port, two stroke, cycle engine providing a check valve is placed between the crank case and carburetor to prevent the crank-case compression from forcing its contents back through the inlet passages. A great many manufacturers make special carburetors for two stroke motors that have the check valve built into the carburetor itself. With three port two stroke cycle engines a check valve is not necessary as the piston in this type of engine performs this duty.

In that class of vaporizers known as mixing valves, the valve that controls the flow of gasoline blocks the air passage in such a way that an additional check valve is not necessary.

(108) Kingston Carburetors.

The Kingston Carburetor shown by Fig. 117 differs from the Schebler in many details, the principal difference being in the construction of the spray nozzle and the construction of the auxiliary air valve. The throttle valve E controls the exit of the mixture through the engine connection C which is an extension of the mixing chamber. The spray nozzle J which is surrounded by a hood or tube is controlled by the needle valve A which is threaded into the top of the mixing chamber, this latter adjustment being locked into place by a button head screw and a slot in the casting.

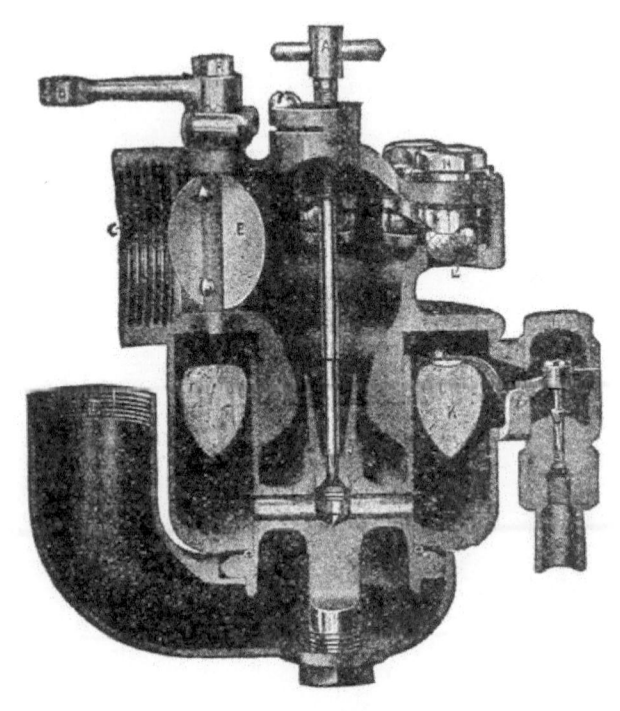

Fig. 117. Cross-Section Through Kingston Carburetor Showing Balls Used for Auxiliary Air Valves.

Surrounding the nozzle tube or hood is a curved restriction in the air intake passage, is known as a Venturi tube, which insures a constant relation between the air and fuel supplies. As the action of the Venturi tube is rather complicated, it will not be taken up in detail. Air is supplied to the Venturi passage through the intake (D). An annular float (K) surrounds the mixing chamber that acts on the gasoline supply valve (I) through a short lever arm. This valve is accessible for cleaning on the removal of the cap H that covers the valve chamber. Gasoline enters the float chamber through the fuel pipe G, and enters the spray nozzle through the two ports in the base of the mixing chamber.

The auxiliary air valve is a particularly novel feature of this carburetor, as no springs nor disc valves are used in its construction. Five balls (M) of different weights and sizes act as air valves, the balls covering the inlet ports (L) under normal operation. As the speed increases, the balls are lifted off their seats in order of their weight or size by the increase in suction. With a slight increase of suction, the lightest ball covering the smallest hole is lifted first, a further increase in suction lifts the next largest ball which still further increases the auxiliary air intake, and so on until at the highest speed all of the balls are off their seats. Access to the ball valves is had through the valve caps (N). The constant supply inlet is circular and may be set at any desired angle, as can the float chamber and gasoline supply connection. Control and adjustment are entirely by the needle valve.

(109) The Feps Carburetor.

The Feps carburetor has the main needle valve surrounded by a Venturi chamber as in the preceding case, the needle valve adjustment being made through a lever on the left of the mixing chamber. An auxiliary nozzle directly under the auxiliary air valve at the right, connects with the float chamber and furnishes an additional mixture of gasoline and air for hill climbing and high speed work when the leather faced auxiliary air valve lifts from its seat. The adjustment for this auxiliary jet is shown at the right of the air valve chamber.

For intermediate speeds, the air valve alone is in action. No controlling springs are used on the air valve which insures positive action and sensitive control of the air. A float surrounding the Venturi tube controls the fuel valve through the usual lever arm. A wire gauze strainer placed in the fuel chamber to the left prevents dirt and water from being drawn into the nozzle, and as this strainer easily removed it is a simple matter to clean and prevent the troubles due to dirty fuel.

By closing the upper valve in the vertical engine connection the vacuum is increased in the manifold when starting the engine. This increase of vacuum draws gasoline from the float chamber and primes the engine making the engine easy to start in cold weather. The tube through which the gasoline is drawn for priming is the small crooked tube bending over the float and terminating above the starting valve. Below this valve is the throttle valve which controls the mixture in the ordinary manner. The adjustment for intermediate speeds is made by the center knurled thumbscrew shown over the air valve chamber which controls the travel of auxiliary air valve. In effect this is a double carburetor, one jet for high speed and one for low.

(111) Gasoline Strainers.

Much trouble is caused in carburetors by dirt, water and sediment, collecting in the small passages and obstructing the flow of the gasoline.

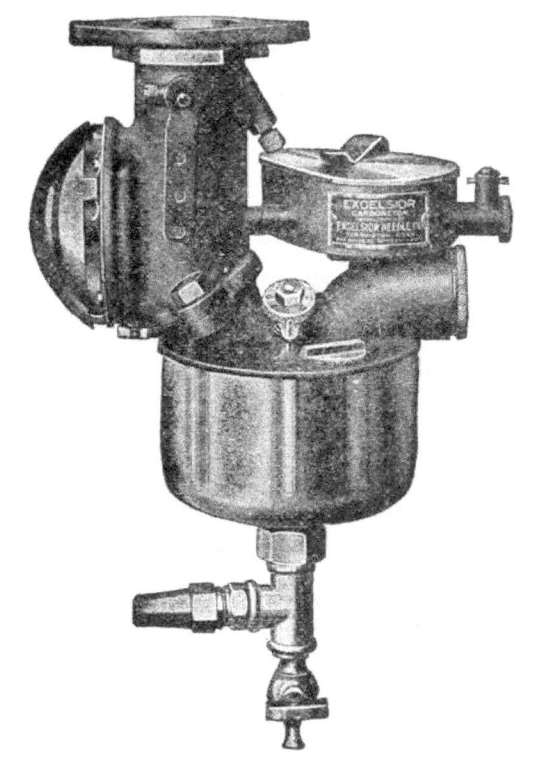

Fig. 119. The Excelsior Carburetor in Which the Air is Regulated by a Ball which Lies in the Tapering Venturi Tube. An Increase of Suction Lifts the Ball and Allows More Air to Pass.

The purpose of the gasoline strainer is to prevent any water or foreign matter from being carried into the carburetor, and this device should be used on every engine if the owner wishes to be free from carburetor troubles.

(112) Installing Gasoline Carburetors.

(1) Use brass or copper pipe from the tank to carburetor if possible to avoid trouble from dirt and flakes of rust.

(2) When installing a gasoline tank be sure that the bottom of the tank is at least six inches above the carburetor to insure a good flow.

(3) The tank should be provided with an air vent hole, or the gasoline will not flow because of the vacuum in the top of the tank.

(4) All tanks should be provided with a drain cock at the lowest point so that water and dirt may be easily removed.

(5) Clean out the tank thoroughly before filling with gasoline to avoid clogged carburetors.

(6) Pipes from the tank to carburetor should never be placed near exhaust pipes or hot surfaces for the gasoline vapor may prevent the feeding of gasoline.

(7) Clean out pipes before using.

(8) If common threaded pipe joints are used on the gasoline piping, use common soap in place of red lead.

(113) Installing the Carburetor.

The carburetor should be placed as near to the cylinder as possible, the shorter the pipe, the less the amount of vapor condensed in the manifold. With multi-cylinder engines the carburetor should be so situated, that is, an equal distance from each cylinder, so that each cylinder will inhale an equal amount of vapor.

The intake opening of the pipe should be placed near one of the cylinders, or draw warm air off the surface of the exhaust pipe in order that gasoline will evaporate readily in cold weather, and form a uniform mixture at varying temperatures.

Great care should be taken to prevent any air leaks in the carburetor, or intake manifold connections, as a small leak will greatly reduce the strength of the mixture and cause irregular running. Always use a gasket between the valves of a flanged connection and keep the bolts tight. If a brazed sheet brass manifold is used, look out for cracks in the brazing.

Leaks may be detected in the connections by spurting a little water on the joints, and turning the engine over on the suction stroke. If the water is sucked in the leaks should be repaired at once. Make sure when placing gaskets, that the gasket does not obstruct the opening in the pipe, and that it is securely fastened so that it is not drawn in by the suction.

Never allow the carburetor to support any weight, as the shell is easily sprung which will result in leaking needle valves.

CARBURETOR ADJUSTMENT. When adjusting the carburetor of multiple cylinder engine, it is advisable to open the muffler cutout in order that the character of the exhaust may be seen or heard. With the muffler open, the color of the exhaust should be noted. With a **PURPLE** flame you may be sure that the adjustment is nearly correct for that load and speed; a yellow flame indicates too much air; a thin blue flame too much gasoline, and is not the best for power.

Before starting for the adjustment test, try the compression, and the spark. If the compression is poor, try the effects of a little oil on the piston, which may be introduced into the cylinder through the priming cup. It will be well to dilute the oil to about one-half with kerosene. After all trouble with all the parts are clear, you may start the engine.

Turn on the gasoline at the tank, and after standing a moment see whether there is any dripping at the carburetor, if there is, the trouble will probably be due to a leaky float, dirt in the float valve, or to poor float adjustment. Locate the leak and remedy it before proceeding further. Dirt on the seat of the needle valve may sometimes be removed by "flooding" the carburetor, which is done by holding down the "tickler" lever for a few seconds, causing the gasoline to overflow, and wash out the dirt.

If the motor has been standing for a time it would be well to "prime" the motor by admitting a little gasoline into the cylinder through the priming cup, or by pushing the tickler a couple of times so as to slightly flood the carburetor.

Now turn on the spark and turn over the engine for the start, taking care that the throttle is just a little farther open than its fully closed position. If the engine takes a few explosions and stops, you will find the nozzle, or that some part of the fuel piping is clogged which will stop the engine. If the motor gradually slows down, and stops, with **BLACK SMOKE** issuing from the end of the exhaust pipe, or **MISFIRES** badly, the mixture is **TOO RICH**, and should be reduced by cutting down the gasoline supply by means of the needle valve adjusting screw. If it stops quickly, with a **BACKFIRE**, or explosion at the supply of gasoline should be **INCREASED** by adjusting the mouth of the carburetor, the mixture is **TOO LEAN**, and the needle valve.

In all cases be sure that the auxiliary valves are closed when the engine is running slowly, with the throttle closed, as in the above test. If they are open at low speed, the mixture will be weakened and the test will be of no avail.

After adjusting the needle valve as above until the engine is running (with throttle in the same partially closed position), turn the valve slowly in one direction or the other until the motor seems to be running at its best. During the above tests the spark should be left retarded throughout the adjustment, and the throttle should not be moved.

The carburetor should now be tested for high speed adjustment, by opening the throttle wide (spark ¼ advanced), and observing the action of the motor. If the engine back-fires through the carburetor at high speed, it indicates that the mixture is too weak which may be due to the auxiliary air valve spring tension being too weak and allowing an excess of air to be admitted. Increase the tension of the spring, and if this does not remedy matters, admit a little more fuel to strengthen the mixture by means of the needle valve adjustment. Do not touch the needle valve if you can possibly avoid it, or the high-speed adjustment, as the fuel adjustment will be disturbed for low speed.

If the engine misfires, with loud reports at the exhaust, does not run smoothly, or emits clouds of black smoke at high speed, the engine is not receiving enough air in the auxiliary air valve, consequently the tension of the spring should be reduced.

Back firing through the carburetor denotes a weak mixture.

Trouble in cold weather may be caused either by slow evaporation of the gasoline, or by water in the fuel that freezes and obstructs the piping or nozzle. In cold weather a higher gravity of gasoline should be used than in summer, as it evaporates more readily, and therefore forms a combustible gas the rate at lower temperatures.

To increase the rate of evaporation of the gasoline, it should be placed in a bottle and held in hot water for a time before pouring it into the carburetor or tank, or the air inlet warmed with a torch.

The cylinder water jacket should always be filled with hot water before trying to start the engine, and will prevent the gas from condensing on the cold walls of the cylinder. Often good results may be had by wrapping a cloth or towel around the carburetor, that has been dipped in hot water.

The cylinder of an air-cooled engine may be warmed by **gently** applying the heat of a torch to the ribs, or by wrapping hot cloths about it.

The tank, piping, and carburetor should be drained more frequently in cold weather than in hot, to prevent any accumulation of water from freezing, and stopping the fuel supply. A gasoline strainer should always be supplied on the fuel line, and should be regularly drained.

The motor may often be made to start in cold weather by cutting out the spark, and cranking the engine two or three revolutions with the throttle wide open. The throttle should now be closed within $1/8$ of its fully closed position, the ignition current turned on, and the engine cranked for starting. This system will very seldom fail of success at the first attempt.

Carburetor flooding is shown by the dripping of gasoline from the carburetor, and which results in too much gasoline in the mixture. Flooding may be caused by dirt accumulating under float valve, by a leaking float (Copper Float), by Water Logged Float (Shellac worn off Cork Float), by float adjustment causing too high a level of gasoline, by leaking float valve, by cutting out ignition when engine is running full speed, by rust or corrosion sticking float valve lever, by float binding in chamber, by float being out of the horizontal, by float valve binding in guide, by excessive pressure on gasoline, or by tickler lever held against float continuously.

Dirt accumulated under float valve may sometimes be flushed out by depressing tickler lever several times; if this does not suffice, the cap over the valve must be removed, and the orifice cleaned by wiping with a cloth.

LEAKING FLOAT VALVES should be reground with ground glass or very fine sand; never use emery as the particles will become imbedded in the metal, which will be the cause of worse leaks.

Should the shellac be worn off of a cork float allowing the gasoline to penetrate the pores of the cork, a new float should be installed, as it is a doubtful policy for owner to give the float an additional coat of shellac.

MISFIRING AT LOW SPEED. If the carburetor cannot be adjusted to run evenly on low speed after making all possible adjustments with the needle valve, the trouble is probably due to air leaks between the carburetor and engine, caused by broken gaskets, cracked brazing in the intake manifold, or by leaks around the valve stem diluting the mixture.

INCORRECT VALVE TIMING will cause missing, especially on multiple cylinder engines, as the carburetor cannot furnish mixture to several cylinders that have different individual timing. Look for air leaks around the spark edge openings, and be sure that all valves seat gas tight. Always be sure that the auxiliary air valve remains closed at low speeds, as a valve that opens at too low a speed will surely cause misfiring as it dilutes the mixture.

MISSING in one cylinder may be caused by an air leak in that cylinder.

WATER in gasoline will cause misfiring, especially in freezing weather, as it obstructs the flow of fuel to the carburetor. The carburetor and tank should be drained at regular intervals, and if possible, a strainer should be introduced in the gasoline line.

CLOGGED NOZZLE. Particles of loose dirt in the nozzle will occasion an intermittent flow of gasoline that will result in misfiring. The nozzle should be cleaned with a small wire run back and forth throughout the opening.

CLOGGED AIR VENT in the float chamber will change the level of the fuel, and will either "starve" the engine, or flood the carburetor. The air in the float chamber is a very small hole, and is likely to clog.

HOT FUEL PIPE. If the fuel pipe that connects the tank with the carburetor, becomes hot, due to its proximity to the exhaust pipe of cylinders, vapor will be formed in the pipe that will interfere with the flow of fuel.

DIRT UNDER AUXILIARY AIR VALVE will prevent the valve from seating properly, causing the engine to misfire at low speed.

CRACKS OR LEAKS in intake pipe or gaskets will cause intermittent leaks of air and spasms of misfiring. Old cracks that have been brazed will sometimes open and close alternately causing baffling cases of spasmodic misfiring.

DIRT IN AIR INTAKE will change the air ratio, and the increased suction will cause a greater flow of gasoline. Do not place the end of the inlet pipe in a dusty place, nor where oil can be splashed into it by the engine. Clean out periodically.

"**LOADING UP**" of the inlet piping in cold weather on light load is caused by the mixture condensing in the intake pipe. The only remedy is to keep the piping warm, or to heat the inlet air.

CLOGGED OVERFLOW PIPE, with engines equipped with pump supply will cause flooding, as the fuel does not return rapidly enough to the tank.

(114) Kerosene Vaporizer for Motorcycles.

An ingenious vaporizing device has been designed for the use of kerosene as a fuel for motorcycle engines, by the M. G. and G. Motor Patents Syndicate, Ltd., England, is described in Motor Cycling. The device consists of a comminuter, or vaporizer, which screws into the sparkling-plug hole in the cylinder, the plug being transferred to an aperture in the vaporizer, a feeder for regulating the supply of fuel to the vaporizer, and a throttle and air barrel, or mixing chamber, for the purpose of proportioning the amount of air and gas supplied to the engine, and for controlling the speed of the machine as in an ordinary carburetor.

The feeder receives the fuel—in this case kerosene—although any heavy oil can be used with almost equally good results. The feeder answers a purpose similar to the ordinary float chamber of the carburetor, i. e., to regulate the amount of kerosene it is required to pass through the vaporizer. It consists of a small chamber mounted upon the end of a pipe leading to the vaporizer. Kerosene is fed to this device by a copper pipe from the tank, and enters at the lowest point through a 316-inch hole or jet. This is covered by a small valve, operated by engine suction. The lift of this valve can be adjusted by the insertion of washers to suit any particular size of engine, just as one would use various size jets to suit either a large or small engine. One of the greatest advantages of the device lies in the size of this aperture or jet, inasmuch as it cannot possibly choke up with grit, and even water will pass through and not stop the operation of the carburetor. At the top of the feeder is an air hole, which admits just sufficient air to pass the kerosene through the vaporizer, the reason for this being that the heat of the vaporizer shall only act upon the fuel, the mixture afterwards being balanced by air being admitted through the mixing chamber.

After the kerosene leaves the feeder it passes through a pipe to the vaporizer. This consists of a gunmetal body with cooling ribs cast on the outside, whilst through the center runs a thin copper tube of $5/8$-inch diameter and only 20 gauge, which would really melt during the heat of combustion were it not for the fact of the fuel passing through it. The heat derived from this formation of vaporizer is approximately 1,000 degrees Fahr. Inside the central tube is a strip-steel spiral, which serves the double purpose of giving a centrifugal motion to the fuel, and at the same time forming a supporter for the tube, preventing it crushing under the force of the explosions. It is, of course, understood that the inside of the feeding tube is entirely isolated from the combustion chamber. The sparking plug is screwed into the wall of the vaporizer, which is now really an extension of the combustion chamber.

Obviously this slightly reduces the compression of the engine, which, however, is a necessary feature when kerosene is used as a fuel. After passing through this device the kerosene is thoroughly vaporized, and the vapor is led through a flexible pipe to the throttle chamber; this taking the place of an ordinary carburetor and being fitted to the induction pipe.

There are two slides, operated by Bowden levers from the handle-bar, one being for the main air intake and the other for the gas.

Fig. 121-a. The English Aster Electric Lighting Unit.

Undoubtedly the greatest claim for this vaporizer is the fact that practically no carbon deposit forms upon the inside of the cylinder or on the piston. What little deposit is formed takes the shape of small, soft flakes, which, instead of adhering to the cylinder walls, break away before they have attained any size and are blown through the exhaust valve. Altogether, this device seems to have finally solved the problem of using kerosene as a fuel on air-cooled engines, especially if the carbon deposit difficulty has been finally overcome.

The device was fitted to a 3½ h.p. Matchless with a White and Poppe engine. In order to start up, a small gasoline tank, holding about one half-pint of gasoline, is fitted under the main tank and communicates with the feeder. Half a minute is all that is necessary running on gasoline, when the kerosene can be turned on. The machine would fire at a walking pace, and could also be accelerated up to 55 m.p.h.

CHAPTER X
LUBRICATION
(116) General Notes on Lubrication.

No matter how carefully the surface of a shaft or bearing may be finished, there always remains a slight roughness or burr of metal, which although of microscopic proportions is productive of friction or wear. Each minute projection of metal on a dry shaft acts exactly as a lathe tool, when the shaft revolves in cutting a groove in the stationary bearing. Since there are a multitude of these projections in a journal, the wear would be very rapid, and would in a short time completely destroy either the shaft or bearing, no matter how highly finished in the beginning.

When lubricating oil is introduced into a bearing it immediately covers the rubbing surface, and as the oil has a considerable resistance to being deformed, or is "stiff," it separates the surface of the shaft from that of the bearing for a distance equal to the thickness of the oil film. With ordinary lubricants this distance is more than enough to raise the irregularities of the shaft out of engagement with those of the bearing. This property of "stiffness" in the oil is known as "viscosity." The value of viscosity varies greatly with different grades of oil, and also with the temperature with the result that the allowable pressure on the oil per square inch also varies. With oils of low viscosity a small pressure per square inch on the bearing will squeeze it out, and allow the two metallic surfaces to come against into contact, causing wear and friction, while an oil of greater viscosity will successfully resist the pressure.

The life and satisfactory operation of the engine depends almost entirely upon the lubricant and the devices that apply it to the bearings. Excessive wear and change in the adjustments are nearly always the result of defective lubricating devices or a poor lubricant. The principal lubricants are:

(1) Solid lubricants such as graphite, soapstone, or mica.

(2) Semi-solid lubricants such as vaseline, tallow, and soap emulsions, or greases compounded of animal fats, vegetable and mineral oils; and

(3) Liquid lubricants, such as sperm oil, or one of the products of petroleum, the latter medium being the class of lubricant most suitable for internal combustion engines, owing to its combining the qualities of a high flash-point with a comparative freedom from either acidity or causticity.

Oils of animal or vegetable origin should never be used with gas engine as the high temperatures encountered will char and render them useless. Tallow and lard oil are especially to be avoided, at least in a pure state.

In the cylinder only the best grade of **GAS ENGINE** cylinder oil should be used, which according to different makers has a flash point ranging from 500 to 700 degrees. Using cheap oil in the cylinder is an expensive luxury. In general, the oils having the highest flash points have also the objectionable tendency of causing carbon deposits in the combustion chamber and rings which is productive of preignition and compression leakage. The lower flash oils have a tendency to vaporize and to carry off with the exhaust which will leave the walls insufficiently lubricated unless an excessive amount is fed to the cylinder. By starting with samples of well known brands recommended by the builder of the engine it will be an easy matter to find which is the cheapest and gives the best results. In figuring the cost of oil do not take the cost per gallon as a basis, but the cost for so many hours of running, or better yet the number of horse-power hours. Unless you are fond of buying replacements and new parts do not stint on the oil supply.

On the other hand, an excess of oil should be avoided as this means not only a waste of oil through the exhaust pipe, but trouble with carbon deposits and ignition troubles as well. Foul igniters, misfiring, and stuck piston rings are the inevitable result of a flood of lubricating oil. When a whitish yellow cloud of smoke appears at the end of the exhaust pipe, cut down the oil feed. The exhaust should be colorless and practically odorless.

Too much oil cannot be fed to the main bearings of the crank shaft if the waste oil is caught, filtered and returned to the bearings by a circulating system, for the flood of oil not only insures ample lubrication but removes the heat generated as well. The bearings require a much lighter oil, of a lower fire test than the cylinder oil. It is evident that its viscosity is a most important element, as it determines the allowable pressure on the shaft. The viscosity of an oil varies with the temperature and is greatly reduced at cylinder heat. A comparative test of the viscosity or load bearing qualities of an oil may be made by making bubbles with it by means of a clay pipe; the larger the bubble, the higher the viscosity of the oil.

Different sizes of bearings, and bearing pressures, call for oils of different viscosities, and consequently an oil that would be suitable for one engine would not answer for another; heavy bodied oils being used for heavy bearing pressures, and light thin oil for small high speed bearings. The best way to determine the value of an oil for a particular shaft bearing is by experiment, attention being paid to its adaptability for the feeding devices used.

The compression attained in a gas engine cylinder depends to a certain extent upon the body of the cylinder oil, for many engines that leak compression past the rings with thin oil will work satisfactorily with a heavy viscous oil that clings tightly to the surfaces. An engine will often lose compression when an oil of poor quality is used.

Air cooled engine cylinders require an oil of heavier body than water cooled because of the higher temperature of the cylinder walls. Gum and sticky residue are usually formed by animal oils or adulterants added to the numeral oil base. Oils containing free acids should be avoided as they not only corrode and etch the bearing, but also clog the oil pipes or feeds with the products of the corrosion.

Free acid is left from the refining process, and may be determined by means of litmus paper inserted into the oil. If the litmus paper turns red after coming into contact with the oil, acid is present, and the oil should be rejected.

The following are the characteristics of an oil suitable for use on an engine:

(a) The oil must be viscous enough to properly support the bearings or to prevent leakage past the piston rings.

(b) It should be thin enough so that it can be properly handled by the oil pumps, or drip freely in the oil cups.

(c) It should not form heavy deposits of oil in the cylinder and cause the formation of "gum."

(d) It should contain no free acid.

Ordinarily a good grade of fairly heavy machine oil will be suitable for use on the bearings of the average engine, such as the cam-shaft and crank-shaft bearings.

Only very light clean oil, or vaseline should be used on ball-bearings, as heavy greases and solid lubricants pack in the races and cause binding or breakages.

Flake graphite is much used as lubricant, and too much cannot be said in its favor, as it furnishes a smooth, even coat over the shaft, fills up small scores and depressions, and makes the use of light oil possible under heavy bearing pressures. With graphite, less oil is used, as the graphite is practically permanent, and should the oil fail for a time, the graphite coat will provide the necessary lubrication until the feed is resumed without danger of a scoring or cutting. In fact, when graphite is used, the oil simply acts as medium by which the graphite is carried to the bearings.

If graphite is injected into the cylinder in small quantities it greatly improves the compression, as it fills up all small cuts and abrasions in the cylinder walls.

A good mixture to use for bearings is about 1½ teaspoonsful of graphite, to a pint of light machine oil, thoroughly mixed.

Graphite can be placed in the crank chamber of a splash feed engine, by means of an insect powder gun.

Trouble with oil cups is always in evidence during cold weather, as the oil congeals, and does not drip properly into the bearings. The fluidity of the oil can be increased in cold weather by the addition of about ten per cent of kerosene to the oil.

If too much oil is fed to the cylinders, the piston rings will be clogged with gum, and a loss of compression, or a tight piston will be the result. An excess of oil will short-circuit the igniter or sharp plugs, and will form a thick deposit in the combustion chamber that will eventually result in preignition or back-firing. Deposits and gum formed in the cylinder will cause leaky valves and a loss of compression. Feed enough oil to insure perfect lubrication, but not enough to cause light colored smoke at the exhaust.

Lubricating systems may be divided into three principal classes: Sight-feed, splash system, and the force feed system. Sight feeding by means of dripping oil cups is too common to require description, and is used on many stationary engines, both large and small.

The splash system is in general use on small high speed engines both stationary, and of the automobile type.

The force feed system in which oil is fed under pressure by a pump is by far the most desirable as the amount of oil fed is given in positive quantities proportional to the engine speed, and with sufficient pressure to force it past any ordinary obstructions that may exist in the oil pipe.

Another system that is half splash, and half force feed, is the pump circulated system much used in automobiles.

THE SPLASH FEED SYSTEM is the simplest of all, as the bearings are lubricated by the oil spray caused by the connecting rod end splashing through an oil puddle located in the bottom of the closed crank case. The piston and cylinder are lubricated by the spray, as well as the bearings, as the lower end of the piston projects into the crank chamber at the moment that the connecting rod end strikes the oil puddle.

To maintain constant lubrication, it is necessary that the oil in the puddle be kept at a constant height, or as in some cases be varied in such a way that the surface of the puddle is raised and lowered in proportion to the load on the engine. In the average engine the oil level is maintained by overflow pipes or openings that allow any excess of oil over the fixed level to flow back to the pump. In the Knight engine the puddles are formed in movable cups which are connected with the throttle in such a way that the opening of the throttle raises the oil level and supplies more oil to the engine at the greater load, or speed.

Oil in splash systems is supplied by a low pressure pump, usually of the rotary type, in the base of the engine. Oil from the pump passes to the bearings, drops into the puddle, overflows through the overflow opening, and returns to the pump through a filter, the same oil being used over and over again until exhausted. This strainer should be removed occasionally and the dirt removed, for should it be allowed to collect it is likely to obstruct the oil supply. The oil should be replaced before it becomes too black or foul, the crank case and bearings thoroughly cleaned with kerosene, and new oil replaced. The supply may be interrupted by the failure of the pump, caused by sheared keys or leakage of air in the suction line due to cracks. It would be well to run the engine for a few minutes with the kerosene in the crank case, in order that all of the oil may be removed. See that the drain cock is closed at the bottom of the cylinder or all of the oil will be lost. Lock the valve handle carefully so that it cannot jar open. If light colored smoke appears in intermittent puffs with a multiple cylinder engine, it indicates that one cylinder is receiving too much oil.

(117) Force Feed Lubricating System.

The force feed system is by far the most reliable of all oiling systems, as it feeds uniformly and continuously at almost any temperature, and against the pressure of practically any obstruction in the pipe.

The oil is supplied by a small pump driven from the engine, the pump being incased in the oil tank housing. Frequently a hand pump is used in combination with the power pump when starting the engine, or at times when the power pump is out of service. A single pump is used with any number of leads, each lead, or feed, having an independent regulating valve and sight feed, or a pump unit may be provided for each lead, depending on the size of the engine.

(118) Bosch Force Feed Oiler.

The force feed of the Bosch Oiler is so positive in character, that the flow of oil is not affected by heavy back-pressure due to elbows and the diameter of the conducting pipes. Springs, valves and other devices, which would check the flow of oil, are fundamentally eliminated. The amount of oil fed may be accurately and permanently regulated. Glands and other packings and bushings are eliminated. Connecting rods and all links are eliminated by the direct application of the movements of the oscillating cam disks to the pump plungers and piston valves.

Each feed of this oiler is provided with a separate pump element consisting of a pump body plunger and a piston valve, the suction and feed ducts connecting directly with the pump body of their respective elements. With this construction, pump elements may be replaced or added. The oiler requires no attention other than to be supplied with oil; and the opening and closing of the valves, pet cocks, etc., on starting and stopping the machine is rendered unnecessary. The correct and regular operation of the elements may be verified by observation of the reciprocating movements of the regulating screws.

Each pump plunger is provided with an adjusting screw through which the feed may be regulated from 0 to 0.2 cubic centimeters for each stroke.

The Bosch Oiler (Fig. 121) being positively driven by the machine that it supplies, the oil fed is in all cases proportional to the engine speed; overloads are thus automatically taken care of.

The circular arrangement of the elements of the Bosch Oiler permits the device to be driven by a single shaft, and the oil is forced through the feeds from a single reservoir to the required points of application. A pump element consists of a pump body 1, a pump plunger 2 and a piston valve 3, and is supported on the base plate 13. The elements are arranged concentrically about the drive shaft in such a manner that the pump plungers form a circle around the circle formed by the piston valves.

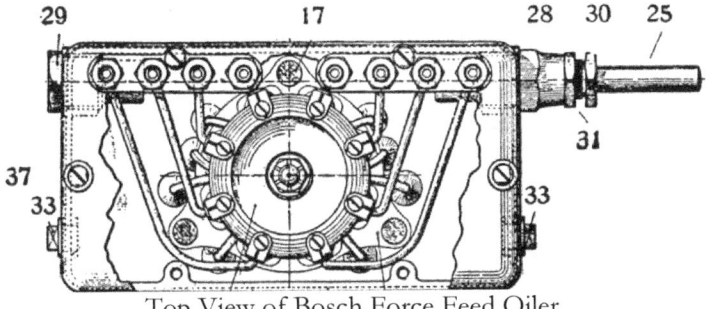

Top View of Bosch Force Feed Oiler.

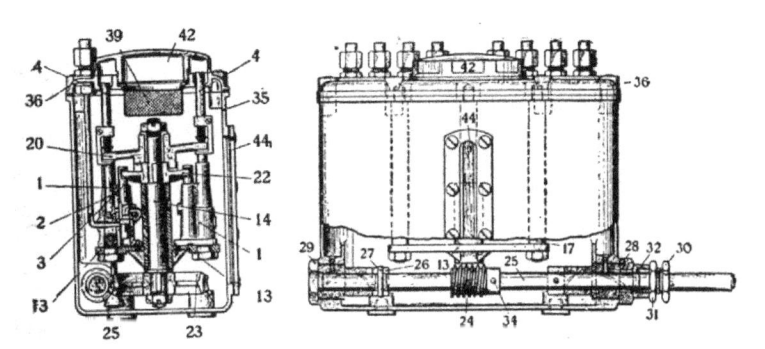

Fig. 121. Cross-Section Bosch Oiler.

The pump cam disk 20 and the valve cam disk 22 are set on the drive shaft at other than a right angle with its axis, and the rims of the disks are gripped by slots formed in the heads of the pump plungers and piston valves. The relation of these cam disks is such that the valve cam disk is 90° in advance of the plunger cam disk. The valve cam disk is solid on the drive shaft, but the pump cam shaft is loose and driven through a lug on the valve cam disk. When the drive of the pump is reversed, the lug on the valve cam disk frees itself and again takes up the drive of the pump cam disk, after the drive shaft has made a half revolution.

Regulating screws 4 are set in the slotted heads of the pump plunger, and by means of this the back-lash or play of the cam disk may be regulated. The regulating screws are provided with lock nuts, and project through the cover of the oil tank housing, being exposed by the removal of the filler cover 42. The filler opening is provided with a removable strainer to prevent the entrance of foreign particles into the oil tank.

Pump shaft 14 is driven through worm gear 23 which meshes with worm 24 on drive shaft 25; drive shaft 25 projects from the oiler housing, and is coupled with the driving shaft of the machine to be lubricated.

Base plate 13 is attached to the oiler cover by three stud bolts, thus permitting the removal of the entire oiler mechanism from the housing.

The quantity of oil in the oil tank is shown by gauge glass 44.

On the starting of the machine to which the oiler is attached, the pump shaft and the cam disks that it supports are set in motion through worm 24 and worm gear 23. A direct reciprocating motion is given to the pump plunger and to the piston valve by the rotation of the cam disks which have a movement similar to that of the "wobble saw." The relation of the cam disk is such that the piston valve movements are 90° in advance of the movements of the pump plungers. The pump will run in either direction without alteration.

To secure this effect a play of 90° is provided between the cam disk. When cam 22 is driven clockwise, cam disk 20 is driven by the lug which meshes with a lug on disk 22. The cams are then in such a relation that the cam valve disk is 90° in advance of the pump cam disk. When reversed, cam 20 remains at rest until cam 22 catches the lug and cam 20, when the drive continues as before. The cams are then in the same relation as previously for as the valve disk 22 has traveled through 180° it is evident that it is 90° in advance of the pump disk.

(119) Castor Oil for Aero Engines.

Castor oil is used almost exclusively in the Gnome and other rotary engines of the same type, but has not been particularly successful on stationary cylinders.

Chemically, castor oil differs from all other vegetable or animal oils in containing neither palmitin or olein. It is soluble in absolute alcohol, but practically insoluble in gasoline. On the other hand, the castor oil is capable of dissolving small quantities of mineral oil, the more fluid they are the less it absorbs of them. But the insolubility of castor oil in mineral oil disappears completely when it is mixed with even a very small quantity of another vegetable or animal oil, such as colza or lard oil. An adulteration may thus result in a serious reversal of the oil's best qualities; in fact, in serious seizures, Castor oil does not attack rubber, but it contains 1 to 2 per cent of acid fats; sometimes more.

"In my opinion says a writer in '*Autocar*' castor oil can only be used in fixed cylinders with impunity for short distances and then with repeated cleanings between runs, but on rotary engines of the Gnome type cleaning is almost unnecessary. The reason is that one cannot consistently use castor oil over and over again, for the fact is indisputable that it has a far greater tendency than mineral oils to absorb oxygen, and so gradually to increase in body and finally to gum. When once it commences to gum the carbonization becomes more rapid, because the thickened and pitch-like oil acts as an insulating covering on the top of the pistons and of the cylinder, and cannot get away with sufficient rapidity to avoid decomposition and baking to a coke. Therefore if castor oil is to be used on the ordinary stationary cylinder type of engine, it is necessary to wash out the crank chamber and to replace with fresh oil at frequent intervals. On a rotary engine such as the Gnome this cleaning is unnecessary, because there is a continuous stream of fresh castor oil brought into the crank chamber and then thrown by centrifugal force past the pistons and through the cylinder into the exhaust. Thus the stream of oil never has sufficient time to oxidize fully, gum or decompose. This action of centrifugal force accounts for the large consumption of oil on the rotary engine, and also for the fact that the pistons and cylinders keep comparatively clean.

"In thus criticizing the use of castor oil I do not wish it to be inferred that it is not an excellent lubricant. What I wish to suggest is that in the case of an internal combustion engine it must be made with discretion. A point in favor of castor oil is the fact that it maintains its viscosity in a remarkable manner at high temperatures, and that at those high temperatures it has a peculiar creeping or capillary action which enables it to spread uniformly over the whole of the metallic surfaces, whereas under the same conditions a similarly bodied mineral oil would be unevenly distributed in patches. Another point is that the specific heat of castor oil is considerably higher than that of a pure mineral oil. This is in its favor, insomuch that it shows castor oil to be a better heat remover than a mineral oil.

"Motorists and aviators have from time to time informed me that they are using castor oil, but have apparently been under some misapprehension. I find that they have been using a brand of prepared oil under the impression that it is a specially refined castor oil, or that it is a blend of castor oil."

Producer Gas Engine Plant at Göttingen, Germany, Consisting of Four 3,500 Horse-Power Units.

A simple method for testing the purity of castor oil is at the disposal of all. It is known as the Finkener test. Ten cubic centimeters of castor oil is placed in a graduate. Five times as much alcohol, 90 per cent, is added and stirred in. The solution should remain clear and brilliant at 15 to 20 degrees C. An admixture of foreign oils, even if only 5 per cent, riles the solution at this temperature, though not above it.

(120) Force Feed Troubles.

The most common trouble with force feed systems is the failure of the operator to remove the dirt collected by the strainer. The oil piping should be cleaned out at least once every year by means of a wire and gasoline, to remove any gum that may have been deposited. Driving belts should be kept tight to prevent slipping, and belts that are soaked with oil should be cleaned with gasoline and readjusted.

Leaking pump valves generally of the ball type are a common cause of failure. They may leak because of wear or by an accumulation of grit and dirt on their seats, which prevents the valves from seating properly. If the valves leak, the oil will be forced back into the tank, or will not be drawn into the pump cylinder at all, depending on whether the inlet or discharge valve is the offender. Plunger leakage which is rare will cause oil failure.

If the oil pipes that lead to the bearings rub against any moving part, or against a sharp edge, a hole will be worn in the pipe, a leak caused which will prevent the oil from reaching the bearing. A dented or "squashed" pipe will prevent the flow of oil.

The set screw or pin holding the pulley to the pump shaft may loosen and cause it to run idly on the shaft without turning the pump. This will of course, prevent the circulation of oil.

The worm and worm wheel may wear so that the pump is no longer driven by the pulley shaft, or a poor pipe connection may leak all that the pump delivers.

The amount of oil required by each lead or bearing should be carefully determined by experiment, and kept constantly at the right number of drops per minute.

The feed adjustments jar loose, and should be inspected frequently.

(121) Oil Cup Failure.

Oil cups should be cleaned out frequently with gasoline or kerosene, as any gum or lint will interfere seriously with the feed. They should be adjusted and filled frequently to prevent any possible chance of a hot bearing.

Oil cups should be as large as possible in order that they may be left for considerable periods without danger of a hot box.

Cold weather affects the oil feed to a considerable extent, especially with small oil cups, and they should be kept as warm as possible. When heavy oils are used a cold draft will stop the feed.

Oils may be made more fluid in cold weather by the addition of about ten per cent of kerosene.

(122) Hot Bearings.

A hot bearing is almost a sure sign of insufficient oil, and the trouble should be located and remedied immediately. Oil pumps stopping, clogged oil pipes or holes, frozen oil, or oil leaks are common causes of hot bearings.

Never allow an engine to run with a hot bearing for any length of time, as the bearing or piston may seize tight and wreck the engine. Inspect the journals frequently to see if they are above normal temperature. A hot, binding bearing often causes the effect of an overload on the engine, slowing it down, and increasing the governor and fuel feed, this is followed in a short time by the bearing seizing.

(123) Cold Weather Lubrication.

It is by no means uncommon trouble in cold weather to find excessive fluctuations in pressure as the engine speed and temperature of the oil varies. Thus, if the pressure be set correctly with the engine running fast, and when just started up, it will be found, after half-an-hour's running, that, with the engine turning slowly, the pressure is far too low, owing to the oil having become thin. If the pressure be then reset, it may be found on next starting up from cold that the gauge goes hard over, and may very easily be burst if the engine is run fast.

The point is one to which many designers of engines pay far too little attention, though the difficulty may be very easily gotten over. The secret lies in having the by-pass outlet of most ample proportions, so that the excess of oil, however thick, can get away quite easily. If there is any throttling of the by-pass, back pressure must result with consequent increase of the pressure at which the by-pass valve comes into operation. In other words, the pressure of the main supply to the bearings will be increased.

A writer to "The Motor," London solved this problem in the following manner:

"Originally, the by-passage was somewhat small, little larger than the oil delivery pipe to the engine, which was about 316 inch bore, and the result was that the pressure when starting with the oil cold rose to about 25 pounds per square inch, and fell to about one pound per square inch with the oil hot and the engine running slow. It was possible, however, to bore out the by-pass passage and fit a larger pipe, about three times the area of the main delivery pipe, with the result that the oil, when cold, never rose above about 15 pounds per square inch, however fast the engine run. When thoroughly heated, the normal running pressure was about 6 pounds per square inch, falling to 2 pounds per square inch with the engine only just turning over, which brings up the question of the correct working pressure. This will vary very largely with the design of the engine, but, broadly speaking, the higher the pressure the better for the bearings. The limiting figure is determined by the tendency of the engine to throw out oil at the end of crankshaft bearings, and by the amount that gets past the piston rings. Obviously, an engine with new, tight bearings and new piston rings will stand a higher pressure without undue waste of oil or excess deposit in the cylinder head than will an old engine with worn bearings and slack rings. And, again, the question will be affected by the design of the pistons. For instance, where the trunk of the piston is bored for lightness, much more oil will get past the rings than in cases where a 'solid' trunk is employed. Roughly speaking, 8 to 15 pounds per square inch is a good figure for a new, high-speed engine. An old and worn engine, particularly if not of a high-speed type, may require no more than 2 to 6 pounds per square inch."

Brookes Gasoline-Electric Generating Units for Operating Search Lights. An Independent Unit is Used for Each Light.

The writer recently encountered a rather curious difficulty in connection with obtaining a free by-pass. The return pipe from the by-pass led into the case carrying the gearwheels of the camshaft and magneto drive, and oil continually flooded out from the end of the camshaft and other bearings. The waste and mess were sufficiently serious to warrant investigation, and the cover plate over the gears was accordingly taken off. It was then noticed that the oil delivered to the gearwheel case had only two small holes by which to drain away to the crankcase. The flow from the by-pass was beyond the proper capacity of these holes, and so the whole gearwheel case became filled with oil under considerable pressure, quite possibly 2 or 3 pounds per square inch, and it was not surprising that oil exuded from the ends of the bearing. A few extra limber-holes, if one may borrow a nautical expression, were drilled through to the crankcase, and no further trouble was experienced.

(124) Plug Oil Holes When Painting.

When the chassis of the car is repainted it is well to see that all exposed oil holes are stuffed with waste to prevent them from being choked. Failure to observe this precaution may result in the holes being clogged with paint, which if not removed before the car is started, will prevent oil reaching the bearings.

(125) Oiling the Magneto.

Never oil the circuit breaker or circuit breaker mechanism, unless for a drop of sperm oil that may be applied to the cam roller by means of a toothpick. If oil gets on the circuit breaker contact points, it will cause them to spark badly, resulting in pitting or destruction of the points. If the oil is occasionally applied to the cam roller or should oil accumulate on breaker points, the breaker should be rinsed out with gasoline to remove the surplus.

Pitted or carbonized contact points are capable of causing much trouble, and gummy oil or dirt will develop this trouble quicker than any other cause. Use only the best grade of thin sperm oil on the ball bearings.

In the course of time the circuit breaker contact points will wear or burn, causing imperfect contact, and too great a separation between the points. The contacts should be examined from time to time, and if rough or pitted, should be dressed down to a flat even bearing by means of a dead smooth file, and the distance readjusted. The contacts should not bear on a corner or edge, but should bear evenly over their entire surface to insure a maximum primary current and spark.

CHAPTER XI
COOLING SYSTEMS

The object of the cooling system is not to keep the cylinder cold, but to prevent the heat of the successive explosions from heating the cylinder walls to a degree that would vaporize the lubricating oil and prevent satisfactory lubrication of the cylinder and piston. The hotter the cylinder can be kept without interfering with the lubricating oil, the higher will be the efficiency of the engine and the greater the output of power.

To obtain the greatest power from an engine, the heat developed by the combustion should be confined to the gas in order that the pressure and expansion be at a maximum, it is evident that the pressure and power will be reduced by over-cooling as the heat of the expanding gas will be taken from the cylinder and transferred to the cooling medium. The temperature of the cylinder, and therefore the efficiency of the engine is determined principally by the vaporizing point of the lubricating oil, and consequently the higher the grade of the oil, the higher the allowable temperature of the cylinder.

If cold water from a hydrant or well be forced around the water jacket rapidly, the power will be greatly reduced owing to the chilling effect on the expanding gas. There is not much danger in keeping the cylinder of an air cooled engine too cool, in fact the great difficulty with this type of engine is to keep it cool enough to prevent an excessive loss of lubricating oil.

The valves, particularly the exhaust valves, should be surrounded with sufficient water to keep them cool as they are subjected to more heat than any other part of the engine, and are liable to wrap or pit. The water leaving the jacket of a gasoline engine should not exceed 160° F., as temperatures in excess of this amount cause deposits of lime scale.

When possible, a portion of the cooling water should be run into the exhaust pipe immediately after it has completed its flow around the valves and cylinders, as the water cools the gas so suddenly that the exhaust to atmosphere is rendered almost noiseless, and the exhaust pipe is kept much cooler and less liable to cause fire by coming into contact with combustible objects.

On some engines the exhaust pipe is water jacketed for some distance to prevent dirty rusty pipes in the vicinity of the engine mechanism and also to prevent injury to the operator should he come into contact with the pipe.

Small engines and medium size vertical engines usually have the water jacket cast in one piece with the cylinder casting and others have a separate head that is bolted to the cylinder.

In the latter type the water flows from the cylinder to the head through ports or slots cut in the end of the cylinder water jacket that register with similar slots in the jacket of the head.

Thus in this construction we have not only to pack the joint to prevent leakage of gas from the cylinder, but also to prevent the leakage of cooling water from the jacket into the cylinder, or outside. Thus there is always a chance of water leaking into the cylinder bore and causing trouble unless the packing is very carefully installed and looked after.

In large horizontal engines the gas and water joints are never made at the same point, as it would be practically impossible to prevent leakage into the cylinders of such engines.

When the cylinder and cylinder water jackets are cast in one piece without a water joint at the junction of the cylinder and the head, the water connection between the head and the cylinder being made by pipes external to the castings.

Small, portable, stationary engines are sometimes **"HOPPER COOLED,"** or cooled by means of the evaporation of the water contained in an open water jacket that surrounds the cylinder.

The hopper is merely an extension of the water jacket such as used on all water cooled engines, the only difference being that the top of the hopper is open permitting the free escape of water vapor or steam to the atmosphere. The water level should be carried within two inches from the top of the hopper.

Water when converted into vapor or steam absorbs a great quantity of heat, and of course the steam carries the heat of evaporation with it when it escapes to the atmosphere.

As the hopper is open to the air, the temperature of the cylinder cannot exceed 212° F. (temperature of boiling water) as long as there is sufficient water left to cover the cylinder.

The hoppers contain sufficient water for runs of several hours' duration, and as the water boils away or evaporates, it may be replenished by simply pouring more water in the top of the hopper. Hopper cooling is used principally for small portable engines where the weight of a water tank or other cooling device would be objectionable and also where there is danger of freezing the pipes and connections of other systems.

The loss of water by evaporization is from .3 to .6 of a gallon per horsepower hour; that is, for a 5 hp. engine the loss would be from 1.5 to 3 gals. for every hour that the engine was operated under full load.

The cylinder and the water jacket are cast in one integral piece, with no joints of any kind in either the combustion chamber or in the water jacket.

Fig. 124. Air Cooled "Grey Eagle" Aeronautical Motor. Note the Depth of Cooling Ribs.

A system of cooling by which the heat of the walls is radiated to the air directly without the medium of water is often used on small high speed engines, and is known as "**AIR COOLING.**"

This type of cylinder is surrounded with radiating ribs or spires which increases the radiating surface of the cylinder to the extent that the required amount of heat is lost to allow of economical lubrication. This system is desirable where the weight of radiators and water would be a drawback, where it would be inconvenient to obtain water, or where there would be trouble from freezing. An air cooled motor generally is provided with a fan that increases the efficiency of the radiating surface by changing the air between the ribs. With aeronautical motors such as the Gnome, and Gray Eagle, shown by Fig. 124, the circulation of the air due to the propeller and the rush of the aeroplane is sufficient to thoroughly cool the machine.

As a rule, the air cooled motor is made more efficient in fuel consumption than the water cooled type because of the high temperature of the cylinder walls. In fact all engines are air cooled eventually, whether the heat is radiated at a high temperature by the fires, or at a lower temperature through the circulating water and radiator.

When the engines are of the portable type, and likely to be used out of convenient reach of water, the hopper or **EVAPORATOR TANK** system is used, the tank system being used for the larger engines. In effect, the tank system is the same as the hopper cooler, the heat being dissipated principally by evaporation, although some heat is radiated from the surface of the tank itself. The difference between the two systems is merely one of size, the tank offering a greater area for the emission of heat than the hopper.

A tank-cooled engine has one pipe running from the top of the cylinder to a point near the top of the tank, the bottoms of the cylinder and tank being connected together by another pipe.

When the water becomes heated in the cylinder, it expands and becomes lighter than the cold water in the tank and consequently rises to the surface of the water in the tank through the upper pipe. As the warm water flows into the tank, it is immediately replaced by the heavier cold water that flows into the cylinder from the bottom of the tank through the lower pipe. This successive discharge of the heated water from the cylinder to the tank sets up a continuous flow of water through the water jacket of the cylinder, which transfers the excess heat of the cylinder to the tank where it is dissipated to the atmosphere by evaporation and radiation.

The circulation of the cooling water set up by the action of heat or the expansion of the water is called Natural or Thermo Syphon circulation.

Cooling tanks may be used profitably with stationary engines if the tank can be located so that vapor and steam produced will not be objectionable. If the tank is used inside of a building, the vapor should be conveyed to the outside air by means of a stack or chimney, or by means of a small ventilating fan driven by the engine.

The water consumption of a cooling tank is from .3 to .6 gallons per hour, the exact quantity varying with the atmospheric conditions and temperature.

Fig. 124-a. De Dion Bonton "V" Type, Air Cooled Aero Motor. The Cooling Air is Furnished by a Blower Mounted on the Crank Shaft at the Rear of the Motor. The Propeller is Driven from the Cam Shaft. Courtesy of Aero.

For engines of from 10 to 50 horsepower a battery of cooling tanks may be used, the number depending on the size of the engine. For natural circulation, the tank should be installed so that bottom of the tank is above the bottom of the cylinder for maximum results, if placed much lower the engine should be provided with a circulating pump.

If water is used from the city mains from 10 to 15 gallons will be required per horsepower hour, the exact quantity varies with the temperature of the supply.

The water from very large stationary engines is cooled by allowing it to trickle down through a cooling tower, which is built somewhat like the screen cooler only on a larger scale. The object of the cooling tower is to present the greatest possible surface of water to the air, this is accomplished by screens or baffles that turn the water over and over as it falls. The water, well cooled, finally collects in a cistern at the base of the tower from which it is pumped back to the engine and thus is used over and over again. This is an ideal system when water is expensive and when engines of considerable power are used.

(126) Cooling System Troubles.

Overheating caused by deposits of scale or lime in the jacket is one of the most common causes of an excessively hot cylinder. When hard water containing much lime is heated, the lime is deposited as a solid on the walls of the vessel forming a hard, dense, non-conducting sheet. When scale is deposited on the outside of the cylinder walls it prevents the transfer of the heat from the cylinder to the cooling water and consequently is the cause of the cylinder overheating. Besides acting as an insulator or heat, the deposit also causes trouble by obstructing the pipes and water passages, diminishing the water supply and aggravating the trouble.

Scale interferes with the action of the thermo syphon system more than with a pump, as the pressure tending to circulate the water is much lower. Whatever system is used, the scale should be removed as often as possible, the number of removals depending, of course, on the "hardness" of the water.

Large horizontal engines are usually provided with hand holes in the jacket, through which access may be had to the interior surfaces on which the scale collects. Under these conditions the scale may be removed by means of a hammer and chisel.

The scale may be softened by emptying half the water from the jacket and pouring in a quantity of kerosene oil, the inlet and outlet pipes being stopped to prevent the escape of the oil. The engine should now be started and run for a few minutes with the mixture of kerosene and water in the jacket; no fresh water being admitted during this time. After the mixture has become boiling hot, stop the engine and allow it to cool; it will be found that the scale has softened to the consistency of mud, and may easily be washed out of the jacket.

The work of removing the scale can be reduced to a minimum by filling the jacket with a solution of 1 part of Sulphuric Acid and 10 parts of water, allowing it to stand over night. The scale will be precipitated to the bottom of the jacket in the form of a fine powder and may be easily washed out in the morning.

If the jacket water is kept at a temperature above 185° F. the amount of scale deposited will be nearly doubled over that deposited at 160° F.

Wash out sand and dirt occasionally, a strainer located in the pump line will help to keep the jacket clear and free from foreign matter.

If a solution of carbonate of soda, or lye, and water are allowed to stand in the cylinder over night, the deposit will be softened and the work with the chisel will be made much easier.

If a radiator is used (automobile or aero engine) the deposit can be removed with soda, never use acid, lye, or kerosene in a radiator or with an engine with a sheet metal water jacket.

Obstructions in Water Pipes. Poor water circulation may be caused by sand, particles of scale, etc., clogging the water pipes, or by the deterioration of the inner walls of the rubber hose connections. Sometimes a layer of the rubber, or fabric of the hose may loosen from the rest and the ragged end may obstruct the passage.

A sharp bend in a rubber hose may result in a "kink" and entirely close the opening.

The packing in a joint may swell, or a washer may not have the opening cut large enough, either case will result in a poor circulation.

Sediment is particularly liable to collect or form in a pocket, pipe elbow, or in the jacket opposite the pipe opening. Oil should be kept off of rubber hose connections as it will cause them to deteriorate rapidly, this may finally result in water circulation troubles. Rubber pipe joints between the engine and the radiator or tanks are advisable as they do not transmit the vibration of the engine, and hence reduce the strain on the piping. A strainer should be provided in order to reduce the amount of foreign material in the water.

Radiators. A clogged radiator will give the same results as a clogged jacket with the exception that steam will issue from the radiator if the circulation is not perfect.

If the radiator becomes warm over its entire surface it is evident that the water is circulating, the temperature being a rough index of the freedom of the water, or the interior condition of the surfaces. A leaking radiator may be temporarily repaired with a piece of chewing gum.

Should the radiator be hot and steaming at the top and remain cold at the bottom for a time, it shows that the water is not circulating and that the jackets on the cylinders are full of steam. Such a condition usually is indicative of clogging between the bottom of the radiator and pump, between the pump and bottom of cylinders, or of a defective pump.

Natural Gas Plant at Independence, Kansas. Used for Pumping Gas From the Wells to Various Distributing Points.

Thermo-syphon radiators are more susceptible to the effects of sediment and clogging than those circulated by pumps.

A radiator may fail to cool an engine because of a slipping or broken belt driving the fan, or on account of a loose pulley or defective belt tension adjuster. Keep the belt tight. The fan may stick on account of defective bearings.

Radiator may be **AIR BOUND**, due to pockets or bends in the piping holding the air.

Rotary Pump Defects. A defective circulating pump will cause overheating, as it will supply little if any water to the jackets.

Examine the clutch or coupling that drives the pump and see that the key or pin that fastens it to the shaft is in place. Next see that the driving pinion and gear are in mesh and properly keyed to their respective shafts.

In some cases the shaft has been twisted off, or the coupling pin sheared through by reason of the shaft rusting to the pump casing. Worn gears or impellers **IN THE PUMP** reduce the output and cause heating, as will a sheared driving pin in the impeller. Wear and bad impeller fits reduce the capacity of the pump.

Scale or sediment collecting in the pump sometimes strips the pins or impeller teeth. Note the condition of the gaskets or whether the pump shaft is receiving the proper amount of grease. Put a strainer in pump intake. See that no leak occurs on pump intake pipe.

To avoid the trouble and expense due to cracked water jackets, never neglect to drain the cylinders and piping from all water in freezing weather. Drain cocks should be provided at the lowest points in the water circulating system for this purpose. It would be well to provide an air cock at the highest point in the line in order that all of the water can drain out as soon as the drain cock is opened.

With automobile or portable engines it is not always convenient or possible to drain the engine every time that it is stopped and consequently we must resort to a "non-freezing" mixture or at least a solution that will not solidify under ordinary winter temperatures. Such a solution should be chosen with care, as many will cause the corrosion and destruction of the jackets and piping; **NEVER USE COMMON SALT** and water under any conditions.

Wood alcohol and water in equal parts, is often used for automobiles, but is rather expensive for portable engines having a comparatively great amount of water in circulation.

Unless the circulating system is absolutely air tight, as it is when radiators are used, alcohol will be lost by evaporation and must be replaced frequently.

The most practical solution for the average engine used, is made up by dissolving about five pounds of **CALCIUM CHLORIDE** in one gallon of water. This mixture will stand a temperature of about 15° F below zero, and if diluted to half the strength will not freeze above zero.

Use **CALCIUM CHLORIDE**, not ordinary Salt (Sodium Chloride).

CHAPTER XII
GOVERNORS AND VALVE GEAR
(127) Hit and Miss Governing.

When the speed of an engine is held constant for varying loads by missing explosions on the light loads and increasing the number for heavy loads, the governing system is said to be of the "hit and miss type." The mixture remains constant in quantity and quality in this type of engine. A hit and miss governor allows only enough charges to be fired to keep the speed constant.

When the load falls off, with a natural tendency on the part of the engine to increase its speed, the governor cuts out the next explosion by holding the exhaust valve open and the inlet closed, thus preventing fresh mixture from being drawn into the cylinder. With an increase in load, the governor allows the valves to follow their regular cycle with the result that a greater or less number are fired in succession. Hit and miss governing is very economical for only full charges of the most perfect mixture are fired, and with short exhaust pipes the scavenging is much better than with other forms of governing. The principal difficulty with this system is that the regulation is not as perfect as with some other types.

(128) The Throttling System.

Unlike the hit and miss system of governing, the throttling type of governor allows the engine to take an explosion on every working stroke, the speed being held constant by either regulating the quality or quantity of the mixture, or both. Throttle governor permits of close speed regulation as the impulses are more frequent and not so violent as with the hit and miss system.

The governor acts directly on the throttle valve, and at no time is the operating mechanism disengaged from the driving cam. The throttle governor engine is particularly well adapted for driving dynamos, supply electric light, as the uniform speed gives a smooth, steady light without the objectionable flickering so likely with the hit and miss engine. To obtain the best fuel economy with a throttling engine, it should be run close to its rated capacity, as the poor and imperfect mixture admitted at light loads considerably increases the fuel consumption.

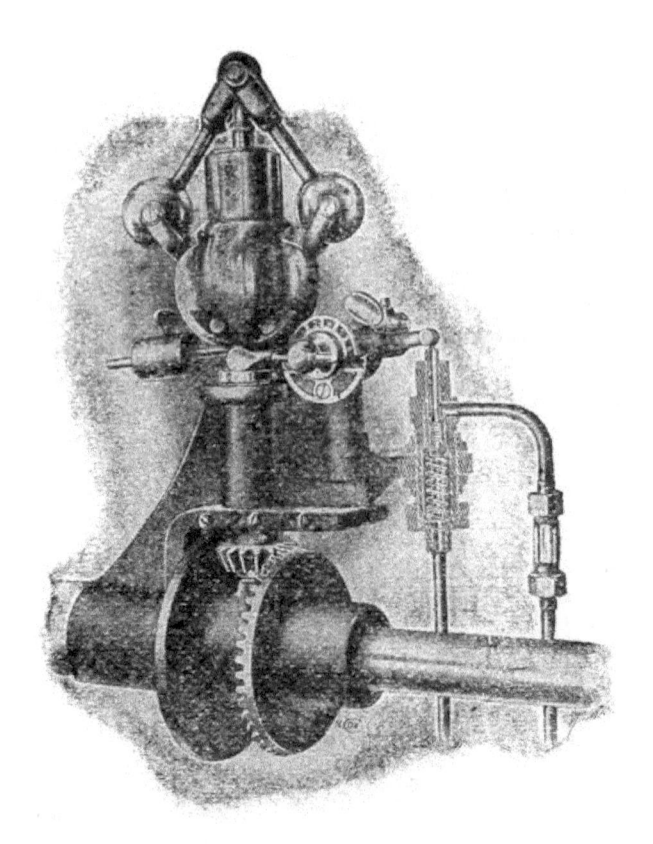

Fig. 76-d. De La Vergne Governor.

Practically all motors of the variable speed type such as are used on automobiles and motor boats are controlled manually by the throttle; although marine motors are often fitted with governors to prevent racing when the screw is lifted out of the water in a heavy sea.

(129) The Controlling Governor.

The governor proper depends upon centrifugal force for its action, and generally consists of two weights which are pivoted at one end to a rotating shaft driven by the engine. When these weights are rotated rapidly the bottoms are thrown outwardly by the centrifugal force and tend to assume a horizontal position. The faster the weights are rotated, the greater will be the tendency for the bottoms of the weights to come into the horizontal, and the greater will be the pressure exerted by them on the controlling levers connected to the throttle. It is evident that the centrifugal pull on the weights varies directly with the speed of rotation and consequently with the speed of the engine. The exact relation between the travel of the weights and the speed of the engine is controlled by a spring that acts between arms cast on the weights and the spindle. If a heavy spring is used, greater speed must be attained to move the weights a given distance than with a weak spring, as the centrifugal force must be greater.

Fig. 124-d. Governor and Governor Mechanism of Fairbanks-Morse Type "R E" Engine. The Fly-Balls, Springs, and Control Rods Are Shown on the Governor Staff. The Upper End of the Bell Crank Goes to the Throttle.

The throttle valve of the engine is connected by a rod to the governor through a sliding collar in such a way that the movement of the governor weights due to an **INCREASE** of speed partially closes the valve until the speed of the engine is reduced. Should the speed of the engines **DECREASE**, owing to a heavy load coming on, the spring will force the balls to occupy a lower position which will increase the valve opening until the engine again reaches the normal speed for which the tension of the spring is adjusted.

Thus the speed of the engine is kept practically constant by the action of the governor in opening and closing the throttle, which in turn, varies the **QUANTITY** of mixture admitted to the cylinder. The **QUALITY** of the mixture is varied by hand, in the engine by means of cocks in both the air and gas pipes. The **GOVERNOR PROPER** is of practically the same construction in the hit and miss engine, the difference of the two types lying in the method of connecting it to the controlling system. In one case (hit and miss) the governor controls the exhaust valve, and in the other (throttling) it controls the quantity of gas admitted by the throttle valve. The speed of the engine may be varied within certain limits by a lever connected to the valve controlling rod.

(130) Types of Governors.

The types of governors used on the leading makes of engines will be found described and illustrated in Chapter V which treats of each engine in detail.

(131) Governor Troubles.

Hit and miss governor troubles may be due to the following defects:

BINDING GOVERNOR COLLAR, stuck with dirt or gummy oil, will cause the engine to die under load, and overspeed on light load.

INLET VALVE LOCK may be worn in such a manner as to prevent the valve from seating during the idle strokes and lose fuel, or cause overspeeding.

DETENT LEVER KNIFE EDGE may be worn, or rounded off, so that the exhaust valve is not held open for the idle stroke. This defect will cause overspeeding.

SPEED CHANGING LEVER may work loose and cause the speed to vary erratically.

GOVERNOR WEIGHTS may be stuck on pins with dirt or gummy oil causing engine to overspeed.

LOST MOTION IN GOVERNOR GEAR such as loose pins and bushings, worn rollers, or bearing surfaces will cause the speed to vary continuously. **LOST MOTION** on portable engines will cause the engine to run normally in one position, and overspeed in another.

WEAK OR BROKEN SPRINGS ON GOVERNOR will cause engine to lose speed or die down altogether. Springs may be stiffened by pulling out the coils.

DRY GOVERNOR BEARINGS or joints will cause binding and cause governor to act sluggishly. Use plenty of lubricant.

WORN ROLLERS may cause a speed variation. Keep the governor well oiled, clean, and free from gum.

If the knife edges are allowed to slip over one another, much wear is caused on the cams and if allowed to continue, sooner or later the engine will run away. Springs will weaken with age and hard usage. With belt driven governors see that the belt is tight and that the lacing is in good condition for a slack belt may allow the engine to overspeed.

I advise that every purchaser of an agricultural motor read his instruction book with care, that is, locate all oil holes and note the action and purpose of every part. If in doubt as to any part of its use write the manufacturer of the motor.

(132) Throttling Governor Troubles.

STICKING GOVERNOR VALVE will cause the engine to overspeed; remove the gum and dirt.

LOOSE PINS OR BUSHINGS, or lost motion in any part of the governor mechanism will cause irregular motion or running; be sure that the bearings and joints are well oiled.

STUCK PINS will cause the engine to overspeed on light loads, and fall down on the normal load, or cause racing.

WEAK OR BROKEN SPRINGS will cause the engine to lose speed or to lie down altogether even on light loads.

STIFF GOVERNOR SPRINGS cause the engine to speed up.

SLIDING COLLAR stuck will cause racing or a fluctuation in the speed. Keep the governor well oiled, clean, and free from gum.

The governing valve should be removed from its care frequently and thoroughly cleaned with kerosene. Deposits of carbon and gummed oil at this point are dangerous because of the likelihood of their causing overspeeding.

(133) Valve Gear Arrangement.

The valve operating mechanism lay-out depends upon the cylinder and valve arrangement, and consequently varies in detail with different engines.

Overhead Valve Arrangement of the Fairbanks-Morse "R E" Engine.

Fig. F-14-15 in Chapter V, shows the valve gear of an upright engine having the inlet and the exhaust valves located in pockets placed at one side of the cylinder. The inlet valve is operated by a valve rod that is actuated by the cam. The exhaust valve stem is raised and lowered, directly, through a cam on the same shaft. The method of driving the valves in this engine is practically standard for all vertical engines having the valves located in pockets. This system is used in a greater proportion of automobile engines.

The opposed engine has the cylinders arranged on opposite side of the crank case, and makes an exceedingly well balanced and quiet running engine; as there is no point in the revolution where either the crank throws or connecting rods have an unequal angularity, or differ in velocity.

While this type of two cylinder engine is common in automobile practice, it is not often met with in stationary work, the cam-box and the cam being directly in the center of the crank case.

The opposed type of engine is particularly well adapted for aeroplane service as a steady, quiet running engine is an absolute necessity because of the frail construction of the aeroplane frame.

(134) Cam Shaft Speeds.

The valves of the gas engine are opened and closed by means of cams or eccentrics, that are geared to the crankshaft, and which also control the timing.

As a four stroke cycle engine performs all of the events, or a complete cycle in two revolutions of the crankshaft, it is evident that the cam must go through the routine in one revolution or must revolve at **ONE-HALF OF THE CRANKSHAFT SPEED**.

Therefore the cam gear ratio must be as one is to two, the smaller gear being placed on the crankshaft, the gears being known as the "half time gears."

As a two stroke cycle engine goes through the routine of events in every revolution, the cam-shaft must run at crankshaft speed so that the cam outline makes one revolution in the same time as the crank. The cam shaft speeds given here apply to all engines of the corresponding cycle no matter whether the valves are of the poppet, rotary or slide-sleeve types.

(135) Valve Gear Troubles.

The valve gear mechanism causes trouble principally through the wear of the various parts which results in a change in the valve timing, or in the lift of the valves. Loss of power, **MISFIRING**, and overheating are the result of such derangements.

Often trouble is caused in reassembling the valve mechanism after the engine has been torn down for repairs, which trouble may generally be traced to incorrect gear meshing.

The following list will give the principal defects due to the wear of the valve mechanism.

(a) **WORN CAM GEARS** change timing because of play, or "back lash" in the teeth, or cause a howling or grinding noise, that will cause the owner to believe that the end of the engine is near. **MISFIRING** and **LOSS** of power are probable results of a change in the timing. If any of the teeth are stripped from the gear you may be sure that the timing is changed. Replacement with a new gear is the only cure for a worn or broken gear.

(b) **GEARS NOT IN PROPER MESH** due to an error in assembling the gears, will prevent the engine from being started, or cause misfiring and loss of power.

The maker of the engine generally marks the teeth that go together, but if no such marks appear, the owner should center punch or scratch them before taking down the engine.

(c) **A GEAR SLIPPING ON THE SHAFT**, due to a missing key in the gear, or to a loose set-screw will cause all of the troubles due to a change in the timing. Examine the key carefully, for dirt often collects in the key-way to such an extent that it is liable to be mistaken for the key. Keys and pins have sheared in two, allowing the shaft to slip in the gear.

(d) **WORN CAM-SHAFT BEARINGS** are the cause of trouble, as they will change both the timing and the lift of the valves. If much play exists in the bearing, it will prevent the valves from lifting at the proper time, and will also reduce the lift by the amount of the play, which sometimes has a considerable effect on the free passage of the gases. If the cam-shaft bearings are of the bushing type they should be replaced with new paying attention at the same time to the condition of the shaft. If rough or shouldered the shaft should be machined to a dead smooth surface. If on a large engine and of the adjustable type, the shims should be removed as required or the wedges adjusted.

(e) **LOOSE CAMS OR ECCENTRICS** will change the timing because of lost or sheared keys. If your cams are not integral with the shaft, look them over occasionally and be sure that the keys are tight. Loose cams will produce thumping and grinding and may often be located by the sound. See that the key-way is not worn when fitting keys.

If the cams are fitted with taper pins it would be well to ream the hole before placing new pins, as there is a liability of the hole being worn oval.

(f) **A TWISTED OR SPRUNG CAM-SHAFT** will change the positions of the cams relative to one another, and not only will change the time of all cylinders, but will change their time relatively causing the engine to run out of balance, or produce an unusual vibration.

(g) **WORN CAMS** are causes of a change of timing on all types of engines, and are the most frequent cause of reduced valve lift with its consequent trouble of overheating.

If the outline or contour of a cam is changed with wear it should be replaced, if keyed to the shaft, as it will be a constant source of trouble. If the cams and cam-shaft are in one integral piece, it will be necessary to replace the entire shaft.

(h) **WORN CAM ROLLERS AND ROLLER PINS** will reduce the lift of the valves, and in the case of a broken or sheared pin will prevent the valve from lifting at all. Always replace loose pins or loose rattling roller.

(i) **PUSH ROD DEFECTS.** Too much clearance between the push rod and valve stem will reduce the lift of the valves and change the timing. The clearance for small engines should be equal to the thickness of a visiting card, and for large engines is somewhat larger, say 1–16". The increase of clearance is due principally to wear.

Too small a clearance should be avoided for the reason that the valve stems expand with the heat and will lift the valves too soon, or even permanently until readjusted. Broken valve springs will cause trouble, or lost keys that retain the valve spring washers. Loose adjusting screws on the push rods or stripped threads will delay the valve opening.

(j) **TAPPET LEVER DEFECTS** are generally caused by wear or poor adjustment. Loose pins or bushings, too much clearance between the tappet and valve stem or broken valve springs, or loose adjusting screws will produce changes in the timing or valve lift.

(k) **BENT VALVE ROD.** A bent valve rod will shorten the travel of the valves, and change the timing.

(l) **CAM LEVER OR PIN** will cause timing troubles if the pin or bushing are loose or worn, by reducing the travel of the valves.

When occasion arises for the removal of valves, the opportunity should be taken to clean the stems and guides, which may be more or less gummed with ancient oil. Freedom of valve movement is of extreme importance, and for this reason neither the cleaning nor the lubrication of the stems and guides should be neglected. The occasional use of a little kerosene will prevent gummy accumulations, but care should be taken not to allow the kerosene to wash out all of the oil and thereby leave the surfaces dry.

A broken valve spring, though not a common occurrence, is not an unknown possibility. If no spare spring is at hand, a plan that can be recommended is to turn the broken spring end for end, thus bringing the finished ends up together; this will prevent the spring from shortening by overlapping, and winding itself together.

(136) Valve Timing.

The exact time at which the valves of a four stroke cycle engine open and close depends to a great extent upon the speed of the engine, the fuel used, the compression pressure, and the relation of the bore to the stroke.

As these items vary in nearly every make of engine there has appeared in the technical press, a great mass of seemingly conflicting data. Engine speed is the principal factor in determining the timing.

Correct valve timing plays a considerable part in the output and efficiency of an engine, for if the inlet valve, for example, opens too late, the cylinder will not receive a full charge. If it opens too early the hot gases in the cylinder will ignite the gas in the carburetor and cause back-firing. Should the exhaust open too late, the retention of the hot gas in the cylinder is likely to cause overheating.

The timing of the valves is usually expressed in degrees of the circle described by the crank-pin, or the angle formed by the crank with the center line of the cylinder at the time the valve is to open or close.

(137) Valve Setting on Stationary Engines.

The exhaust should open when the crank lacks 30° of completing the outer end of the power stroke, that is, the crank should make an angle of 30° with the center line of the cylinder when the exhaust valve begins to open, and should be inclined **AWAY** from the cylinder. Some makers have the exhaust open a little later in the stroke, but little is to be gained with a later opening as the retention of the charge beyond 30° heats the cylinder and does very little towards developing power. The only advantage of the late opening is that the valve opens against a lower pressure and causes slightly less wear on the parts.

The exhaust valve should close 5° **AFTER** the crank has passed the **INNER** dead center on the exhaust or scavenging stroke, although some makers close the valve exactly on the dead center. The 5° should be given to allow the gas all possible chance of escape. The piston is said to be on the inner dead center when it is in the cylinder as far as it will go, and on the outer dead center when it is on the center nearest the crankshaft.

The **INTAKE** valve should open about 5° **AFTER** the exhaust valve closes, or 10° after the crank passes the inner dead center. The inlet valve should **NEVER** open before the exhaust valve closes on a low speed engine. The above timing is for engines running 150–600 R.P.M. The automatic type of inlet valve, of course, cannot be timed, but attention should be paid to the strength and tension of the spring and the condition of the valve stem guides.

The inlet valve should close 10° **AFTER** the crank passes the outer dead center in order that the cylinder be filled to the fullest possible extent. If the valve closed exactly on the dead center a partial vacuum will exist and the charge retained in the cylinder will be comparatively small, but if the valve remains open past this point the air would have time to completely fill the cylinder and develop the capacity of the engine. The longer the inlet pipe, the longer the inlet valve opening.

(138) High Speed Engine Valve Timing.

The faster a motor turns, all other things being equal, the greater the amount of advance necessary with the valves, as the higher the speed the less the time required to fill or empty the cylinder. In a short stroke high speed motor the exhaust should close and the intake open as early as possible in order to admit the full charge. The exhaust should open early to allow of the full escape of the gases, as the time allowed for expulsion is extremely short when an engine runs 1,000 R.P.M. and the back pressure is liable to be considerable.

The inlet valve of high speed engines should remain open for a considerable period after the crank passes the outer dead center on the suction stroke, owing to the inertia of the gases which tends to fill the cylinder. Lengthening the period of opening of the inlet valve in multiple cylinder engines produces better carbureting conditions and reduces the variations of pressure in the manifold.

EXHAUST VALVES. The exhaust valve should begin to open 40° **BEFORE** the crank reaches the **OUTER** dead center on the working stroke, and should close 10° **AFTER** the crank has passed the inner dead center.

INLET VALVES. The inlet valve should open 15° **AFTER** the crank passes the inner dead center on the suction stroke, and should close 35° after the crank passes the outer dead center.

The inlet valve should never open before the exhaust valve closes, although this is done on several types of high speed aeronautical engines. The makers of these engines claim that this practice scavenges the combustion chamber more thoroughly and makes the mixture more effective owing to the inertia of the burnt gases forming a partial vacuum in the combustion chamber. The writer has never been able to get satisfactory results with this timing and doubts whether it can be accomplished successfully.

In timing an engine great care should be taken to get the crank exactly on the dead center.

(139) Timing Offset Cylinders.

The only difference in timing engines with offset cylinders and timing those with the center line of the cylinder in direct line with the crank shaft, is in the locating of the dead center. With no offset, the center of the cylinder, the crank pin and the crank shaft are all in one direct line when the engine is on the dead center.

With offset cylinders the crank pin lies to one side of the cylinder center line when on the dead center, on either the inner, or the outer center. To find the center on an offset engine proceed as follows:

Turn the engine over slowly until the crank-pin reaches either the extreme top or bottom point of the crank circle, depending on which center is to be determined, and then turn very slowly until the centers of the piston-pin, crank-pin, and crank-shaft are in line. With the average engine this will be found a difficult and tedious job, and it will be well to mark the dead center on the flywheel or other convenient point to prevent a repetition of the job. The quickest method of accomplishing the feat is to remove the spark plug or relief cock to gain access to the piston, and insert a rod or pointer in the opening thus provided.

Draw the piston back a short distance from the end of the stroke with the pointer resting on the head of the piston, and mark this position of the piston both on the pointer, and on the flywheel, using some stationary part of the engine as a reference point.

Now turn the crank over the center line until the piston is moving in the opposite direction, and is the same distance from the end of the stroke as shown by the mark on the pointer. Mark this position on the flywheel using the same reference mark as before. We now have two marks on the flywheel, and will bisect the distance between them, using the dividing mark to obtain the center.

Place the bisection mark even with the reference point used for obtaining the two previous marks on the flywheel, and the engine will be on the true dead center, as the flywheel is now midway between two points of equal stroke.

(140) Auxiliary Exhaust Ports.

To decrease the amount of hot gas and flame passing over the exhaust valve some makers provide their engines with auxiliary exhaust ports, which are similar to the exhaust ports used on two stroke cycle engines.

The auxiliary exhaust consists of a series of holes drilled or cored through a rib on the cylinder wall, the holes being so situated that they are covered by the piston until it is at the extreme end of its outward stroke. The holes are not uncovered until the burning charge has been expanded and cooled to the greatest extent possible in the cylinder. As soon as the piston uncovers the ports the greater portion of the dead gas escapes instantly to the atmosphere, carrying with them the greater percentage of the heat and flame. The small amount of residual gas that remains is forced out through the exhaust valve in the usual manner, thus no flame ever reaches the exhaust valve.

The use of auxiliary exhaust ports produces a cooler cylinder as the gas passes over the cylinder wall only once, and consequently is in contact with the walls only one-half of the time usual with the ordinary system. The cool cylinder lessens the liability of **PREIGNITION** and decreases the consumption of cooling water and lubricating oil. Auxiliary exhaust ports are particularly desirable on air cooled engines.

(141) Valves and Compression Leaks—Misfiring.

Owing to the intense heat in the cylinder, and the action of the gases on the valves the seating surfaces become **ROUGH** and **PITTED** which causes leakage and loss of compression. Exhaust valves cause the most trouble in this respect as they are surrounded by the hot gases during the exhaust stroke and are much hotter than the inlet valves.

To determine the value of the compression, turn the engine over slowly by hand.

Leaking inlet valves usually are productive of **BACK FIRING** or **EXPLOSIONS IN THE CARBURETOR** intake passages, or in the mixing valves, as flame from the cylinder leaks through the valve and fires the fresh gas in the intake.

MISFIRING OR LOUD EXPLOSIONS at the end of the **EXHAUST PIPE** are indicative of leaky exhaust valves, if the mixture is correct and the ignition system above suspicion. Misfiring caused by leaky exhaust valves is due to combustible mixture escaping from the cylinder to the exhaust pipe and being ignited by the succeeding exhaust of the engine.

If the engine has more than one cylinder, test one cylinder at a time, opening the relief valves on the other cylinders. Now take a wrench and **ROTATE** the inlet valve on its seat, for it may be that some particles of carbon or dirt have been deposited on surface of the valve seat which prevents the valve from closing properly. Rotating the valve will usually dislodge the deposit.

Try the compression again; if there is no improvement, rotate the exhaust valve on its seat in the same manner, and repeat the test for compression. **ROTATING THE VALVES IN THIS MANNER WILL OFTEN MAKE THE REMOVAL OF THE VALVES UNNECESSARY.** When the valves are closed the end of the valve stem should **NOT** be in contact with the **PUSH ROD**, or cam lever. Suitable **CLEARANCE** should be allowed between the end of the valve stem and the operating mechanism when the valve is closed; this clearance varies from the thickness of a visiting card on small engines to $1/8$ of an inch on the large. If the valve stem is continually in contact with the push rod it cannot seat properly and consequently will leak. Wear on the valve seats and regrinding reduces this clearance, wear on the ends of valve stems and push rods from continuous thumping increases it. Keep the clearance constant and equal to that when the engine was new. On many engines this clearance is adjustable to allow for wear by lock nuts on the ends of the valve stems or push rods.

If the above attempts have proved unsuccessful remove the exhaust valve from the cylinder, if the valve is in a cage, remove the entire cage; this may easily be done on most types of engines. Always remove the exhaust valve first as the inlet valve rarely requires attention. With small engines, and engines having the valves mounted directly in the cylinder head it will be necessary to remove the cylinder head to gain access to the valves. In such a case use care when opening the packed joint between the cylinder and head, to avoid damaging the gasket.

The exhaust valves should be lubricated with Gas Engine Cylinder Oil, never with common machine oil on account of gumming and sticking, or with gas engine cylinder oil thickened with **FLAKE GRAPHITE**. Powdered graphite may be used with success without the addition of oil, but oil makes the application of the graphite much easier.

A cracked valve seat, due to expansion strains or to the hammering of the valve, is a common cause of compression leakage, and is rather difficult to locate as the leakage only occurs under comparatively high pressure. Leakage may also occur between the valve cage and the cylinder casting unless pains are taken to thoroughly clean the cage and the bore before fastening into place.

Warped valves are caused by overheating, the head of pallet of the valve becoming out of square with the stem, or by twisting on the valve seat. If warped valves are suspected the high point of the seat may be determined by means of the following test and should be carefully filed down until it is close to a bearing after which it may be ground down as described under pitted valves.

If the stems are now in good condition examine the seating surfaces of the valve pallets and cage or rings.

The seats should be bright and free from pits, depressions, or streaky blue discolorations. If the seats are deeply grooved from long continued leaks it is best to discard them and replace with new.

Pitted valves, and those slightly grooved or streaked should be reground by the use of a little emery flour and tripoli which operation is performed as follows:

Lift the valve from its seat and apply lubricating oil to the seating surface, then sprinkle a little flour or emery on the oiled surface and drop the valve back on the seat. Do not use coarse emery nor too much of the abrasive, a pinch is enough and will grind as rapidly as a pound. Take care to drop the emery only where required, do not sprinkle it over the engine or working parts as it will cause cutting and the destruction of the bearings.

Now turn the valve around in one direction for about a half dozen turns and then in the other direction for the same length of time, alternately, at the same time applying a moderate pressure on the valve. Small valves may be rotated with a large screw driver entered in the slot found on the valve plate, but the handiest method is with a carpenter's brace in which is inserted a screw-driver bit.

Never turn the valve around and around in one direction continuously as this movement is liable to cause grooving, alternate the direction of rotation frequently with occasional back and forth movements made in a semi-circle.

Do not press heavily on the valve, use only enough pressure to insure contact between the two seating surfaces.

The valve should be lifted occasionally from the seat to prevent grooving, and to redistribute the abrasive, and then dropped back, after which the grinding should proceed as before. Remove the valve after it turns without friction, wipe it clean, apply fresh oil and emery and grind once more. When the grinding has removed all pits and ridges, and presents a smooth even surface, the grinding is complete. To test for accuracy of grinding place a little Prussian Blue on the seat, if the valve is ground to a perfect surface the blue will show uniformly spread over the seat, if the grinding is incomplete bare places showing high spots will be seen. It is a good plan to finish the grinding by using a little Tripoli with oil after the emery has removed the pits and high spots, as Tripoli is finer than emery and will smooth down scratches made by the emery.

After the grinding has been performed to your satisfaction, wash the valve, valve stem, and guides thoroughly with gasoline and kerosene to remove the smaller traces of emery, to prevent wear and cutting.

When the valves are ground in place on the engine stuff up all openings or parts of the cylinder to prevent the emery from gaining access to the bore. After grinding is complete wipe off surfaces thoroughly and remove waste used for stuffing.

CHAPTER XIII
TRACTORS AND FARM POWER

Because of our increased population, which results in a greater planted acreage, and the scarcity and increased cost of farm labor, farming has rapidly developed into an industrial science. Where formerly the farmer was content to perform certain parts of his work by hand, he today employs machinery for the same task, and is far more particular as to the working of his soil and the cost of production per acre. By the use of machinery his crop is marketed at less expense, in a shorter time, and he has more time in which to enjoy life than ever before.

The modern gasoline and oil engine has been the greatest factor contributing to the farmer's ease and prosperity for it has eliminated the terrors and drudgery of plowing, churning, watering stock, sawing wood, threshing, and has besides given him many of the conveniences of city life, such as running water and electric light. The benefits of power are not only conferred on the farmer but his wife as well for the small domestic engines have saved the back of the house wife during the strenuous period of harvest time.

One of the difficulties of farming is the necessity of doing certain work in a limited time or else suffering a heavy loss. The breaking, the plowing, the harvesting, and the threshing each must be done at a certain time, often within a few days of each other in order to obtain the benefits of the best weather conditions. Threshing starts as soon as the grain is ready, and if rain interferes with the threshing, the farmer can start plowing immediately if provided with a tractor and thereby gain the undoubted benefits of fall plowing. Plowing at harvest time has much to do with eliminating weed seeds for the weeds are turned under while green, the seeds sprout and commence their growth and are winter killed before they reach maturity. In this way the field is practically freed from weeds in the spring. When the weather again becomes suitable, the threshing may be resumed and when completed he can again turn to his plowing.

Operator's View of the "Big Four" Tractor, Showing the Four Cylinder Engine in Place.

Gas power is not to be considered merely as a substitute for animal power for the engine not only performs the work of the horses but also performs work that no horse can do, and does it with far less expense. In the hottest weather when horses are dropping in the broiling sun, the tractor moves tirelessly through the fields. Every farmer knows the expense attached to keeping a horse in the idle winter period for it must be fed, watered, and cared for, work or no work. When the engine is idle it costs nothing except for the interest on the investment, while animals grow old and are subject to disease whether they work or not.

The time of plowing and harvest is short and requires quick work, and continuous work. Horses cannot be driven at plow faster than one mile per hour, and cannot be worked more than 10 hours per day, while the tractor under suitable conditions can travel 2 to 3 miles per hour, and keep at it twenty-four hours per day. An ordinary tractor can break from 20 to 40 acres of ordinary loam per day and will plow in cultivated land from 40 to 50 acres per day.

The same factors govern the fuel consumption of a tractor that govern the rate of plowing, that is, the character of the soil and the depth of plowing. On an average, 1½ to 2½ gallons of gasoline will be used in breaking an acre of sod, and 1 to 1½ gallons of gasoline in plowing stubble. As kerosene contains about 18 per cent more heat per gallon than gasoline, the quantity of fuel used by an oil tractor is correspondingly less. When used for pulling wagons on the road at about 3 miles per hour the fuel consumption will approximate 4 gallons per hour, this consumption varying of course with grades, etc.

Thirty horse-power, at the speed given above represents a draw bar pull of about 9,000 pounds, which is equivalent to the tractive effort of from 30 to 40 horses, were it possible to concentrate the pull of so many horses at a single point, at one time. It would of course be impossible for the horses to maintain this effort for as long a time as the tractor. On a level road it will take about 100 pounds tractive effort for each 2,000 pounds of weight in the form of road wagons (including the weight of the wagon). The number of wagons that can be drawn with a given draw bar pull can be easily figured. When pulling on a grade, the effective draw bar pull will be reduced in proportion to the extent of the grade. While no fixed rule can be given regarding the number of plows that can be handled by a tractor, the average machine can pull six to eight breaking plows and from eight to twelve stubble plows, depending on the character of the soil and the depth of plowing. When the conditions permit the use of a greater number of plows, than specified above the amount of work done will of course be greater.

A tractor can haul four ten foot seeders and two twenty foot harrows and cover 7 or 8 acres per hour at a cost of from 12 to 15 cents per acre. At harvest time the tractor will also effect a great saving in time and expense for the average machine will handle five or six eight foot binders, making a cut of nearly 50 feet wide, and this can be kept up for 24 hours at a stretch.

A tractor of the average output can handle any separator, and with a 44" cylinder machine can turn out from 2,000 to 3,000 bushels of wheat and 5,000 bushels of oats in a ten hour run. It will also handle any of the largest shredders. For irrigation work, silo filling, and wood cutting it is equally efficient.

(142) The Gas Tractor.

The tractor of the internal combustion type using gasoline or oil as a fuel is much more successful than the steam machine, both from the standpoints of convenience and cost of operation. There is absolutely no danger of fire whatever around a gas tractor for this reason the engine can be placed in any position regardless of the direction of the wind, which would be impracticable with a steam engine. This is a great advantage for if the wind is allowed to blow directly from the engine to the separator, it will be of great assistance to the pitchers who feed the separator.

When threshing or plowing in a remote field considerable difficulty is always experienced in supplying the steam tractor with the enormous amount of water that it consumes. To supply the water requires a team, tank wagon and drivers which is a considerable item in the running expense. The small amount of water used for cooling the gas engine is renewed once, or at the most, twice a day. Steam coal is bulky and requires the continuous service of a man and team to keep things moving, and this expense is greatly increased by the expense of the coal.

A gas tractor can be started in a very few moments while the engineer of a steam rig has to start in an hour or more before the crew to get steam up, etc. In addition to this there is the usual tedious routine of "oiling up," cleaning the flues, etc. There is absolutely no danger of explosions with the gas engine which have proved so disastrous in the past with steam threshing engines.

With the gasoline, the operator is left free to work on the separator as he has no firing to do and does not have to concentrate his attention on keeping the water level at the correct point in the gauge glass. The engine is automatically lubricated in all cases so no attention is demanded on this score for it will run smoothly hour after hour without the least attention. This feature eliminates one high priced man from the job. On heavy loads the problem of keeping up the steam pressure is often a vexatious one, especially if a poor grade of coal is used. With a lower priced man as operator tending both the separator and the gas engine the crew need only consist of two pitchers to feed the machine, with a man and team for each pitcher. This small crew is easily accommodated at the farmers house, and does not require the services of a separate cook and camp equipment.

With a gasoline rig the expenses will be approximately as follows:

Engineer, wages and expenses	$ 5.00
Two pitchers, at $3.00	6.00

Four men and teams		20.00
60 gallons of gasoline at 15c		9.00
Lubricating oil		1.00
Cost per day		$41.00

Taking 1,500 bushels (wheat) as a day's work, the cost of threshing figures out at 2¾ cents per bushel.

According to data furnished by the M. Rumely Company, which is based on an actual test, the total cost of plowing, seeding, cutting and threshing, including ground rental and depreciation, amounted to $8.65 with horses and $6.55 with their oil tractor. These figures will of course vary in individual cases, but are principally of interest in showing the comparative cost of horse and tractor operation.

With a gasoline or oil tractor equipped with engine plows one man can tend to both the plows and the engine, although some operators prefer to have two men, one relieving the other consequently plowing more acres per day and reducing the cost per acre. In some cases one man is placed on the plows and the other on the engine. By running the tractor twenty-four hours per day, with two shifts of men, a much better showing is made by the tractor when compared with horse plowing, for with the latter method it would be necessary to supply twice the number of horses.

To show the relative merits of various grades of fuel we will print the data kindly furnished by Fairbanks Morse for a ten hour day.

ITEM	Fuel Oil 3c	Kerosene 6½c	Gasoline 15c
60 Gallons Fuel	$1.80	$3.90	$ 9.00
Lubricants	.40	.40	.40
Engineer	3.50	3.50	3.50
Plowman	2.00	2.00	2.00
Repairs	.12	.12	.12
Cost to Plow 24 Acres	7.82	9.92	15.02
Cost per Acre	.32	.41	.63

Plowing at the rate of 20 acres per day, and kerosene at 6⅔ cents per gallon, the Rumely Company obtain the cost of plowing one acre as $0.66. In the latter figure the interest and depreciation are included which will increase the figures over those given by Fairbanks Morse. It should be understood that these costs are approximate and will vary considerably in different localities and under various conditions.

Oil Injection Engines.

Engines using low grade fuels such as kerosene, usually experience much trouble in obtaining a proper mixture when the fuel is vaporized in an external carbureter even when the carbureter is specially designed for the heavy oil. This leads to fuel waste, starting troubles and cylinder carbonization, to say nothing of the objections of an odorous, dirty exhaust. To overcome the objections of carbureting the heavy oils it has been common practice to inject or aspirate a small amount of water, the water vapor tending to prevent the fuel from cracking and to distribute the temperature more uniformly through the stroke. The injection of water is not a particularly desirable feature, since its use involves one more adjustment and possible source of trouble when running on variable loads.

In the semi-Diesel engine the fuel is sprayed directly into the combustion chamber by mechanical means, thus making the fuel supply to a certain extent independent of atmospheric and temperature conditions. After the injection the spray is vaporized both by the hot walls of the combustion chamber and the heat of compression, the latter being principally instrumental in causing the ignition of the gas. In this case no electrical ignition devices are required, thus at one stroke overcoming one of the principal objections to a gas engine.

Until recently the semi-Diesel engines were confined to units of rather large size, the smallest being much larger than the engines usually used on the farm. It is now possible, however, to obtain oil engines of the fuel injection type in very small sizes, built especially for portable or semi-portable service. Not only is it possible to use a cheaper grade of fuel with this type of engine, but the fuel consumption is also less than with the carbureting type. To this may be added the advantages of an engine free from the troubles incident to the ignition and carbureting systems.

Good results may be obtained with small injection engines on oils running from kerosene (48 gravity) down to 28 gravity, the combustion in all cases being complete and without excessive carbon deposits. Little trouble is caused by variable loads as long as the speed is kept constant. Compared with gasoline, the heavier fuels are much safer to store and handle, owing to their high flash points.

The compression of the injection engine is much higher than the old carbureting kerosene engine as the compression heat is used in a great part to ignite the oil vapor. Usually the pressure is in excess of 150 pounds per square inch, the exact value being determined by the form of the combustion chamber, whether a hot bulb is used, etc. The high compression assists in increasing the economy of the engine.

Usually the piston either draws in a complete volume of pure air or draws in pure air through the greater part of the induction stroke, the spray either starting near the end of the suction stroke or during the early part of the compression. When a hot bulb is used the oil spray strikes the bulb forming vapor, the increasing compression caused by the advancing piston furnishing the air for combustion and forces the mixture into contact with the hot walls. Another type has no hot bulb, the lighter constituents of the fuel being vaporized and ignited by the compression alone, their inflammation serving to kindle the main, heavy body of the oil. In some engines, the combustion of the light constituents serves to spray the heavy oil through the valve and into the combustion chamber. Details of several of the most prominent makes of oil engines are described in an early chapter of this book.

As a rule, this class of oil engine does not run well when the speed is varied through any great range, nor when governed by a throttling type governor, since both of these conditions affect the compression. They may be either of the two or four stroke cycle type, and when of the latter they are much more successful than a two stroke cycle engine using a carbureter.

On small engines the fuel consumption will run about 0.7 pint per brake horsepower hour, this consumption decreasing on large engines to about 0.6 pint per brake horsepower hour or even less.

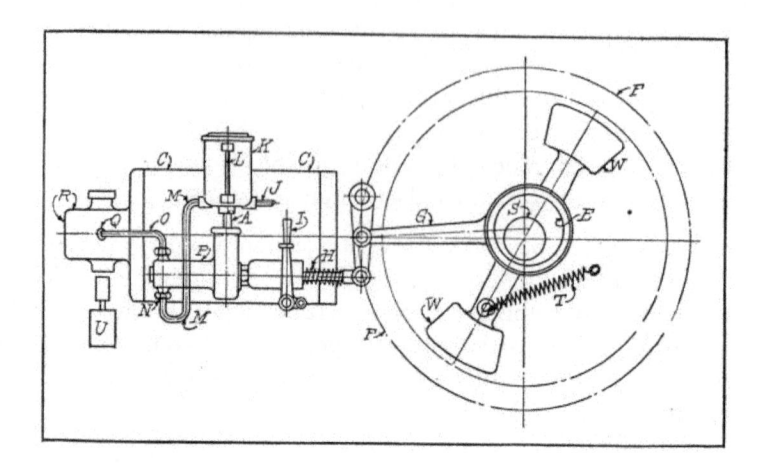

Oil Injection Type. Injection pump P driven by eccentric E through rods G-H draws oil from tank K through M-N and sprays it into combustion chamber R through O-Q. Amount of oil sprayed is controlled by fly-wheel governor W-W shifting E on shaft S, thus varying stroke of P. Engine is started by heating R with torch U and injecting first oil with hand lever I. A second pump supplies constant level of oil to K, level being observed in glass L. C-C is the cylinder, and F is the fly-wheel.

The accompanying diagram shows a diagram of a typical oil engine of the injection type, a pump P supplying the oil from auxiliary tank to the hot, extended combustion chamber R, this chamber being an extension of the cylinder C-C. Oil is kept at a constant level in K by an overflow pipe, the oil entering from the supply pump through pipe J, and entering the pump through M at N. By gauge glass L, the operator can tell whether he has a sufficient supply of oil.

The injection pump P is driven from the eccentric E (mounted on the main shaft S) through the eccentric rod G and the rod H. The governor weights W-W alter the amount of fuel supplied by changing the stroke of the pump, thus keeping the speed constant under varying loads. The governor acts by shifting E in relation to the shaft S, a spring T controlling the throw of the governor. The entire governor mechanism is contained in the fly-wheel F.

To start, the combustion chamber R is heated by the torch U, and after thoroughly heated, the starting fuel is injected by means of the hand lever I. This engine is of the two cycle type with scavenging air furnished by crank-case compression.

(143) Construction of Gas Tractors.

A gas tractor may be considered as being simply a special application of the gasoline or oil engine in which the engine drives the road wheels through a train of gearing instead of driving its load by a belt from the pulley. Four intermediate mechanisms must be provided between the engine and the road wheels in order that the tractor may properly perform its work. These devices are known as the "clutch," the driving gears, reverse gear and the "differential" gear. It should be understood that these mechanisms do not change the construction or operation of the engine in the slightest, and that the principles that apply to the engines described in the previous chapters apply also to the engine of the tractor. The following will briefly describe the functions of the intermediate trains in their proper order, starting at the engine.

The Clutch.

A tractor is arranged to pull its load in two different ways, first by the draw bar, as when pulling plows, and secondly by a belt from the engine pulley as in driving a threshing machine or circular saw. In the first case it is necessary to drive the road wheels through the gear train, and in the second case it is necessary to disconnect the road wheels while driving the thresher or saw. As the engine cannot be started while under load it is also necessary to disconnect the road wheels to free the engine while turning it over to get the first explosion.

The device that connects and disconnects the engine from the road wheels is known as the **CLUTCH**. This usually consists of two or more friction surfaces that form a part of the driving gear, which may be brought into frictional contact with the engine pulley, when it is necessary to drive the road wheels. When the two members of the clutch are brought into contact they revolve together, thus connecting the engine with the driving gear.

Reverse Gears.

The Reverse Gear of the "Big Four" Tractor.

As it is not practicable to reverse the direction of rotation of the gas engine, the rotation of the road wheels is reversed by means of gears contained in the driving train. In some tractors the reverse gears are similar to those in an automobile, being located in the transmission. In other tractors two bevel pinions are provided that fit loosely on the engine shaft and engage with a large bevel wheel that forms part of the driving gear. A sliding jaw clutch that revolves on the engine shaft is arranged so that it can connect with either of the bevel pinions causing them to rotate with the engine shaft and drive the main wheel. As the two pinions are on opposite sides of the large bevel wheel, they run in opposite directions in regard to it, so that it is possible to reverse the large wheel by engaging the clutch with either one or the other of the bevel pinions.

The Differential Gear.

The differential gear makes it possible to apply the same amount of power to each of the road wheels, and also allows one wheel to rotate faster than the other when turning around a corner. If both road wheels were rigidly fastened to a single rotating axle it would be practically impossible to turn a corner for it would be necessary for the engine to slip one or the other of the wheels because of their difference in speed, as the outer wheels turn faster than the inner.

Differential Gear of the "Big Four" Tractor.

The Driving Gear.

The driving gear consists of a series of spur gears arranged for the purpose of reducing the high speed and small "pull" of the motor into the low speed and heavy pull of the road wheels. This reduction in speed is generally brought about by a double system of shafts, the second shaft from the motor carrying the differential gear and meshes directly with the master gear on the bull wheel. The first shaft is an idler.

Fig. 125. Fairbanks-Morse Oil Tractor, Showing General Layout.

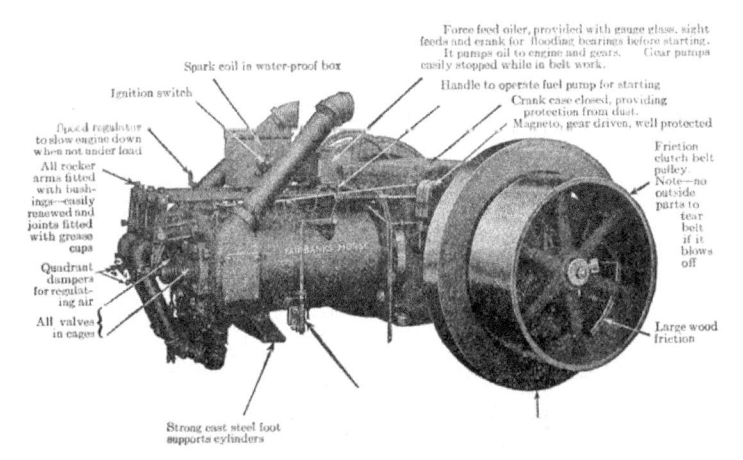

Fig. 126. Two Cylinder Engine of Fairbanks-Morse Oil Tractor.

(144) Fairbanks-Morse Oil Tractor.

The Fairbanks-Morse 30–60 Horse-power Oil Tractor gives an effective draw bar pull of 9,000 pounds and develops over 60 horse-power at the belt pulley which is more than sufficient to drive any farm machinery. It will operate equally well on kerosene, distillate oils, and gasoline, any of which will develop the rated horse-power. Two forward speeds and one reverse are obtained by a gear transmission of the automobile type, the forward speeds being 1¾ and 2½ miles per hour and the reverse 1¾ miles. Combined with the governor variation, it is possible to get the proper speed for any kind of work.

The fuel is sprayed directly into the cylinder with a spray of water, the proportion of water to oil being nearly equal at full load. As explained in Chapter VII, the water spray aids in the combustion of the heavier oils, eliminates soot and tarry deposits, and makes the engine run more smoothly because of the reduction of the explosion pressure. The spray also reduces the temperature of the cylinder and minimizes the dangers of preignition. The engine is of the slow speed type running at a normal speed of 375 revolutions per minute, and the two cylinders have a bore and stroke of 10½ × 12 inches. The speed regulator supplied with the engine gives an extreme variation of 300 to 375 R.P.M.

Fig. 127. Fairbanks-Morse Tractor Transmission with Two Forward Speeds and One Reverse.

The cylinders are cast two in a block which arrangement permits of the bores being brought close together and gives an easy circulation of cooling water. The value of this practice has been proved in automobile work where a simple and rigid structure is absolutely necessary.

All of the valves are in the heads of the cylinder which eliminates heat radiating pockets in the combustion chamber. Both the inlet and exhaust valve are mechanically operated through substantial push rods and valve rockers, and are completely surrounded by water. Large clean out holes are provided in the separately cast cylinder head making it accessible for the removal of scale and sediment. A single cylinder head serves for both cylinders which contributes to easy cooling passages and a single arrangement of exhaust and inlet piping. The valves are in cages bolted to the cylinder head in such a way that they are easily removed for inspection without disturbing the piping or connections.

The pistons are easily removed without taking the heads out of the cylinder or taking down any shafting. The valve rocker arms are provided with easily renewed bushings and grease cups. As the engine is of the four stroke cycle type with both cylinders on the same side of the crank-shaft, only a single throw crank shaft is used, which is without intermediate bearings.

Dual ignition is used, the high tension magneto and the two unit spark coils shown in Fig. 126 being independent of one another so that either the magneto or battery can be used for starting or for continuous operation. The magneto is mounted directly on the engine bed and is gear driven from the crank shaft. The ignition advance and retard lever and ignition switch are mounted on the engine in an accessible position. As the coil is mounted on the engine the leads are short and the vibrators are directly under the supervision of the operator.

Close speed regulation is maintained by a throttling type governor. The voluntary speed variation used to slow the engine down to meet certain conditions encountered in plowing is accomplished by a small lever located at the end of the cylinders. The cooling water is circulated through the cylinders by a gear driven centrifugal pump. From the cylinders the water enters a closed radiator of the automobile type located at the front of the traction where it is cooled without loss. A nine feed, forced type oiler is used which supplies oil to the cylinders and bearings, and also to the transmission gears. External bearings which are subjected to dust are equipped with grease cups. The fuel pump which takes its supply from an 80 gallon tank is in an accessible position near the operator and is provided with a handle by which it is operated when starting the engine.

The clutch which is located in the flywheel at the right of the engine is operated by a lever on the footboard. All of the friction faces and levers are arranged inside of the pulley so that they are not only protected from injury but are prevented from tearing the belt should it slip from the pulley face.

A powerful foot with a drum on the differential gear will hold the outfit on a grade independent of the engine.

The transmission is of the shifting gear type with hardened steel gears. The transmission gears are enclosed in a practically dust proof case, this being connected with enclosed crank case and better providing for air displacement of the pistons. Power is transmitted to the truck through the clutch on the left hand side of the engine, which is operated by combined clutch and shifting lever on the footboard. This lever has an interlocking device, arranged so that it is impossible for the operator to shift the gears before the clutch is disengaged, or to engage the clutch until the gears are completely in mesh. It is also impossible to get two sets of gearing in mesh at one time and prevents any possibility of stripping gears by applying the load on the corners of the teeth.

The drive wheels are 78" diameter, 30" face. These give a very large bearing on the ground which is particularly desirable when using the engine for cultivating or seeding on plowed ground. The front wheels are 48" in diameter, 14" face. The wheel base is long and engine is easy to guide. The drive wheels are covered by a metal housing which protects the operator and the working parts of the engine from dust and mud.

This engine gives a drawbar pull on low gear of 9,000 lbs., which will haul from 8 to 12–14" plows, according to the character of the plowing. The hitch is placed about 18" above the ground and consists of a heavy bar extending approximately to the middle of the bull wheels on each side, thus providing for hitching the load most satisfactorily.

(145) The Rumely "Oil Pull" Tractor.

The Rumely oil-pull tractor is driven by a two cylinder, four stroke cycle oil engine, having a bore and stroke of 10 × 12 inches giving 30 tractive horse-power and 60 horse-power at the pulley. The cylinders are cast single and are provided with independent heads. The pistons are easily removed by unbolting the cylinder heads and the crank end of the connecting rod, after which operation they may be pulled out upon the platform. The exhaust and inlet valves are in easily removable cages placed on either side of the cylinder. The stems of the valves are at right angles to the bore of the cylinder and open directly into the combustion chamber without pockets or extensions to the chamber.

A bell crank rocker arm acts on the valve stems which in turn is actuated through a push rod that extends from the cam-shaft in the crank chamber. The cam-shaft, rocker arms, valves, and half time gears are clearly shown by Fig. 128. The housings of the inlet valves connect directly with the special kerosene carburetor made by the Rumely Company. The Higgins carburetor used on these engines is very simple and effective in vaporizing the heavier fuels and has no springs nor internal mechanism to get out of order. The carburetor is controlled directly from the governor which regulates the air, water and kerosene required for the combustion, and has no manual adjustments that need attention from the operator. A constant flow of kerosene and water is maintained through the carburetor by means of force pumps, the level in the device being kept constant by overflow pipes through which the excess returns to the supply tanks.

Fig. 128. Phantom View of the Rumely "Oil Pull" Engine.

As in nearly all types of low compression, or carbureting oil engines, the Rumely engine receives an injection of water in the cylinder to aid the combustion and cooling, and to reduce the initial pressure of the explosion. While the initial pressure is reduced by the water vapor, and with it the strain on the engine, the mean effective pressure is increased because of the absorption of heat from the walls and the more perfect combustion. The only moving part in the carburetor is a single plate controlled by the governor which is produced with one or more air passages. The governor that operates this valve is driven by gears and regulates the speed by throttling the charge. The speed of the engine can be varied from 300 to 400 revolutions per minute while the engine is running.

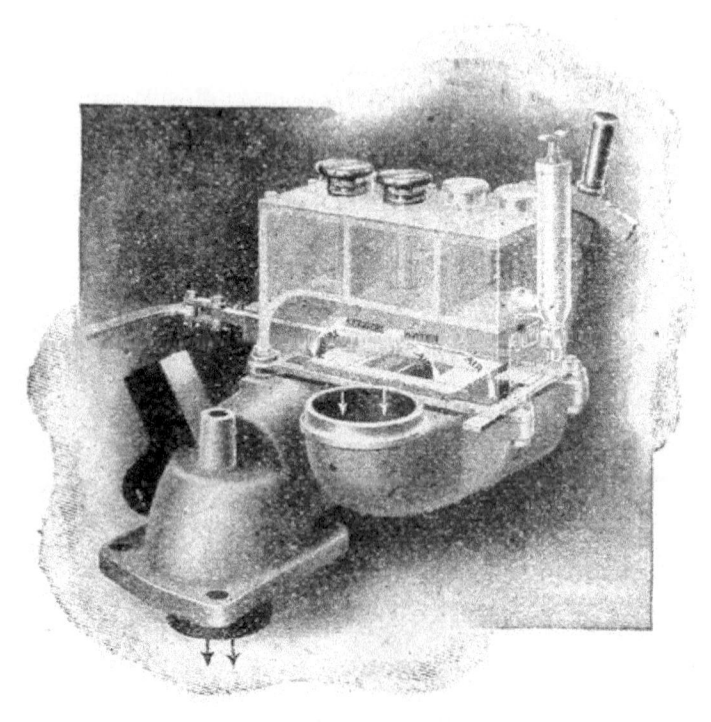

Fig. 129. Higgins Oil Carburetor.

In this engine it is a very simple matter to remove the crank-case cover and the cylinder heads and expose the whole of the working mechanism of the engine.

After removing the cylinder heads and without changing his position, the operator can examine, clean, and, if necessary, regrind the valves. Also without changing position the operator can control his reverse transmission gears, friction clutch for starting the tractor. He is also in reach of the ignition apparatus, governor carburetor and oiler.

The crank case is cast in one piece. The bearings are cast integral with the crank case, and are fitted with interchangeable, adjustable, babbitted shells. Binder caps hold the bearings together and keep the babbitted shells securely in position. The design permits removal of binder caps for examination of crank shaft bearings without distributing the adjustment. The crank case is secured to tractor frame by well fitted bolts, thereby avoiding annoyance from loose bolts and nuts.

The crank case is covered with a sheet steel lid that shuts out all dust and dirt. This cover can easily be removed at any time by simply unscrewing the bolts that hold it in place. It is constructed with this cover on top instead of on the side or end, which permits of easy access to any working parts in the crank case.

Fig. 130. Rumely Oil Pull Tractor.

To further facilitate the accessibility to working parts in the crank case, a secondary cover is provided which can be removed in a couple of minutes. This opening is large enough to allow the operator to reach any point within the crank case.

All cams are key-seated upon the cam shaft with double key-seats, which absolutely prevent any possibility of slipping or alteration in the timing of the engine. The exhaust and intake valves are mechanically operated. The valves are constructed with steel stem, nickel-steel heads, the whole being highly finished.

Valve cages are oil cooled, thereby eliminating all possibility of the valves overheating or warping. The valves themselves can be removed by simply unscrewing the connection. The engine is provided with a set of relief cams by which the compression can be relieved—this greatly facilitates the starting of the engine.

The piston is equipped with five self-expanding rings. Connecting rod is of drop-forged steel construction. Crank-pin bearings are made in halves and lined with shells of special metal.

A combination of mechanical force feed and splash lubrication is employed. Six force feed tubes enter the crank case, on to each bearing, and two tubes force oil into the cylinder. The crank case contains two gallons of oil and is arranged so that any surplus may be drawn off immediately. The lubricator has a gauge glass that shows the quantity of oil supplied at all times, and which is always in view of the operator.

A make and break system (low tension) furnishes the ignition spark, which is supplied with current by a Bosch low tension magneto under normal running conditions, and a battery for starting and for use when the magneto fails. The magneto is of course gear driven so that its armature has a fixed relation with the piston position. The igniters of either cylinder may be easily removed for examination by simply unscrewing two nuts.

Oil is used as a medium for carrying heat from the cylinder walls to the radiator. In the construction of the cooler the company have followed new principles, thus accomplishing the desired result with a minimum amount of metal and liquid. There is no surplus of liquid, just enough oil being used to fill the cylinder jackets, radiator and circulation pipes. The oil is kept in a constant flow from the cylinders to the radiator and back to the cylinders by a large pump which is driven by a chain direct from the crank shaft. The radiator is self-contained and will hold the oil for an indefinite period.

The radiator is composed of a number of sections of pressed galvanized steel. Oil circulates freely within the sections and the air is drawn round the outside. There is a constant flow of oil inside and a constant current of air outside.

The engine is provided with a smooth-working, efficient friction clutch, which is easily handled by a platform lever and with little exertion on the part of the operator. The toggle bolts are adjustable so that any wear in the blocks can be taken up.

The clutch and brake are so connected that when the clutch is thrown out the brake is immediately applied and when thrown in the brake is released.

The various movements of the valves and the ignition mechanism on the face of the flywheel, are marked so that one can check up the timing of the engine. By bringing any one of these marks to coincide with the stationary pointer attached to the side of the crank case, one can easily ascertain whether the adjustments and the timing are exact.

The crank shaft is supported by two end, and one intermediate bearing, the latter bearing being placed between the two throws of the crank shaft. As the two cylinders are placed on the same side of the crank shaft, the two throws are also on the same side of the shaft and to balance these throws cast iron counter weights are bolted on the bottom of the crank arms. The bearings are exceptionally long, the total length of the three bearings amounting to more than half the length between the outer ends of the bearings.

The frame of the tractor consists of four twelve inch "I" beams securely riveted together with intermediate channel stiffeners. The cast steel bearings are riveted to the frame so that the whole construction is one unyielding mass. The bearings are in halves which makes the removal of the shafts an easy task.

With the exception of the differential and master gears all of the gears are cut out of semi steel blanks. The fly wheel has a face of 11 inches, and a diameter of 36 inches.

(146) The "Big Four" Tractor.

The Big Four tractor differs from the majority of tractors in having a four cylinder vertical type motor of 30 tractive and 60 brake horse-power capacity. The cylinders have a bore of 6½ inches and a stroke of 8 inches. The engine runs at the comparatively high speed of 450 revolutions per minute. Gasoline is used for fuel, and is vaporized in a conventional type of jet carburetor.

Both the inlet and the exhaust valves are placed in a pocket at one side of the cylinder making what is known as an "L" engine. The cylinders and the heads are cast in one piece, doing away with points between the cylinders and heads. The pistons and connecting rods may be removed without disturbing the cylinders or their connections by pulling them out through hand holes in the base of the crank case.

The four throw crank shaft is provided with five bearings, these intermediate bearings between the throws and two end bearings in the case. The interior working parts of the motor are lubricated by the splash system with a positive forced feed oiler. The splash pools can be adjusted at a minute's notice so that any desired oil level can be obtained. Grease cups provide the lubrication for all bearings outside of the motor.

Water is circulated by a direct driven centrifugal pump, and as the cooling water is in a closed system the same water is used over and over again without much loss, a bucketful or so a day being an ample supply. The tubular radiator situated in the front of the tractor is provided with a cooling fan that is driven from the engine in a manner similar to automobile practice. A high tension magneto is gear driven from the cam shaft, and is mounted on a rocking bracket so that the armature is advanced and retarded as well as the circuit breaker.

Fig. 131. Views of the Four Cylinder Motor of the "Big Four" Tractor. Note the Massive Construction Compared with Automobile Practice.

An internal expanding clutch connects the motor with the driving gear by operating on the inner run of the fly-wheel. The motion of the engine is transmitted to an intermediate reversing device through bevel gears, this being necessary for the reason that the crank-shaft runs "fore and aft," or parallel to the length of the tractor. A double acting jaw clutch engages with either one or the other of a pair of bevel pinions that run in opposite directions. Motion from the reverse gear is transmitted directly to the different shaft, and from there it is transmitted to the master gears on the bull wheels. The differential shaft is in one piece.

"Big Four," Four Cylinder Tractor Motor.

Showing the Position of Engine on "Big Four" Tractor.

The main driving wheels are very large when compared with the wheels of an ordinary tractor, for they are eight feet in diameter and are proportionately broad. This no doubt gives splendid tractive effect in soft and uneven fields and must save the machine from "stalling" under adverse conditions. Another unusual feature is the automatic steering device used in plowing. This device consists of a long tubular boom that is fastened to the swiveled front axle of the tractor and a small wheel fastened to the outer end of the boom. The small wheel rolls in the next furrow and compels the tractor to plow in a line parallel to it. This steers the tractor more accurately than would be possible by hand and at the same time enables one man to operate both the engine and the plows.

The "Case" Gas Tractor.

Cost of Gas Engine Operation (American).

		GAS PRODUCER PLANT.			NATURAL-GAS ENGINE.			LOW-PRESSURE OIL ENGINE.			DIESEL ENGINE.		
		Load.	Three-quarter Load.	Half Load.	Load.	Three-quarter Load.	Half Load.	Load.	Three-quarter Load.	Half Load.	Load.	Three-quarter Load.	Half Load.
1	Fuel per hp-hour	1.25 lb.	1.5	1.8 lb.	10 cu. ft.	12 cu. ft.	15 cu ft.	1 lb.	1.25 lb.	1.60 lb.	0.50 lb.	0.55 lb.	0.60 lb.
2	Fuel per hp-year (4,500 hours)	2.5 tons	3 tons	3.6 tons	45,000 cu. ft.	54,000 cu. ft.	67,500 cu. ft.	643 gals.	80 3.5 gals.	10 28.5 gal s.	32 1.5 gals.	35 3.5 gals.	38 6 gals.
3	Cost of	$4.00 per ton			30 cents per 1,000			3 cents per gallon			3 cents per gallon		

- 399 -

	fuel				cu. ft.							
4	Cost of fuel per year	$10.00	$12.00	$14.40	$13.50	$16.26	$20.25	$19.30	$24.10	$30.85	$9.65	$10.60	$11.58
5	Cost of attendance per hp-hour	0.40 cent			0.25 cent			0.25 cent			0.30 cent		
6	Cost of attendance per year	$18.00			$11.25			$11.25			$13.50		
7	Lubricating oil per hp-hour	0.006 pint			0.006 pint			0.006 pint			0.007 pint		
8	Cost of oil per year at 25 cents per gal.	$0.84			$0.84			$0.84			$0.98		
9	Scrubber and cooling water per hp-hour	8 gals.			5 gals.			5 gals.			4 gals.		
10	Cost of water per year at 30	$1.44			0.90			0.90			0.72		

	cents per 1,000 cubic feet												
11	Operating expenses; items 4, 6, 8 and 10	$30.28	$32.28	$34.68	$26.49	$29.19	$34.24	$32.29	$37.09	$43.84	$24.85	$25.80	$26.78
12	Saving by Diesel engine	5.43	6.47	7.90	1.64	3.39	6.56	7.44	11.29	17.06	...	...	...
13	Interest, depreciation and maintenance respectively in per cent of investment	6 + 7 + 2 = 15%			6 + 7 + 2 = 15%			6 + 7 + 2 = 15%			6 + 10 + 3 = 19%		
14	Assuming $80 initial cost per hp. the yearly fixed charges will be	$12.00			$12.00			$12.00			$15.00		

From a Paper Read Before the American Institute of Electrical Engineers.

CHAPTER XIV
THE STEAM TRACTOR
(147) The Steam Tractor.

The steam tractor consists of the following elements, which will take up in detail under separate headings.

(1) Engine proper, consisting of the cylinder, piston, valve motion, guides, crank, fly wheel, etc.

(2) Boiler—with the grates, burners, etc.

(3) Feed pump or injector.

(4) Feed water heater.

(5) Driving gear, differential, clutch, etc.

As in the case of the gas tractor, the machine consists simply of a steam engine and its boiler that drive the road wheels of the tractor through a gear train. With the steam tractor the gearing is simplified as the reverse is performed by the engine's valve motion, and not through gearing. There is no need of speed changing transmission gears in the steam tractor as the engine is sufficiently flexible to provide an innumerable number of speeds by simple throttle control.

While the fuel most commonly used is coal, straw and wood, crude oil is often used, the fuel being determined principally by the location of the engine, and by its cost on the job. The matter of fuel should be taken into consideration when the engine is purchased as the different grades demand different fire box and boiler construction. When it is possible to obtain crude oil at a reasonable figure, it certainly should be used in preference to all others as liquid fuel is the most compact, most easily controlled, and efficient of any. The subject of oil burners is taken up later in this chapter, a number of types of which are clearly illustrated.

(148) The Cylinder and Slide Valve.

The steam engine cylinder consists essentially of a smoothly bored iron casting in which a plunger called the "piston" slides to and fro, the cylinder acting not only as a container for the steam acting on the piston but as a guide and support as well. Needless to say, the contact or fit between the piston and cylinder walls must be as perfect as possible, tight enough to prevent steam passing the piston, and free enough to allow the piston to slide without unnecessary friction. The reciprocating piston is connected to the crank through a connecting rod by which the pressure on the piston is communicated to the crank arm.

The pressure exerted on the crank pin by the piston depends on the area of the piston (in square inches) and the pressure of the steam on each square inch of the area. With a given steam pressure, the greater the area, the greater the force tending to turn the crank. As power is the rate or distance through which the force acts in a unit of time it is obvious that the power developed by the engine is equal (in foot pounds) to the force in pounds multiplied by the velocity of the piston in feet per minute. Since there are 33,000 foot pound minutes in a horse-power, the power developed by such a cylinder is equal to the force multiplied by the piston velocity, divided by 33,000.

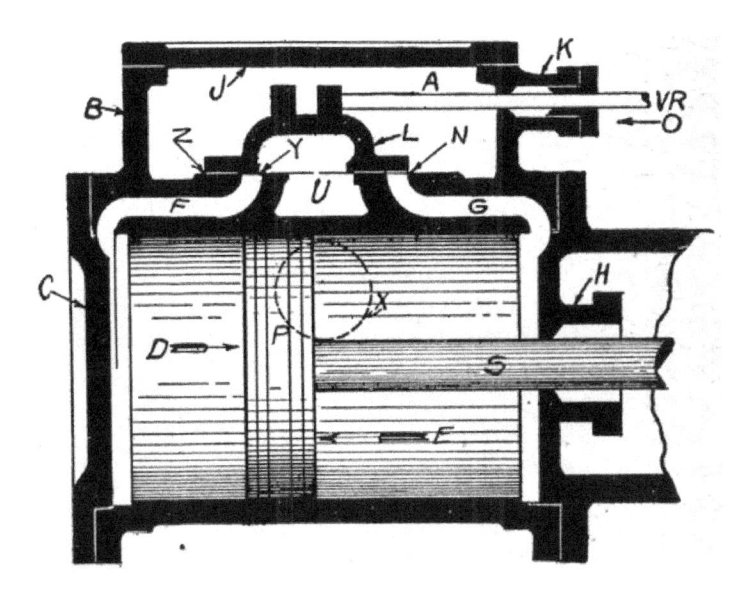

As the cylinder is necessarily limited in length it is evident that the piston cannot travel in one direction continuously but must be reversed in direction when it travels the length of the cylinder bore thereby traveling the next distance in the opposite direction. This reversal of the piston is accomplished by admitting the steam in one end of the cylinder and then into the other, this causing the steam to act on the opposite sides of the piston alternately. To establish a difference of pressure on the two piston forces, the steam pressure is relieved on one side while the steam acts on the other.

A typical cylinder furnished with the ordinary steam tractor is shown by Fig. 133, in which T is the cylinder, P the piston and R is the piston rod. When the steam in the cylinder end E acts in the direction shown by arrow E, the piston pulls the rod R in the direction shown by arrow S, the pressure in the cylinder end D being relieved to atmospheric at this time. The steam is admitted and relieved by the valve L which slides back and forth on its seat actuated by the valve rod VR.

In the position shown, the valve L is moving to the left as shown by arrow O. The edge of the valve N is just opening the steam port G through which the cylinder end F is placed in communication with the steam filled valve chest A. Steam at boiler pressure fills the space A, which flows into E past N and through G when the valve opens and establishes pressure against P, which, through the piston and connecting rods turns the crank.

The steam is exhausted from the cylinder end D, through the port F, through the exhaust port U, and out of the exhaust pipe X. As will be seen from the figure, the inside valve edge Y has moved to the left so that the port F is fully opened. When the piston reaches the left hand end of the cylinder, the valve L moves to the right so that the end of the cylinder E is connected to the exhaust port V through the cylinder port G, thus allowing the steam in the space E to pass out of the exhaust pipe X. A further movement of the valve to the right causes the left edge Z of the valve to uncover the cylinder port F which allows the steam to flow into the cylinder space D and push the piston to the right. This motion is carried on continuously, the valve moving in a fixed relation to the piston, and admits the steam and releases it first on one side of the piston and then on the other. The valve shown is known as a "D" valve and is one of a variety of valves furnished with steam engines, which, however perform exactly the same functions as the valve shown.

An "eccentric" which is really a form of crank, drives the valve to and fro, the eccentric being fastened to the crankshaft. The full pressure of the steam forces the D valve down on its seat, and as the valve is of considerable size, this pressure causes much friction and power loss. In some engines a "balanced" valve is used in which the pressure on the valve is balanced by an equal pressure that acts on the under side of the valve face. Balanced or unbalanced, the function of the slide is to alternate the flow of steam in the two ends of the cylinder.

Steam is prevented from passing the piston into the opposite end of the cylinder by elastic rings placed in grooves on the piston which are known as "piston rings." Being thin and elastic these rings instantly conform with any irregularity of the piston bore and effectually stop the flow of steam past them. At the point where the reciprocating piston rod R passes through the cylinder, a steam tight joint is made by the "stuffing box" or gland H. The space between the inner walls of the stuffing box and the piston rod are either filled with some description of fibrous packing or a metallic packing that fits around the rod in the same manner that the piston rings fit in the bore of the cylinder. The packing is arranged around the valve rod VR in the same manner.

As the piston, piston rod, and valve slide on their respective surfaces with considerable pressure it is absolutely necessary that these parts receive ample lubrication. In practically all engines the oil is taken into the cylinder with the steam in the form of drops, the oil being measured out by a sight feed lubricator that is tapped into the steam supply pipe. In this device, the oil from the lubricator reservoir is fed through a regulating needle valve, drop by drop, up through a gauge glass so that the engineer can tell the amount of oil that he is feeding. The body of the lubricator is filled with condensed water up to the level of the outlet through which the oil passes into the cylinder, and the entire lubricator, reservoir and all is under boiler pressure at all points. The oil regulating valve is placed at the bottom of the lubricator, and as oil is lighter than water, it floats up from the valve to the level of the outlet, through the gauge glass, and from the outlet level floats out into the steam pipe and mixes with the steam. By floating the oil in this manner, the engineer can see every drop that is fed.

(149) Expansion of Steam.

In order to reduce the amount of steam used, the valve does not allow the steam to follow the piston at full boiler pressure through the entire stroke, but cuts it off at a certain point after the piston has started on its travel. As the volume of the steam is increased by the further travel of the piston after the point of cut-off, the steam expands in volume until the end of the stroke is reached, at which point the pressure is naturally much below the initial or boiler pressure. This reduction in temperature and pressure results in a wider working temperature range than would be the case with the steam following the piston throughout the stroke, and as the steam is exhausted to atmosphere at a temperature much lower than that of the boiler steam, much less heat is carried out through the exhaust. As a general rule, the most economical point of cut-off is at ¼ of the stroke. Engines requiring more steam than is supplied at ¼ cut-off in order to carry the load, are too highly taxed for efficient results. Since the most efficient point of cut-off is only ¼ of the possible steam travel it is evident that an engine can carry a load much greater than that for which it is rated, but it is also evident that this increased capacity is gained at the expense operating economy. Wear and tear on the engine parts are also duly increased.

Fig. 134. Case Steam Tractor.

(150) Speed Regulation.

On steam tractors a constant speed is maintained by "throttling" the steam, to meet the demands of the load by partially restricting the flow of steam at light loads and opening the inlet at full load. The valve that controls the steam for the different loads is controlled by a "governor" which depends on the centrifugal force exerted by two fly-balls. The balls, or weights are hinged to a revolving spindle, driven by the engine, in such manner that an increase of speed tends to straighten out and revolve in a more nearly horizontal plane. The amount of travel of the balls for a given speed increase, is governed by a spring, which returns them to a vertical position when the speed decreases. By means of a simple system of levers, the valve is closed when the balls fly out, due to an increase of speed, and is opened when the speed decreases, so that the engine will receive the steam at a higher pressure and again build up its speed to normal. As the load fluctuates, the balls are constantly moving up and down, seeking a valve position that will keep the engine at a constant speed.

Speed variation is generally accomplished by increasing or decreasing the tension of the spring that controls the travel of the governor fly balls, and in the majority of engines this may be done without stopping the engine.

Another form of governor used extensively on stationary engines controls the speed by increasing or decreasing the cut-off. Thus with a heavy load the cut-off may occur at $\frac{1}{2}$ the stroke while with a very light load it may be at 110 stroke. This is by far the most sensitive and economical form of governor, but on account of the reverse gear it is difficult to apply it on a tractor.

(151) Reverse Gear.

As explained under "Cylinders" the travel of the valve bears a definite relation to the piston position so that the ports may be opened and closed at the proper times. It may be shown by a rather complicated diagram that this relation of the valve together with that of the eccentric that drives it is only correct for one direction of rotation. For any other direction of rotation the relation of the valve and piston position must be changed. This may be done in several ways but the most common types are the Stevenson Link and the Wolff slotted yoke.

The Stevenson link motion used on the majority of engines, consists of two independent eccentrics, one being fixed in the relation for forward motion and the other for the reverse direction. The ends of the eccentric rods leading from these eccentrics are connected by a slotted bar or link, in which a block is placed that is connected with the valve rod. The block is free to slide in the slot of the links, that is, it may be moved from one end of the slot to the other. When it is desired to have the engine rotate in a right handed direction, for example, the link is lowered so that the rod from the forward eccentric is brought directly in line with the block so that this eccentric alone acts directly on the valve through the valve stem. When the reverse is desired the link is raised until the rod from the reverse eccentric is brought in line with the block and valve stem, drive being by the reverse eccentric.

When the block is on the link in a position between the two points mentioned, the valve has less travel and it cuts off earlier in the stroke than when driven directly by one eccentric, for the motion at an intermediate point on the link is much different than at the ends of the slots. This fact is taken advantage of in operating engines with the idea of economy in view, and is known commonly as "hooking up" the engine. The best point at which to "hook up" the engine is best determined by experiment, and is equivalent in many respects to the problem of advancing and retarding the spark of a gas engine. We earnestly advise an engineer of a traction engine to take up this subject and determine the best point of cut-off for different loads as he will find that different positions make a considerable difference in his coal bill. Of course the proper way is to determine this point with a steam engine indicator, but as few engineers have such an appliance, the work is generally of the cut and try order. Wear and varying adjustment soon change the points marked on the reverse sector, and for economy's sake these points should be checked occasionally.

In the Wolff motion, a single eccentric is used for both directions of rotation, in connection with a slotted link. A single eccentric is securely keyed to the crank shaft. The eccentric strap has an extended arm which is pivoted to a block that slides back and forth in a curved guide. The angle at which the guide stands with the horizontal determines the direction of rotation, the angle being changed by the reverse lever. The degree of the angle made by the block also determines the point of cut-off. This is a very efficient and simple valve gear.

Guides and Cross-Head.

The outer end of the piston rod is supported by a sliding block known as the "cross-head" which in turn is supported by the guides. An oscillating rod called the "connecting rod" connects the reciprocating cross-head with the crank pin, this rod is used in the same way as the connecting rod of the gas engine except that it is connected to the cross-head instead of the piston.

Clutch.

The clutch affords a means of connecting and disconnecting the driving wheels and engine shaft. It is usually of the friction type described under "Gas Tractors." By releasing the clutch the engine is disconnected from the driving gear so that the tractor remains stationary while the engine is driving a load through the belt.

Use of the Exhaust Steam.

The exhaust from the cylinders is used in two ways, first to create a draft for the fire, and second to heat the feed water pumped into the boiler. The draft is increased by exhausting a portion of the steam into a nozzle placed directly under the stack. The friction of the steam on the surrounding air, draws the air with it, forming a partial vacuum over the grate at each puff, and in this way it causes additional air to rush through the fuel and increases the temperature of the combustion. As the load increases the "puffs" increase in intensity due to the greater terminal pressure and the fire is accelerated in proportion. This is a simple but rather expensive way of increasing the draft.

A considerable proportion of the heat in the exhaust steam is saved by using it to heat the feed water supplied to the boiler. Besides the saving in fuel, affected by heating the water from steam that would otherwise be thrown away, the strains on the boiler due to the injection of cold water are greatly decreased as the difference between the temperatures of the boiling water in the boiler and the hot feed water are much less than in the former case.

The feed water heater consists essentially of a series of tubes in a cylindrical shell. The tubes are surrounded on the outside by the feed water, and are filled with the exhaust steam which passes from end to end through the tubes. The hot water is pumped from the heater into the boiler. An efficient feed water heater adds greatly to the steaming capacity of the boiler.

(152) Feed Pump.

A small steam pump is furnished for pumping the water into the boiler. This device consists of a small steam cylinder connected directly with the pump plunger and is absolutely independent of the main engine so that it can be used whether the engine is running or not. The exhaust of the pump should be turned into the feed water heater when the engine is not running so as to heat the water, but should be directed to atmosphere when the main exhaust is passing through the heater. An injector is usually supplied with the engine for feeding the boiler in emergencies.

The injector forces water into the boiler by means of a steam jet which is arranged so that a high velocity is imparted to the water in the injector nozzle by the condensation of the steam furnished by the jet. In this way water is pumped into the boiler against a pressure that is equal to the pressure of the steam acting on the water. Except for a check valve there are no moving parts. No feed water heater connection is made with the injector for this device raises the temperature of the feed to a considerable temperature. The temperature is not as high, however, as the temperature of the water from the feed water heater and pump, and because of the comparatively low temperature coupled with the fact that live steam is used in heating the injector water, it is not an economical method of pumping.

(153) The Boiler.

As the boilers of traction engines sustain the pull and vibration of the engine as well as the stresses due to traveling over rough roads in addition to the steam pressure strains, they must be made very substantially and of the best materials. The service of the boiler on a traction engine is very different from that met with in stationary or locomotive practice for the tractor seldom receives the attention that is given to the other types and as it goes bumping over the fields with the water whacking at every joint and the engine rushing and surging at every little grade, it receives an "endurance" test every moment of its existence.

A boiler should show an inspection pressure considerably in excess of that which it is intended to carry. It should be well stayed and braced, and should be suspended from the road wheels in such a way as to be relieved from as much strain as possible. No transverse seams should be permitted, and the barrel should be well reinforced at the point where the front bolster is attached as well as at points where pipe connections are tapped into the shell. No large bolts should be tapped into the steam or water space. The tubes should be placed so that they may be easily withdrawn or cleaned. The location of the hand holes and washout holes is also an important item, for inaccessible hand-holes are an abomination.

Boiler lagging or covering is intended to reduce the heat loss by radiation, and for this reason it should be of a good insulating material and should be thick enough to be effective. The cost of jacketing is more than covered by the saving in coal, especially in cold weather.

A straw-burning fire box differs from a coal burner in having a fire brick arch and a shorter grate, and in having a special chute on the fire door for feeding the straw into the furnace. After a short time, the fire brick arch becomes incandescent, keeping the firebox temperature constant and producing perfect combustion of the tarry vapors distilled from the straw. A trap door is provided on the straw chute which automatically keeps the outside air from chilling the fire.

(154) Oil-Burning Steam Tractors.

As with the straw-burning furnace, a brick arch is used in burning oil for the purpose of preventing fractional distillation of the oil during the combustion. In some forms of oil furnaces a brick checker-work is used that provides a much greater surface to the gases than the ordinary firebrick arch and therefore keeps a steadier temperature and pressure. Broken firebrick in the furnace placed in heaps with a rather porous formation is also an aid to combustion. With very heavy oils a jet of steam in the firebox is of great assistance in consuming the free carbon of the fuel (soot).

The oil in practically all cases is atomized or is broken up into a very finely subdivided state by the action of a jet of steam. The finer this subdivision the better will be the combustion for the oil particles will be brought into more intimate contact with the air. Provision is also made in the burner for either whirling or stirring the oil vapor with the air so that a rapidly burning mixture is formed. In other respects the oil burning engine is the same as the coal or wood burner.

(155) Care of the Steam Tractor.

During the idle season, the engine should be well housed, all bright parts slushed with grease and the whole engine carefully covered with tarpaulins. A tractor is an expensive machine and should be given care, or it will rapidly depreciate and start giving trouble. When one considers the abuse and neglect given farm machinery it is remarkable that it will work at all, let alone give efficient service.

Small Fairbanks-Morse Motor Driving Binder.

Before starting a new engine or one that has been idle for a considerable time, all of the bearings and lubricating should be thoroughly cleaned with kerosene oil, removing all grit or gum. After cleaning, they should be thoroughly oiled with the proper grade of lubricant and then adjusted for the correct running fit, taking care that the bearings and wedges are not taken up too tight, nor too many shims are taken out. Be sure that the openings in the lubricating cups and oil pipes are not clogged and that oil holes in the bearing bushings register with those in the bearing caps. At points where there are sight feed gauge glasses, the glasses should be cleaned with gasoline and all of the joints repacked with new packing.

Careful attention should be paid to the piston rod and valve rod packing taking care that it is only tight enough to prevent the leakage of steam and no greater. Excessively tight packing burns out rapidly, scores and shoulders the piston rod, making it impossible to keep the joint tight. When rods are badly scored they should be trued up in the lathe taking care not to take off too much metal on the finishing cut. When renewing fibrous packing be sure that all of the old packing is removed before placing the new packing in the box. Keep the packing well lubricated at all times to prevent wear, and in some cases it will be advisable to add an oil cup to the stuffing box to insure sufficient lubrication.

Go over the valve gear and make sure that there is no looseness or play in the eccentrics or pins, and that all of the bolts and keys are tight and in place. Loose connections in the valve gear are not only productive of knocks and wear but also tend to increase the fuel consumption of the engine. When possible, indicator cards should be taken at intervals to make sure that the valves are correctly set. In a test recently made by the author, the indicator cards showed a defective setting due to wear, that when corrected saved the owner of the engine about 600 pounds of coal per day, and as the coal cost $9.50 per ton delivered in the field, the saving soon paid for the expense of the test. Points of adjustment are provided on all valve gears, and as they differ in detail for each engine we cannot give explicit directions for settling the valves, but will leave this point for the direction book of the maker.

The governor and governor belt should now receive attention making sure that there are no loose points or nuts in the mechanism and that the governor belt is in good condition. Defective governor belts are dangerous through the possibility of over speeding. Slipping or oily belts not only increase the chances of fly-wheel explosions, but also cause a fluctuation in the speed which is not desirable especially in threshing, where good results are obtained only by a constant speed. Make sure that the safety lever works properly and shuts off the steam with a loose or broken belt. Test the governor valve stem for sticking or for rough shots that are likely to cause uneven running. Keep the governor well lubricated with light oil, and keep the oil off the belt as much as possible. Governor valve should be carefully tested for tightness and freedom.

The throttle valve must be absolutely steam tight for a leaking valve is a dangerous proposition especially in stopping the engine. It is generally arranged so that it can be reground with pumice stone or crocus powder and oil. If the valve is of bronze or brass do not use emery or carborundum for the particles will become imbedded in the soft metal and put it in a worse condition than ever. Pack the valve stem.

A leaking slide valve is the cause of much loss of power, and waste of coal, and as the leakage mingles directly with the exhaust, it often remains unknown until it has thrown away a considerable quantity of fuel. It is best detected by blocking the engine with the piston at mid-stroke and opening the throttle valve slightly. If the cylinder drain cocks are now opened, the leaking steam that escapes into the cylinder will be seen issuing from the drains. The leakage that passes into the exhaust will be seen escaping from the stack while it is practically impossible to have the valves absolutely tight at all times, the steam should not escape so rapidly that it roars through the openings. Leakage past the piston is another source of loss that can be detected by blocking the engine so that the piston is very near, one end of the stroke, with the valve opening one of the cylinder ports. Any steam that passes the piston will pass out of the exhaust. With an old engine it is likely that the cylinder is worn oval, or that the valve seat is grooved or uneven, in which case it will be necessary to rebore the cylinder and fit new piston rings or reface the valve seat. Broken piston rings are often the source of leakage, and if not replaced with new at an early date, are likely to destroy the cylinder bore as well. Broken rings generally make themselves known by a wheezing click when the engine is running.

The steam feed pump should be well lubricated with a good grade of cylinder oil and should be well packed around the piston rod especially at the water end. To guard against pump troubles a good strainer should be provided on the water suction line to prevent the entrance of sticks and dirt into the cylinder. Great care should be exercised in keeping the suction line air tight, for if any air escapes into this line no water will be lifted. Dirt under the valves is the cause of much pump trouble, as a very small particle of dirt will allow the water to pass in both directions through the valves. Leaking packing will also destroy the vacuum in one end of the cylinder. For the best results the pump should be run slowly but continuously, feeding a small amount of water at one time. This method of feeding allows the feed water heater to bring the water up to the highest possible temperature which reduces the fuel consumption and reduces the strains on the boiler. It is a bad policy to let the water get low in the boiler and then "ram" full of cold water in a couple of minutes. Attention should be paid to the check valve that is located between the pump and boiler. It should be kept clean and the valve kept tight and in good condition.

When the feed water is hard a boiler compound should be used to reduce the amount of scale in the boiler or soften it and make its removal easier. Scale of 1/16 inch thickness will decrease the efficiency of the boiler by 12%, and this loss increases rapidly with a further increase in the thickness of the scale because of its insulating effect on the tubes. Soft sludges such as mud and clay may be removed by-blowing off or by the filtration of the water before it is pumped into the boiled. Lime and magnesia which form flint-hard deposits, require chemical treatment such as the addition of sodium phosphate, etc. In any case, the deposits waste heat and increase the liability of burning out tubes or bagging the sheets.

Buffalo Marine Motor.

A solution that has given good results with waters containing lime, consists of 50 pounds of Sal Soda and 35 pounds of japonica, dissolved in 50 gallons of boiling water. About 140 quart is fed into the boiler for every horse-power in 10 hours, the solution being mixed with the feed water. Kerosene has been used a great deal to soften scale, and gives good results if not fed in quantities to exceed 0.01 quart per horse-power day of 10 hours. An excess of kerosene is to be guarded against for it is likely to accumulate in spots and cause bagged sheets or burn outs.

CHAPTER XV.
OIL BURNERS.
(156) Combustion.

To obtain the full heat value of a liquid fuel it must be provided with sufficient air to complete the combustion, it must be in a very finely subdivided state, or in the form of a vapor at the time of ignition, and it must be thoroughly mixed with the air so that every part of the oil is in contact which its chemical equivalent of oxygen. Failure to comply with any of these conditions will not only result in a waste of fuel but will also be the cause of troublesome carbon deposits and soot, which eventually will interfere with the operation of the burner.

Complete combustion is much more easily attained with the lighter hydrocarbons such as gasoline or naptha than with crude oil or the heavier distillates, for they are more readily vaporized and mix more thoroughly with the oxygen. Only a slight degree of heat and pressure is required with gasoline while with crude oil a high atomizing pressure and high temperature are required to obtain a satisfactory flame. In the majority of cases where heavy oils are used the fuel is not even completely vaporized but enters the combustion chamber in the form of a more or less finely atomized spray. The methods by which the liquid fuel is broken up divides the burners into three primary classes.

(1) LOW PRESSURE BURNERS in which the fuel is atomized by a blast of low pressure air which also supplies a considerable percentage of the air required for combustion.

(2) HIGH PRESSURE BURNER in which a small jet of high pressure air or steam is used to atomize the oil, the air for combustion being supplied from a source external to the burner.

(3) COMBINED HIGH AND LOW PRESSURE BURNER in which the fuel is atomized by high pressure air or steam, and the greater part of the air for combustion is furnished by a blower at a comparatively low pressure.

In class (1) the oil is supplied to the burner under pressure and by means of a specially designed jet is thrown against hot baffle plates or gauze screens where the partially broken up liquid is caught by the high velocity air and reduced to a still finer spray by its impact against other screens or baffles further on in the burner. This system is applicable only to the light and intermediate grades of oils, such as gasoline, naptha or kerosene, unless heat is applied to the external casing to aid in the vaporization. In some cases the projection of the burner into the furnace gives satisfactory results, but with such an arrangement there is a tendency to deposit carbon in the burner and for the flame to "strike back" should the velocity of the air fall below a certain critical point. Better results were had with this type of burner, by the author when the air blast was preheated by passing several long lengths of the intake air pipe over a hot part of the furnace, instead of entering the burner nozzle into the combustion chamber proper.

A well known modification of this type is the gasoline torch used by electricians and plumbers in which the gasoline is sprayed into a perforated hot tube by air pressure in the tank. When the spray formed at the needle valve strikes the surrounding hot tube it is instantly vaporized and is mixed with the air passing through the perforations in the tube. While the air entering the tube is not forced through the openings by external pressure it attains sufficient velocity to aid in the vaporization because of the vacuum established by the jet. This however is only enough for the more volatile fuels—such as gasoline or benzine.

The high pressure which is by far the most commonly used with low grade fuels may be divided into five principal types (a) **ATOMIZER** burner, (b) The **INJECTOR** burner, (c) **DRIP** feed burner, (d) **CHAMBER OR INTERNAL** burner, (e) **EXTERNAL BLAST** burner. All of these burners break up the fuel by high pressure air or steam, the types given being different only in the way that the pressure is applied to the fuel.

The atomizer acts on the same principle as the medical or perfumery atomizer, the high pressure jet playing directly across the open end of the oil passage as shown by Fig. A. As the vacuum created by the blast is very low, and has little effect in lifting the fuel to the burner, the oil either is made to flow by gravity or by a pump. In the figure the oil in the upper passage is shown pouring down in front of the air or steam jet issuing from the lower port. Both ports are supplied by the pipes shown by the circular openings at the right. The steam and oil are controlled by independent valves placed in the two passages.

In practice the oil and steam openings at the end of the burner may be either single or multiple round openings or long thin slots, the former style being the most common. Since only a small amount of air is admitted through the blast nozzle, far too little to completely consume the oil, the air for the combustion is admitted through openings in the combustion chamber proper, this air being supplied by natural draft or by blower. In some cases the burner is entered into the furnace through an opening that is much larger than the burner itself. The atmospheric air enters through baffle plates in this opening which impart a whirling motion to the air that passes over the burner. This is of considerable aid in maintaining complete combustion in the furnace, and also tends to prevent deposits in the burner.

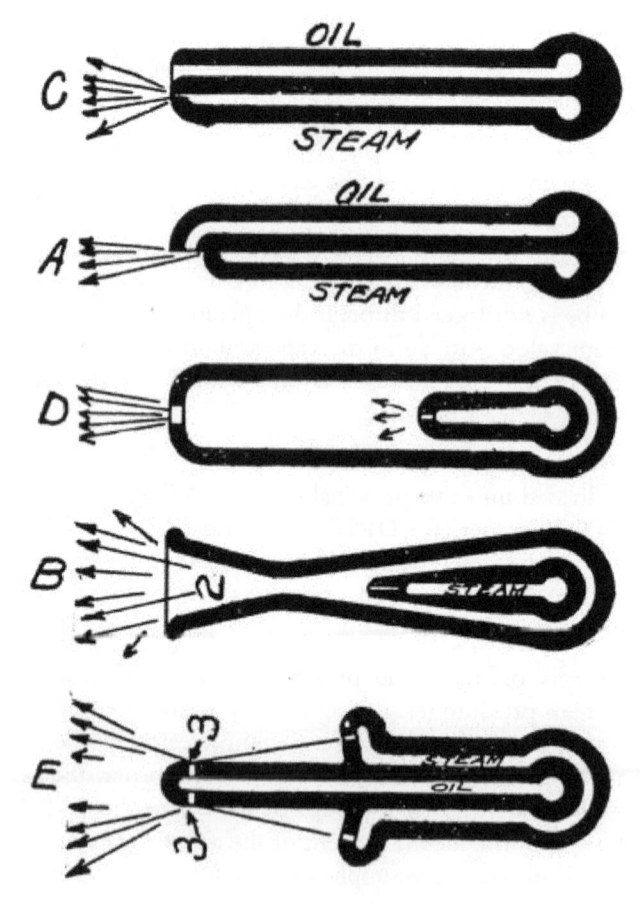

Fig. 135. Showing the Different Classes of Oil Burners.

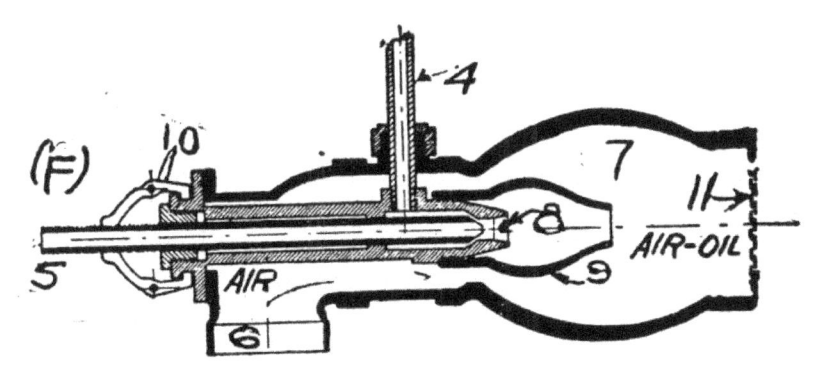

Fig. F. Mixed Pressure Burner, Using Both Steam and Low Pressure Air.

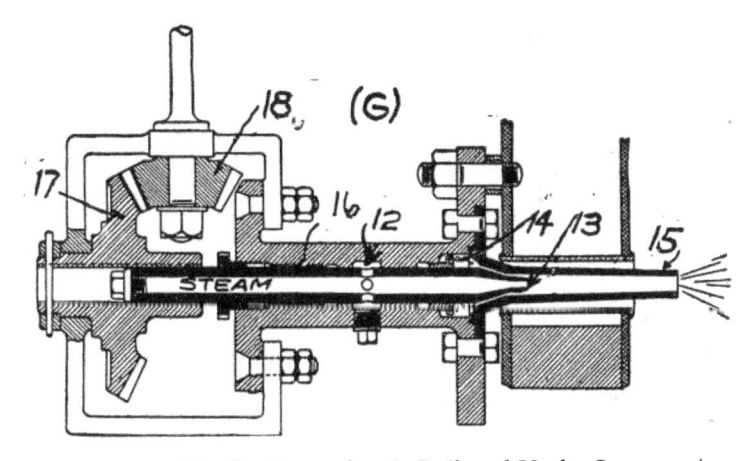

Fig. G. Burner Used by the Pennsylvania Railroad Under Locomotives.

In the injector type of burner shown by Fig. B the air or steam nozzle terminates inside of a shell and is completely surrounded by the oil. A mixture of air and oil issues from main nozzle shown by (2). When the air or steam blows through the inner opening, a partial vacuum is formed in the space (1) which draws the oil into the burner from the supply pipe. On entering this vacuous space the oil comes into contact with the jet and is blown out through the opening (2) in the form of a spray. This vacuum is high enough to lift the fuel for a considerable distance without the aid of a pump and for this reason is the type most commonly met with in practice. A boiler or furnace equipped with this burner will lift the oil directly to the furnace from the reservoir in the same way that a feed water injector will lift water into the boiler. With the commercial injector, the position of the steam jet is made adjustable in relation to the main jet to meet different feed conditions. The steam enters the inner port through the end of the pipe shown at the right. The oil enters the outer port at the right through a port not shown.

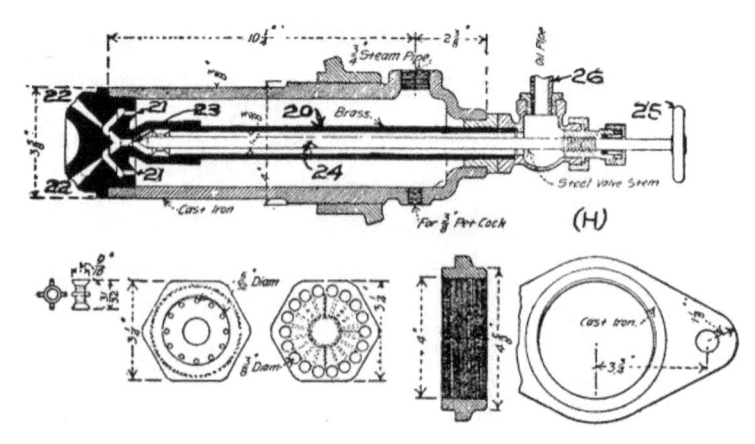

Fig. H. Lassoe-Lovelsin Burner.

Fig. C shows a drip feed or "dribbling" burner in which the oil pours out of the upper port and over the lower port through which the steam or air issues. As would be expected, the atomization is not as perfect with this burner as with the atomizer or injector type.

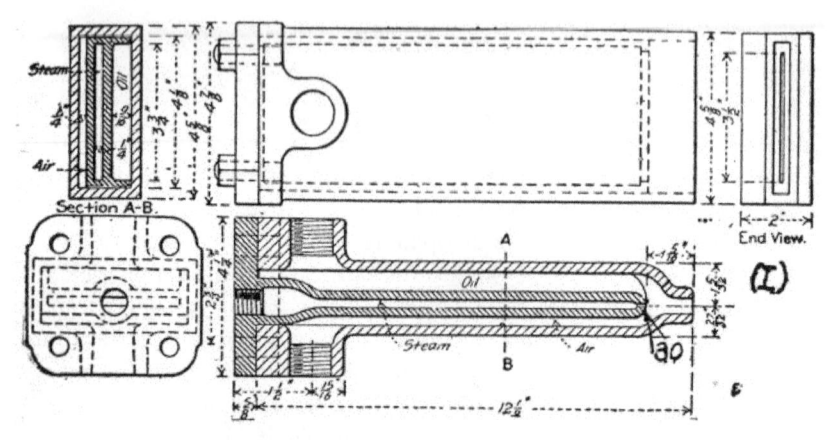

Fig. I. Sheedy Oil Burner, Used for Locomotives.

A burner in which the oil and steam mix before passing out into the furnace through the final opening is known as a "Chamber burner," and is shown by Fig. D. In some respects, at least in construction, it is similar to the injector burner, but it does not possess the lifting abilities of the latter because of the open space in front of the steam nozzle. The atomization takes place largely within the burner because of the eddy currents of air and oil vapor created both by the vapor striking the walls of the outer tube and by the large space in which it has to circulate before passing out of the orifice.

An external blast burner as shown by Fig. E, in which the oil is forced out of the openings (3–3) at the extreme end of the burner atomizes by blowing the oil off of the tube by jets of steam directed by a series of annular openings in a disc. This is really a type of atomizer burner as will be seen by close inspection. This type must be very carefully constructed and the steam jets must be kept very clean in order to have good results for a little variation in the pressure or a small particle of dirt in the openings will deflect the steam and prevent a perfect oil spray. It's one advantage lies in the fact that the oil and air are always separate and therefore minimize the danger of carbonization.

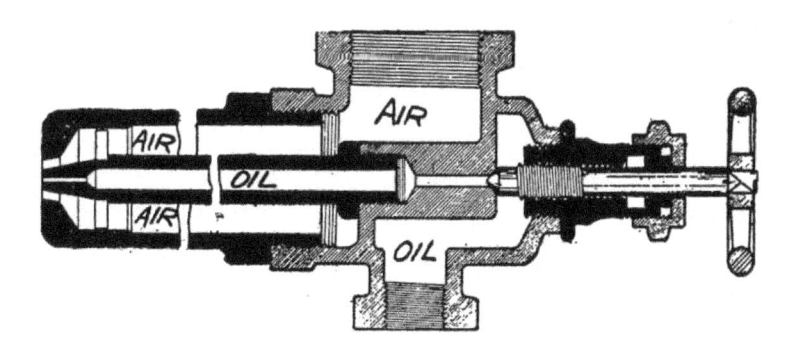

It should be noted that the figures just shown in the illustration of the various classes of burners are diagrammatic only, and that many modifications in detail are made in the practical burner such as regulating valves, sliding steam nozzles, etc.

A burner much used in stationary engine practice and with heating furnaces, where air at two or three ounces pressure is available, is the mixed pressure burner shown by Fig. F. In this burned steam or air compressed, to say 80 pounds per square inch is used for breaking up the fuel oil. A blast of air at low pressure but with considerable volume is used to support combustion in the furnace. The steam or compressed air enters the burner at (5) and meets the oil at the nozzle (8) where it is sprayed into the chamber (9). The oil enters the burner by the pipe (4), flows into the annular passage around the steam nozzle and meets the steam at (8). It will be noted that the steam nozzle (5) is free to slide back and forth in its casing so that the relation between the steam nozzle and spray nozzle may be adjusted to meet different operating conditions. This adjustment is affected by the levers (10) at the end of the burner.

The low pressure air entering through opening (6) from the blower passes around the chamber (9) and mixes with the oil spray from (8) in the mixing chamber (7). This causes a violent swirl in (7) with the result that a comparatively intimate mixture of oil vapor and air is formed before they issue into the furnace. In many burners of this type a gauze screen (11) is placed over the mouth of the final orifice so that back fires are prevented and a still better mixture is formed. Many burners of this type have been built by the author with very satisfactory results, and he knows of only one weak point in the type. This is due to the fact that if a sufficient volume of air is not kept flowing through the low pressure pipe (6), the oil vapor may collect in the piping with the result a back fire will wreck all of the low pressure connections. To prevent this trouble a light galvanized iron weighted damper was placed beneath (6) which closed the pipe when the pressure fell below a certain amount. Since this check valve was placed there were no more pipe fires.

In all cases a sliding damper should be placed in the opening so that the blast can be regulated to suit the amount of oil injected.

As these burners were used in a closed building continuously without smoke or smell and with indifferent grade of oil it will be seen that the combustion was as nearly perfect as could be expected with any type of oil burner.

Several of these burners were made from ordinary steam pipe fittings without steam nozzle adjustment.

While the burners shown are arranged to give a flat flame (with the exception of burner F) they may all be built for a circular flame by surrounding the injection nozzle with a suitable nozzle. A **ROSE** or circular flame is particularly desirable for a vertical boiler where it can be made to conform with the circular shell and apply the heat directly to the tube sheet through suitable fire brick baffles.

A burner of the injector type shown by Fig. G, has been used by the Pennsylvania Railroad with a considerable degree of success. The steam enters the steam nozzle at (12) through the circular openings from which point it passes through the nozzle (13) and carries the oil from the air port

(14). The mixture or spray of steam and oil passes out of the nozzle (15) into the furnace. The steam nozzle is threaded into the casing at (16), and is keyed to the bevel gear (17). Meshing with (17) is the bevel mounted on the vertical stem which terminates in a hand-wheel in the engineer's cab. By turning the bevels, the nozzle turns in the casing threads causing it to move back and forth for the adjustment.

In many types of burners having a nozzle similar to (15) a twisted form of rifling is placed in the bore that gives the escaping gas a rotary motion. This is very effective in mixing the air and oil vapor and spreads the flame very close to the orifice. In burners of the chamber type a spiral vane is sometimes used to gain the same effect, and in one make a rotating fan, is placed near the opening of the outer nozzle which gives a sudden whirl to the gases. While this latter attachment does all that is claimed for it while it is in good repair, it is very likely to stick and put the burner out of commission.

The Lassoe-Lovelsin locomotive burner is shown by Fig. H in which the gas exits through a series of holes in the end of the nozzle (22). The steam enters the outside casing, and unlike the burners just described, entirely surrounds the central oil nozzle, (20). The steam in passing through the openings 21–22 draws the oil through the central opening (23), this oil nozzle being controlled by the needle valve (24) which terminates in the handle (25). Oil enters the oil nozzle through the inlet pipe (26).

The Sheedy oil burner shown by Fig. I has a rectangular nozzle for a flat flame, and has no steam nozzle adjustment. Oil surrounds the steam nozzle and enters the casing through the upper connection. Air enters the lower port through the lower opening as shown in the cross-section of the burner. As the oil flows over the trough formed by the steam nozzle it meets the jet of steam at (30) and is atomized. The air from the lower port aids in bringing the combustion near the tip of the nozzle and therefore prevents carbon deposits from being formed in the burner as well as spreading the flame at a wide angle.